Lecture Notes in Computer Science 14743

Founding Editors

Gerhard Goos
Juris Hartmanis

The series Lecture Notes in Computer Science (LNCS), including its subseries Lecture Notes in Artificial Intelligence (LNAI) and Lecture Notes in Bioinformatics (LNBI), has established itself as a medium for the publication of new developments in computer science and information technology research, teaching, and education.

LNCS enjoys close cooperation with the computer science R & D community, the series counts many renowned academics among its volume editors and paper authors, and collaborates with prestigious societies. Its mission is to serve this international community by providing an invaluable service, mainly focused on the publication of conference and workshop proceedings and postproceedings. LNCS commenced publication in 1973.

Bistra Dilkina
Editor

Integration of Constraint Programming, Artificial Intelligence, and Operations Research

21st International Conference, CPAIOR 2024
Uppsala, Sweden, May 28–31, 2024
Proceedings, Part II

 Springer

Editor
Bistra Dilkina 🆔
University of Southern California
Los Angeles, CA, USA

ISSN 0302-9743 ISSN 1611-3349 (electronic)
Lecture Notes in Computer Science
ISBN 978-3-031-60601-4 ISBN 978-3-031-60599-4 (eBook)
https://doi.org/10.1007/978-3-031-60599-4

Preface

This book constitutes the proceedings of the 21st International Conference on the Integration of Constraint Programming, Artificial Intelligence, and Operations Research (CPAIOR 2024). The conference was held as an in-person event in Uppsala, Sweden at the Uppsala University campus, May 28–31, 2024.

The conference received a total of 104 paper submissions of original unpublished work, which were reviewed by at least three Program Committee members in a single-blind process. The reviewing phase was followed by an author response period and an extensive discussion period carried out by the Program Committee. At the end of the review period, 42 regular papers were accepted for presentation during the conference and were published in this volume. In addition, the conference received 17 extended abstract submissions, containing either original unpublished work or a summary of work, which were reviewed for appropriateness for the conference, and 12 abstracts were accepted for a short presentation and a poster at the conference. Of the regular papers accepted to the conference, the paper "Assessing Group Fairness with Social Welfare Optimization" by Violet Chen, John Hooker and Derek Leben was selected for the Best Paper Award, and the paper "Probabilistic Lookahead Strong Branching via a Stochastic Abstract Branching Model" by Gioni Mexi, Somayeh Shamsi, Mathieu Besançon and Pierre Le Bodic was selected for the Best Student Paper Award. The selection was completed by a 3-member Best Paper Committee.

The conference program also included three keynote talks. Elina Rönnberg (Linköping University, Sweden) gave a keynote on "Decomposition to Tackle Large-Scale Discrete Optimisation Problems". Giacomo Nannicini (University of Southern California, USA) gave a keynote on "Optimization and Machine Learning on Quantum Computers: Yes/No/Maybe?". Tias Guns (KU Leuven, Belgium) gave a keynote on "Decision-Focused Learning: Foundations, State of the Art, Benchmarks and Future Opportunities". The conference program also included a Master Class on "Quantum Computing for CP, AI, and OR, and vice-versa" with invited talks by Carleton Coffrin (Los Alamos National Laboratory, USA), Ashley Montanaro (University of Bristol, UK), Tamás Terlaky (Lehigh University, USA), Harsha Nagarajan (Los Alamos National Laboratory, USA), Andreas Bärtschi (Los Alamos National Laboratory, USA), Zachary Morrell (Los Alamos National Laboratory, USA), Xiaodi Wu (University of Maryland College Park, USA), and David Bernal Neira (Purdue University, USA).

The organization of this conference would not have been possible without the help of many individuals. We would like to thank the Program Committee members and external reviewers for their hard work. We are also very grateful to the Master Class chair Carleton Coffrin (Los Alamos National Laboratory, USA), Sponsorship chair Thiago Serra (Bucknell University, USA), the Publicity chair María Andreína Francisco Rodríguez (Uppsala University), Sweden and Diversity, Equity, and Inclusion (DEI) chair Amira Hijazi (Georgia Institute of Technology, USA). Special thanks go to the conference general chairs Pierre Flener, María Andreína Francisco Rodríguez, and Justin Pearson

(Uppsala University, Sweden), whose support has been instrumental in making this event a success.

Lastly, we want to thank all sponsors for their generous contributions. At the time of writing, these include Artificial Intelligence Journal (AIJ), OptalCP, Quantum Technologies Group at Carnegie Mellon University's Tepper School of Business, Coupa, Association for Constraint Programming, Google, Boeing, Uppsala municipality, nextmv, Gurobi, MERL, The Optimization Firm, SOAF, COSLING, INQA, Uppsala University.

April 2024   Bistra Dilkina

Organization

Program Chair

Bistra Dilkina   University of Southern California, USA

Conference Chairs

Pierre Flener   Uppsala University, Sweden
María Andreína Francisco Rodríguez   Uppsala University, Sweden
Justin Pearson   Uppsala University, Sweden

Master Class Chair

Carleton Coffrin   Los Alamos National Laboratory, USA

Program Committee

Deepak Ajwani   University College Dublin, Ireland
Aliaa Alnaggar   University of Toronto, Canada
J. Christopher Beck   University of Toronto, Canada
Nicolas Beldiceanu   IMT Atlantique (LS2N), France
Mathieu Besançon   Inria, Université Grenoble Alpes, France
Armin Biere   University of Freiburg, Germany
Christian Blum   Spanish National Research Council (CSIC), Spain
Merve Bodur   University of Edinburgh, UK
Quentin Cappart   Polytechnique Montréal, Canada
Carlos Cardonha   University of Connecticut, USA
Mats Carlsson   RISE Research Institutes of Sweden, Sweden
Margarita Castro   Pontificia Universidad Católica de Chile, Chile
Andre Augusto Cire   University of Toronto, Canada
Simon de Givry   INRAE, France
Mathijs De Weerdt   Delft University of Technology, Netherlands
Emir Demirović   Delft University of Technology, Netherlands
Guillaume Derval   University of Liège, Belgium

Bistra Dilkina (Chair)	University of Southern California, USA
Aaron Ferber	Cornell University, USA
Ambros Gleixner	HTW Berlin, Germany
Carla Gomes	Cornell University, USA
Oktay Gunluk	Cornell University, USA
Emmanuel Hebrard	LAAS, CNRS, France
John Hooker	Carnegie Mellon University, USA
Matti Järvisalo	University of Helsinki, Finland
Serdar Kadioglu	Brown University, USA
George Katsirelos	MIA Paris, INRAE, AgroParisTech, France
Joris Kinable	Amazon, USA
Zeynep Kiziltan	University of Bologna, Italy
T. K. Satish Kumar	University of Southern California, USA
Arnaud Lallouet	Huawei Technologies, France
Thi Thai Le	Zuse Institute Berlin, Germany
Pierre Le Bodic	Monash University, Australia
Christophe Lecoutre	CRIL, Université d'Artois, France
Jimmy Lee	Chinese University of Hong Kong, China
Michele Lombardi	DISI, University of Bologna, Italy
Pierre Lopez	LAAS-CNRS, Université de Toulouse, France
Leonardo Lozano	University of Cincinnati, USA
Arnaud Malapert	Université Côte d'Azur, CNRS, I3S, France
Ciaran McCreesh	University of Glasgow, UK
Laurent Michel	University of Connecticut, USA
Michael Morin	Université Laval, Canada
Nysret Musliu	TU Wien, Austria
Barry O'Sullivan	University College Cork, Ireland
Justin Pearson	Uppsala University, Sweden
Laurent Perron	Google France, France
Gilles Pesant	Polytechnique Montréal, Canada
Milena Petkovic	Zuse Institute Berlin, Germany
Claude-Guy Quimper	Université Laval, Canada
Jean-Charles Regin	University Nice-Sophia Antipolis, I3S, CNRS, France
Michael Römer	Bielefeld University, Germany
Elina Rönnberg	Linköping University, Sweden
Louis-Martin Rousseau	Polytechnique Montréal, Canada
Domenico Salvagnin	University of Padova, Italy
Pierre Schaus	UCLouvain, Belgium
Thomas Schiex	INRAE, France
Thiago Serra	Bucknell University, USA
Paul Shaw	IBM, France

Mohamed Siala	INSA Toulouse & LAAS-CNRS, France
Helmut Simonis	University College Cork, Ireland
Kostas Stergiou	University of Western Macedonia, Greece
K. Subramani	West Virginia University, USA
Guido Tack	Monash University, Australia
Kevin Tierney	Bielefeld University, Germany
Christian Tjandraatmadja	Google, USA
Michael Trick	Carnegie Mellon University, USA
Dimosthenis C. Tsouros	KU Leuven, Belgium
Willem-Jan van Hoeve	Carnegie Mellon University, USA
Hélène Verhaeghe	KU Leuven, Belgium
Thierry Vidal	École nationale d'Ingénieurs de Tarbes, France
Petr Vilím	OptalCP, Czech Republic
Mark Wallace	Monash University, Australia
Roland Yap	National University of Singapore, Singapore
Neil Yorke-Smith	Delft University of Technology, Netherlands
Tallys Yunes	University of Miami, USA

Additional Reviewers

Younes Aalian
Mehmet Anil Akbay
Valentin Antuori
Elif Arslan
Christian Artigues
Federico Baldo
Arthur Bit-Monnot
Ignace Bleukx
Camille Bonnin
François Camelin
Jonas Charfreitag
Alexandre Dubray
Aloïs Duguet
Valentin Durante
Suhendry Effendy
Maaike Elgersma
James Fitzpatrick
Mathias Fleury
Maarten Flippo
Marco Foschini
Matteo Francobaldi
Nikolaus Frohner

Ian Gent
Mohammed Ghannam
Luca Giuliani
Bernhard Gstrein
Tias Guns
Matthias Horn
Taoan Huang
Marie-José Huguet
Gabriele Iommazzo
Tanuj Karia
Olivier Lhomme
Haoming Li
Steve Malalel
Yuri Malitsky
Jayanta Mandi
Imko Marijnissen
Valentin Mayer-Eichberger
Matthew McIlree
Gioni Mexi
Yimeng Min
Eleonora Misino
Tobias Paxian

Felipe Pereira

Léon Planken

Zhongdi Qu

Jaume Reixach

Noah Schutte

Konstantin Sidorov

Anand Subramani

Ajdin Sumic

Fabio Tardivo

Charles Thomas

Kevin Tierney

Junhan Wen

Daniel Wetzel

Damien T. Wojtowicz

Contents – Part II

Contents – Part I

Core Boosting in SAT-Based Multi-objective Optimization

Christoph Jabs$^{(\boxtimes)}$, Jeremias Berg , and Matti Järvisalo

HIIT, Department of Computer Science, University of Helsinki, Helsinki, Finland
christoph.jabs@helsinki.fi

Abstract. Maximum satisfiability (MaxSAT) constitutes today a successful approach to solving various real-world optimization problems through propositional encodings. Building on this success, approaches have recently been proposed for finding Pareto-optimal solutions to multi-objective MaxSAT (MO-MaxSAT) instances, i.e., propositional encodings under multiple objective functions. In this work, we propose *core boosting* as a reformulation/preprocessing technique for improving the runtime performance of MO-MaxSAT solvers. Core boosting in the multi-objective setting allows for shrinking the ranges of the multiple objectives at hand, which can be particularly beneficial for MO-MaxSAT relying on search that requires enforcing increasingly tighter objective bounds through propositional encodings. We show that core boosting is effective in improving the runtime performance of SAT-based MO-MaxSAT solvers typically with little overhead.

Keywords: Multi-objective optimization · maximum satisfiability · core boosting · preprocessing

1 Introduction

Maximum satisfiability (MaxSAT) [5], the optimization extension of Boolean satisfiability (SAT) [12], has developed from a theoretical tool into a competitive practical constraint optimization paradigm. This is in particular due to noticeable algorithmic advances in practical MaxSAT algorithms developed in recent years based on the iterative use of SAT solvers. Today, MaxSAT solvers are successfully employed to efficiently solve large instances of various types of real-world NP-hard combinatorial optimization problems via propositional encodings under a single objective function.

Building on advances in (single-objective) MaxSAT solving, algorithmic advances have been recently made towards developing increasingly effective solvers for the more general and challenging realm of MaxSAT under multiple objectives, i.e., multi-objective MaxSAT (MO-MaxSAT) [14,15,23,28,44,45]. Motivated through practical applications that give rise in a natural way to propositional encodings of optimization problems under multiple objectives, the goal in

MO-MaxSAT solving is to efficiently enumerate all Pareto-optimal—as a standard notion of optimality in the multi-objective setting—solutions (or, more precisely, a representative Pareto-optimal solution for each point in the so-called non-dominated set within the search space of all solutions). Pareto-optimal solutions are solution with respect to which no objective can be improved without making the solution worse in terms of another objective, hence intuitively constituting the best possible solutions in general terms under multiple objectives.

A complementary approach to improving constraint solvers by developing more effective algorithms is that of developing preprocessing (or reformulation) techniques to be applied before calling a solver. The aim of (effective) preprocessing is to improve solver runtimes to the extent that the additional time spent in preprocessing is worthwhile in terms of the combined overall time spent in preprocessing and solving, compared to the time required to directly solve the original problem instance. Preprocessing has been highly influential in SAT solving [13], and motivated by this, extensions of SAT preprocessing have been proposed for MaxSAT [7,10,25,42] and most recently for MO-MaxSAT [27]. However, so-far liftings of (Max)SAT preprocessing techniques to the realm of MO-MaxSAT have turned out to provide relatively small runtime improvements for current state-of-the-art MO-MaxSAT solvers. This suggests that more research is called for towards harnessing the full potential of preprocessing for speeding up MO-MaxSAT solving.

In this work, we propose core boosting as an approach to automatically reformulating MO-MaxSAT instances. Core boosting was earlier proposed in the context of single-objective MaxSAT solving [8] with later applications in single-objective core-guided constraint programming [22] and pseudo-Boolean optimization [16]. In the previous works, core boosting was proposed as an any-time algorithm that combines so-called core-guided and upper-bounding search for finding good solutions to single-objective constraint optimization instances within a stringent runtime limit. In contrast, we develop core boosting for the multi-objective setting as a pre-solving phase technique that allows for reformulating an MO-MaxSAT instance by tightening its search space, in particular via detecting inconsistent parts of the search space which can be subsequently ignored by MO-MaxSAT solvers. This is achieved—in short, as we will later on explain in more detail—by increasing objective offsets in a given MO-MaxSAT instance to match the so-called ideal point of the multi-objective search space without removing any Pareto-optimal solutions. As such core boosting can be viewed as a preprocessing technique which significantly differs from more typical (Max)SAT-based preprocessing techniques so-far studied for MO-MaxSAT. Core boosting intuitively leads to enabling more optimized propositional encodings of pseudo-Boolean constraints used within state-of-the-art MO-MaxSAT solvers. We explain in detail how core boosting can be tightly integrated into MO-MaxSAT solvers, and provide an open-source implementation in conjunction with three recently-proposed algorithms for MO-MaxSAT. Empirically, it turns out that core boosting can be highly effective in speeding up overall runtimes of MO-MaxSAT algorithms, having a noticeably greater positive impact on runtime performance than presently available MaxSAT-based preprocessing techniques for MO-MaxSAT.

2   Multi-objective MaxSAT

For Boolean variable x, there are two literals: the positive x and the negative $\neg x$. A clause is a disjunction of literals, and a (CNF) formula a conjunction of clauses. When convenient, we view a clause as the set of literals in the clause, and a formula as the set of its clauses. An assignment τ maps variables to $\{0, 1\}$, i.e., $\tau(x) = 1$ (true) or $\tau(x) = 0$ (false). Assignments extend to literals, clauses, and formulas by $\tau(\neg x) = 1 - \tau(x)$ for negative literal $\neg x$, $\tau(C) = \max\{\tau(l) \mid l \in C\}$ for clause C, and $\tau(F) = \min\{\tau(l) \mid C \in F\}$ for formula F. An assignment for which $\tau(F) = 1$ is a *solution* to F. If a formula has a solution, the formula is satisfiable, otherwise the formula is unsatisfiable.

A pseudo-Boolean (PB) expression $O = (\sum_i c_i \cdot l_i) + o$ is a sum of terms—each consisting of a literal l_i and a positive integer constant c_i—and a non-negative integer constant o referred to as *offset* of O. We denote the set of literals appearing in O by $\text{LITS}(O)$. The value under O of an assignment τ over $\text{LITS}(O)$ is $O(\tau) = (\sum_i c_i \tau(l_i)) + o$. For an integer B, a pseudo-Boolean constraint $O \leq B$ is satisfied by an assignment τ if $O(\tau) \leq B$. Our work makes extensive use of CNF encodings that encode values of PB constraints into literals [6,18,30]. More precisely, $\text{CNF}(O \leq B)$ is a CNF formula that defines a literal $\langle O \leq B \rangle$ such that any solution τ of $\text{CNF}(O \leq B)$ sets $\tau(\langle O \leq B \rangle) = 1$ if and only if τ satisfies $O \leq B^1$. When clear from context, $\langle O \leq B \rangle$ should be understood as $\text{CNF}(O \leq B) \wedge \langle O \leq B \rangle$. For example, the formula $F \wedge \langle O \leq B \rangle$ stands for the formula $F \wedge \text{CNF}(O \leq B) \wedge \langle O \leq B \rangle$, the solutions of which are the solutions τ of F that satisfy $O \leq B$. We also use $\langle O < B \rangle$ as a shorthand for $\langle O \leq B - 1 \rangle$, and $\langle O \geq B \rangle$ as a shorthand for $\neg \langle O < B - 1 \rangle$.

We focus on the following natural extension of maximum satisfiability to the multi-objective setting. An instance $\mathcal{I} = (F, \mathcal{O})$ of multi-objective maximum satisfiability (MO-MaxSAT) consists of a formula F and p linear objective functions $\mathcal{O} = (O_1, \ldots, O_p)$ represented as pseudo-Boolean expressions. Note that this definition covers single-objective MaxSAT with $p = 1$. Any solution τ to F is a solution to $\mathcal{I}$. A solution τ has cost $\mathcal{O}(\tau) = (O_1(\tau), \ldots, O_p(\tau))$ with respect to $\mathcal{I}$, and cost $O_i(\tau)$ with respect to objective O_i. The ideal point [19] $(\gamma_1, \ldots, \gamma_p)$ of $\mathcal{I}$ consists of the smallest value for each objective over the solutions to $\mathcal{I}$, i.e., $\gamma_i = \min\{O_i(\tau) \mid \tau(F) = 1\}$. We focus on the task of computing Pareto-optimal solutions to MO-MaxSAT instances. A solution τ dominates another solution τ' if $O_i(\tau) \leq O_i(\tau')$ for all $i = 1, \ldots, p$ and $O_i(\tau) < O_i(\tau')$ for some i. A solution τ is *Pareto-optimal* if it is not dominated by any solution to $\mathcal{I}$. The costs of Pareto-optimal solutions form the non-dominated set of $\mathcal{I}$.

Example 1. Consider the bi-objective instance on the left in Fig. 1. The infeasible region with respect to the objectives, i.e., the objective values for which no solutions to F exist, is illustrated on the right. The non-dominated set of the instance is $\{(4, 8), (5, 7), (6, 3)\}$ and its ideal point $(4, 3)$.   □

---

[1] In practice, the algorithms considered in this work often employ an implication relationship—$\tau(\langle O \leq B \rangle) = 1$ *if* τ satisfies $O \leq B$—rather than the mentioned equivalence. We assume equivalences for simplicity without loss of generality.

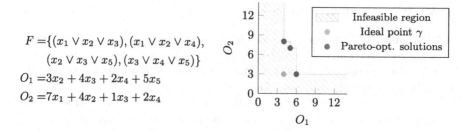

$$F = \{(x_1 \lor x_2 \lor x_3), (x_1 \lor x_2 \lor x_4),$$
$$(x_2 \lor x_3 \lor x_5), (x_3 \lor x_4 \lor x_5)\}$$
$$O_1 = 3x_2 + 4x_3 + 2x_4 + 5x_5$$
$$O_2 = 7x_1 + 4x_2 + 1x_3 + 2x_4$$

Fig. 1. A bi-objective MaxSAT instance and its infeasible region with respect to the objectives.

While we will present core boosting in the context of computing a single representative solution to each element in the non-dominated set, we note that the technique is also applicable when computing *every* Pareto-optimal solution. In particular, unlike other specific preprocessing techniques [27] core boosting maintains all solutions for each element in the non-dominated set.

2.1   SAT-Based MO-MaxSAT Algorithms

Similarly as single-objective MaxSAT algorithms, multi-objective MaxSAT algorithms [14,15,23,28,44,45] make extensive use of *SAT solvers* [36], i.e., decision procedures that either compute a solution of a CNF formula, or determine that the formula is unsatisfiable, i.e., that no such solutions exist. For computing Pareto-optimal solutions, SAT solvers are used to iteratively compute solutions of the instance. When a new solution is found new constraints that rule out dominated solutions from consideration are added until no more solutions remain, at which point the entire non-dominated set has been discovered.

A close analogy in the single-objective case is the Sat-Unsat (LSU) [5] algorithm for minimizing a single objective O subject to a CNF formula F. This algorithm is implemented by various single-objective MaxSAT solvers [11,18,32,46]. Starting from some upper bound UB, LSU minimizes O by incrementally invoking a SAT solver on the formula $F \land \langle O < \text{UB} \rangle$. If the SAT solver finds a new solution τ, we have $O(\tau) < \text{UB}$ and hence the upper bound is improved. If the solver reports unsatisfiability, the latest solution found is optimal and the algorithm terminates. With this intuition, we next detail three state-of-the-art algorithms for computing Pareto-optimal solutions: P-MINIMAL, BIOPTSAT, and LOWERBOUND.

P-MINIMAL [31,44] can be seen as multi-objective LSU. When solving an instance of MO-MaxSAT $(F, (O_1, \ldots, O_p))$, P-MINIMAL iteratively invokes a SAT solver on a working formula consisting of F and additional constraints added in previous iterations. When the SAT solver returns a solution τ, P-MINIMAL (i) blocks all solutions of worse quality, i.e., the ones dominated by τ, and (ii) restricts search in subsequent iterations to solutions dominating τ. Part (i) is achieved by adding the constraints $\bigvee_{i=1}^{p} \langle O_i < O_i(\tau) \rangle$ that enforce subsequent solutions to improve in at least one objective. Part (ii) is achieved

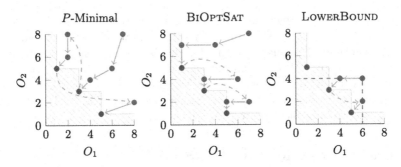

Fig. 2. Search trajectories of P-Minimal (left), BiOptSat (middle), and Lower-Bound (right) in terms of objective values.

by adding the constraints $\bigwedge_{i=1}^{p}\langle O_i \leq O_i(\tau)\rangle$ that enforce subsequent solutions to not worsen in any objective. When the SAT solver reports unsatisfiability, all constraints of type (ii) are removed and search continues. When no more solutions can be found without any constraints of type (ii), all Pareto-optimal solutions have been found. Figure 2 (left) illustrates one possible search trajectory of P-Minimal (with respect to objective values) on a bi-objective instance. Assume search starts at a solution with objective values $(7, 8)$. By iteratively steering search to regions that dominate the current solution, P-Minimal moves through intermediate solutions (marked in blue) until it discovers the (red) non-dominated point at $(3, 3)$, after which the SAT solver reports unsatisfiability. Then, all constraints of type (ii) are removed and P-Minimal starts the minimization procedure again (illustrated by the dashed line) while retaining the blocking constraints (i). After three minimization procedures, the entire non-dominated set in Fig. 2 is discovered and the algorithm terminates since all solutions are blocked.

BiOptSat in its Sat-Unsat variant [28] computes the non-dominated set of a bi-objective MaxSAT instance $(F, (O_1, O_2))$ via the so-called lexicographic method [34]. Assuming the same initial solution with objective values $(7, 8)$, BiOptSat starts by employing single-objective LSU to find a solution τ minimizing O_1, as illustrated in Fig. 2 (middle) by arrows going leftward until the blue solution on the infeasibility boundary is reached. Next, O_2 is minimized while restricting O_1 to at most $O_1(\tau)$, i.e., subject to $F \wedge \langle O_1 \leq O_1(\tau)\rangle$, using LSU, illustrated in the figure by the downward arrows, until the red Pareto-optimal solution τ^p is found. To find the next Pareto-optimal solution, BiOptSat adds $\langle O_2 < O_2(\tau^p)\rangle$ to F and reiterates. This is illustrated by the dashed arrows in Fig. 2. The algorithm terminates when no solutions remain, indicated by the SAT solver reporting unsatisfiability after removing the constraints on O_1.

LowerBound [15], in contrast to P-Minimal and BiOptSat, mainly performs lower-bounding search to compute the non-dominated set. It maintains a *fence*, i.e., a tuple $(\lambda_0, \ldots, \lambda_p)$ of values, initialized to $(0, \ldots, 0)$, that represents the greatest objective values currently considered. During search LowerBound alternates between iteratively loosening the fence until the region bounded by the constraints $\bigwedge_{i=1}^{p} \langle O_i \leq \lambda_i \rangle$ contains feasible solutions, and then employing P-MINIMAL to find all elements of the non-dominated set "inside" the current fence. The search of P-MINIMAL inside a fence is illustrated in Fig. 2 (right) for the fence shown in green. After P-MINIMAL finds all Pareto-optimal solutions within the fence, the fence is loosened further. The algorithm terminates once all solutions have been blocked by P-MINIMAL.

3   Core Boosting for MO-MaxSAT

We now detail core boosting for MO-MaxSAT as our main contribution.

3.1   Effects of Core Boosting

Before describing how core boosting is realized, we explain how core boosting allows for reducing the search space of MO-MaxSAT instances and detail how core boosting reformulates MO-MaxSAT instances.

Core boosting is a technique that through reformulating an MO-MaxSAT instance increases the offsets of the objectives of the instance to match the ideal point without removing any Pareto-optimal solutions. As such core boosting can be viewed as a preprocessing technique which significantly differs from more typical (Max)SAT-based preprocessing techniques recently proposed for MO-MaxSAT [27]. The intuition for the potential usefulness of the core boosting reformulation stems from the fact that MO-MaxSAT algorithms such as P-MINIMAL, BIOPTSAT, and LOWERBOUND search only over the non-constant parts of the objectives in the instance: the range of possible solution costs that the algorithms consider during search is bounded "from below" by the point consisting of the offsets of each objective, and "from above" by the point consisting of the maximum value of each objective. As such, increasing the offsets of the objectives conceptually leads to a smaller search space.

Example 2. Recall the bi-objective MaxSAT instance from Example 1 and Fig. 1. For this instance, the range of solution costs that P-MINIMAL, BIOPTSAT, and LOWERBOUND consider during search is 0 to 12 for both O_1 and O_2, as illustrated on the left in Fig. 3. Applying core boosting on this instance results in a reformulation with the same Pareto-optimal solutions and objectives O_1^{cb}, O_2^{cb} with offsets $o_1^{\text{cb}} = 4$ and $o_2^{\text{cb}} = 3$, respectively. When solving the reformulated instance, MO-MaxSAT algorithms are effectively searching over the costs in the range $4 \ldots 12$ for O_1 and $3 \ldots 12$ for O_2. This search space (depicted on the right in Fig. 3) is smaller than the one that would be considered without core boosting. In particular, after core boosting, the cross-hatched area shown in Fig. 3 does not need to be considered during search.   □

Formally, core boosting transforms an instance $\mathcal{I} = (F, \mathcal{O})$ with an ideal point $(\gamma_1, \ldots, \gamma_p)$ into a reformulated ("core-boosted") instance $\mathcal{I}^{\text{cb}} = (F^{\text{cb}}, \mathcal{O}^{\text{cb}})$ for which the following hold.

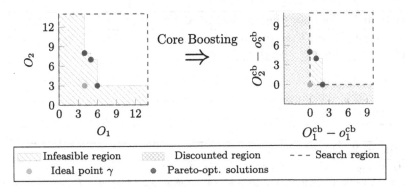

Fig. 3. Illustration on how core boosting shifts the point where search is anchored to the ideal point and reduces the search space.

(i) All solutions of F^{cb} are solutions to F, and any solution to F can be uniquely extended into a solution to F^{cb}.

(ii) $\mathcal{O}(\tau) = \mathcal{O}^{\mathrm{cb}}(\tau)$ for all solutions τ to F^{cb}.

(iii) The offset of objective O_i^{cb} is γ_i.

In other words, core boosting reformulates a given MO-MaxSAT instance in a way that all solutions and their costs are preserved, and the offset of each objective O_i is increased to γ_i, the ith coordinate of its ideal point.

When viewed as a lower-bounding method, the offsets that core boosting derives for each objective are as high as possible while guaranteeing that all Pareto-optimal solutions and the non-dominated set are preserved. More precisely, consider an MO-MaxSAT instance $\mathcal{I} = (F, (O_1, \ldots, O_p))$, its ideal point $(\gamma_1, \ldots, \gamma_p)$ and fix an index i. By definition, there is a Pareto-optimal solution τ for which $O_i(\tau) = \gamma_i$. Since the coefficients of objectives are positive, any reformulation $\mathcal{I}^{\mathrm{ref}} = (F^{\mathrm{ref}}, (O_1^{\mathrm{ref}}, \ldots O_p^{\mathrm{ref}}))$ of $\mathcal{I}$ in which the offset of O_i^{ref} is strictly greater than γ_i will have a different non-dominated set, and specifically the cost of τ will be different. In the context of algorithms computing the entire non-dominated set, core boosting therefore derives the tightest lower bound given by a single point.[2]

3.2 Core Boosting via Single-objective Core-Guided Search

We now detail how the reformulation performed by core boosting can be realized in practice via single-objective lower-bounding search based on so-called unsatisfiable cores, i.e., using core-guided MaxSAT search [1,2,37,40,41]. We detail core boosting in pseudocode as Algorithm 1. Invoked on an MO-MaxSAT instance $(F, \mathcal{O})$, core boosting iteratively invokes single-objective core-guided lower-bounding search (represented in pseudocode by the CoreGuided sub-procedure) on single-objective MaxSAT instances. In the ith iteration, Core

---

[2] Exploring similar ideas from the perspective of so-called lower bound sets [20] constitutes interesting future work beyond the scope of this paper.

Algorithm 1. Core boosting for MO-MaxSAT

Input: An MO-MaxSAT instance $\mathcal{I} = (F, (O_1, \ldots, O_p))$
Output: A reformulated MO-MaxSAT instance $\mathcal{I}^{\text{cb}}$
1: $F^{\text{cb}} \leftarrow F$
2: **for** $i \leftarrow 1$ **to** p **do**
3:    $(F^{\text{cb}}, O_i^{\text{cb}}) \leftarrow \text{CoreGuided}(F^{\text{cb}}, O_i)$
4: **return** $(F^{\text{cb}}, (O_1^{\text{cb}}, \ldots, O_p^{\text{cb}}))$

Algorithm 2. CoreGuided

Input: A single-objective MaxSAT instance $\mathcal{I} = (F, (O))$
Output: An optimal solution τ to $\mathcal{I}$ and a reformulated instance $\mathcal{I}'$
1: $F^{\text{ref}} \leftarrow F, \quad O^{\text{ref}} \leftarrow O$
2: **while** true **do**
3:    $(\text{res}, \kappa, \tau) \leftarrow \text{ExtractCore}(F^{\text{ref}}, O^{\text{ref}})$
4:    **if** res = "unsatisfiable" **then**
5:      $(F^{\text{ref}}, O^{\text{ref}}) \leftarrow \text{Reformulate}(F^{\text{ref}}, O^{\text{ref}}, \kappa)$
6:    **else**
7:      **return** $\tau, (F^{\text{ref}}, (O^{\text{ref}}))$

Guided is invoked on F^{cb} and O_i (line 3), adding new clauses to F^{cb} and reformulating O_i to O_i^{cb}. The formula F^{cb} consists of the clauses of the original instance F and all additional constraints added by CoreGuided in previous iterations.

Algorithm 2 details a generic abstraction of core-guided search under a single objective. The algorithm works by iteratively extracting so-called (unsatisfiable) cores based on which the instance is reformulated. A core κ of a single-objective MaxSAT instance $\mathcal{I} = (F, (O))$ is a subset of objective literals $\kappa \subset \text{LITS}(O)$ out of which at least one literal has to incur cost, i.e., has to be assigned to 1. Such a core can be obtained with a modern off-the-shelf SAT solver by employing its assumption interface [17,36]. This core extraction is done in the ExtractCore subroutine which takes a formula F and an objective O as input and returns a triple $(\text{res}, \kappa, \tau)$ where res indicates whether $F' = F \wedge \bigwedge_{l \in \text{LITS}(O)} \neg l$ is satisfiable. If res = "unsatisfiable", κ contains a core of $(F, (O))$, otherwise τ contains a solution to F'.

When a new core is extracted, the instance is reformulated by the Reformulate subroutine. Existing core-guided algorithms differ mainly in the details of how Reformulate is instantiated. Core boosting makes very lightweight assumptions on the underlying core-guided algorithm. It can be realized with any core-guided algorithm whose instantiation of Reformulate increases the offset of the objective, decreases the sum of coefficients in the objective of the literals in the core, and adds additional clauses and variables to preserve the solutions and their costs in the instance. More specifically, the properties of Reformulate required for core boosting can be summarized as follows. Assume that Reformulate is invoked with formula F, objective O, and core κ, and that it returns a new formula F^{ref} and objective O^{ref}. Then the following must hold for core-boosting to be applicable: (i) Every solution of F^{ref} is a solution of F; (ii) $O(\tau) = O^{\text{ref}}(\tau)$ holds for all solutions of F^{ref}; (iii) the sum of coefficients in O^{ref} is smaller than in O; and (iv) the offset of O^{ref} is greater than the offset of O. It should be noted that these properties are met by practically all modern core-guided algorithms [1,2,21,25,26,37,40,41,43].

As a side-remark, each reformulation performed by all core-guided algorithms we are aware of increases the offset of the objective, and decreases the sum of coefficients, exactly by the minimum coefficient of the literals in the core. This is because at least one literal in the core has to incur cost. Thus, the smallest possible cost incurred due to a core matches the smallest coefficient among the literals in the core.

Example 3. Invoke CoreGuided on the constraints and objective O_1 of the MO-MaxSAT instance from Fig. 1. Let the cores extracted in the first two iterations of executing CoreGuided on $(F, (O_1))$ be $\kappa^1 = \{x_2, x_3, x_5\}$ and $\kappa^2 = \{x_3, x_4, x_5\}$. After reformulating these two cores, the reformulated instance is satisfiable at Line 4. The smallest coefficients in the cores are $c_{\kappa^1} = 3$ and $c_{\kappa^2} = 1$, respectively. The final reformulated objective has a constant offset of $o_1^{\text{cb}} = c_{\kappa^1} + c_{\kappa^2} = 4$ with its coefficient sum reduced by the offset. □

Note that core boosting for MO-MaxSAT, as proposed here, differs from core boosting for (single-objective) MaxSAT [8]. For MaxSAT, CoreGuided is executed under a heuristically determined time limit since core-guided search is complete for MaxSAT. In contrast, in the multi-objective setting running Core Guided without resource limits will *not* fully solve the instance (assuming that the objectives conflict with each other in that their minimum values correspond to different solutions). Instead, the search space is reduced with respect to the individual objectives.

3.3   Realizing Core Boosting

A variety of core-guided single-objective MaxSAT algorithms from the literature [1,2,37,40,41] could be employed for practical implementations of core boosting. We detail here our implementation of core boosting based on the effective core-guided algorithm OLL [1,40].

Informally speaking, OLL instantiates Reformulate by introducing PB constraints over the literals in the extracted cores in a way that systematically allows additional literals to be assigned to 1 in subsequent iterations. More precisely, after obtaining a core κ, OLL (i) decreases the coefficient of each $l \in \kappa$ by c_κ, the smallest coefficient among the literals in κ, removing l from the objective if the new coefficient is 0 (this process is called clause cloning in some references [9]); and (ii) adds new variables $\langle \sum_{l \in \kappa} l \geq k \rangle$ to the reformulated objective with the coefficient c_κ and constraints $\text{CNF}(\sum_{l \in \kappa} l \leq k)$ to the formula for $k = 2, \ldots, |\kappa|$. Conceptually, step (i) relaxes the current objective by removing at least one literal from the objective and allowing at least one literal in κ to be assigned to 1 in subsequent iterations. Step (ii) adds constraints to ensure that at most one literal can be set to 1 without new cores being discovered, thus ensuring the preservation of optimal solutions.

An important intuition for understanding the effects of core boosting is that the changes in the number of literals in the objective depend on the coefficients in the extracted cores. If the coefficients in the variables of a core κ are not equal, not all literals will be removed from the objective in step (i). Since OLL introduces $|\kappa| - 1$ new variables, in those cases the number of literals in the reformulated objective can increase. In contrast, if the coefficients of the variables are all equal, the number of variables in the objective will decrease by one after reformulation as then all variables in the core are removed and $|\kappa| - 1$ variables are introduced.

Example 4. Recall Example 3 where O_1 (see Fig. 1) was reformulated based on the cores κ^1 and κ^2. The detailed objective reformulated by OLL is $O_1^{cb} = x_4 + 4x_5 + 3\langle \kappa_1 \geq 2 \rangle + 3\langle \kappa_1 \geq 3 \rangle + \langle \kappa_2 \geq 2 \rangle + \langle \kappa_2 \geq 3 \rangle$ and the clauses $\text{CNF}(\kappa_1 \leq k) \wedge \text{CNF}(\kappa_2 \leq k)$ are added to the formula.   □

Core boosting can be tightly integrated with SAT-based MO-MaxSAT algorithms in a way that allows for reusing the structure introduced by the core-guided algorithm employed for core boosting in the subsequent MO-MaxSAT search. More specifically, both the implementation of OLL and our implementations of P-MINIMAL, BIOPTSAT, and LOWERBOUND realize their PB constraints using (generalized) totalizers [6,30,39]. For a PB expression O, a totalizer realizes $\text{CNF}(O \leq B)$ by first partitioning O into subsets of size 1 and then iteratively merging partitions by adding extra clauses and variables that count the sum of coefficients of the terms in the partitions to be merged. The merging stops when there is a single partition left, at which point the new variables obtained in the last step correspond to $\langle O \leq k \rangle$ for all $k = 1, \ldots, B$. For an alternative view, the structure of $\text{CNF}(O \leq B)$ created with a totalizer encoding can be visualized as a binary tree. The leaves of the tree correspond to the terms in O. Each internal node corresponds to new variables that count the sum of weights of the terms in the leaves of the subtree rooted at that node set to 1 by satisfying assignments.

When building a totalizer over the reformulated objective O^{ref} obtained after core boosting, our implementation makes use of the fact that some variables already correspond to the roots of other totalizers introduced by OLL. Instead of treating those as leaves in the new totalizer, we instead treat them as internal nodes, thus avoiding redundancy in the encoding which would be incurred by "recounting" the counting variables introduced by OLL in the pseudo-Boolean constraint over O^{ref} used by P-MINIMAL, BIOPTSAT, or LOWERBOUND.

3.4   Core Boosting and MO-MaxSAT Solver Interactions

Finally, in addition to decreasing the range of objectives that algorithms need to search over, we identify two further interactions between core boosting and MO-MaxSAT algorithms.

The first relates to the number of clauses in a PB constraint $\text{CNF}(O \leq B)$ built over an objective $O = \sum_i (c_i \cdot l_i) + o$ by MO-MaxSAT algorithms such as P-MINIMAL, BIOPTSAT, and LOWERBOUND. For CNF encodings used in practice—including the totalizer we use—the number of clauses in $\text{CNF}(O \leq B)$ depends on B and either on the sum of coefficients or the number of unique sums that can be obtained from the coefficients [18,30]. The effect of core boosting on these properties will depend on the specific instance being solved. Due to the properties of typical instantiations of REFORMULATE (recall Sect. 3.2), core boosting is guaranteed to decrease the sum of coefficients of each objective. In contrast (recall Sect. 3.3), the effect of core boosting on the number of variables in the objectives and the subset sums that can be obtained from them will depend on the coefficients of the variables in the extracted cores. Finally, another important contrast between single and multi-objective core boosting is that the intermediate solutions obtained during invocations of core-guided search on the separate objectives can not be used as global upper bound for the objectives by the MO-MaxSAT algorithm. Thus, we expect the effect of core boosting on the number of clauses introduced by the subsequent MO-MaxSAT algorithms to be more limited than in the single-objective case.

As a second interaction, core boosting can alter the structure of the PB constraints built over the reformulated objective by the MO-MaxSAT algorithms. In the case of totalizers this structure is defined by the ordering of the leaves in the totalizer tree and the structure of the tree itself. It is well-known that the structure of PB encodings employed can have a significant impact on the performance of constraint optimization algorithms [3,4]. While fundamental understanding on exactly how the structure of a totalizer affects the performance of constraint optimization algorithms is lacking, conventional wisdom based on empirical evaluation suggests that it is beneficial to place interrelated variables "close" in the tree. Core boosting achieves an approximation of this in the core-guided phase where totalizer tree substructures are built that have variables appearing together in cores as leaves.

4 Empirical Evaluation

We empirically evaluate the impact of core boosting. We integrated core boosting into three recent MO-MaxSAT algorithms, P-MINIMAL [31,44], BIOPTSAT [28], and LOWERBOUND [15] using their implementations in the MO-MaxSAT solver Scuttle [27]. The implementation and benchmarks used in our evaluation, as well as full empirical data, are available in open source (https://bitbucket.org/coreo-group/scuttle). All experiments reported on were run on 2.40-GHz Intel Xeon Gold 6148 CPUs with 381-GB RAM in RHEL under a 1.5-h per-instance time and 32-GB memory limit. Whenever core boosting was applied, the reported runtimes include the time spent in both core boosting and the MO-MaxSAT solver.

We use benchmark instances from seven domains from earlier evaluations of MO-MaxSAT solvers: multi-objective set covering with fixed set cardinality (set-cover-sc) and fixed element probability (set-cover-ep) [28], learning interpretable decision rules (lidr) [33], the flying tourist problem (ftp) [35], package upgradeability (packup) [29], staff shift scheduling (shiftdesign) [38], and satellite photograph scheduling (spot5) [24]. Set-cover-sc and set-cover-ep, instances with two objectives were obtained from [28] and instances with 3–5 objectives were generated similarly following [28]. The lidr instances contain 2 objectives and were also obtained from [28]. The ftp instances containing 2 objectives were obtained from [15] as instances with pseudo-Boolean constraints and encoded with the (generalized) totalizer encoding [6,30]. Package upgradeability instances were obtained from Mancoosi International Solver Competition of years 2011 and 2012 (https://www.mancoosi.org/misc/), and encoded with PackUp [29] with all combinations of 2–5 of the 5 minimization objectives. The shiftdesign and spot5 benchmarks were obtained from MaxSAT Lib (https://www.cs.toronto.edu/maxsat-lib/) and the single objective deconstructed according to [42], resulting in 3 objectives for shiftdesign and 2 for spot5. The packup, shiftdesign, and lidr families have unit coefficients in all objectives, i.e., are unweighted in MaxSAT terminology. For a balanced and meaningful benchmark set, we randomly sampled instances from each benchmark family, discarding instances that were solved in less than five seconds by P-MINIMAL without core boosting, until we obtained 20 instances per number of objectives and benchmark family.

4.1   Impact of Core Boosting on Solver Performance

We turn to the results of the evaluation. Since BIOPTSAT is specific to bi-objective problems, we report on its performance solely on the 120 instances with two objectives. Out of the 20 shiftdesign instances, all configurations of P-MINIMAL and LOWERBOUND solved exactly 1 instance. We therefore exclude shiftdesign from the reported results.

Table 1 shows the number of solved instances and the cumulative runtime (divided by 10^3 seconds) over the solved instances per benchmark family for P-MINIMAL, BIOPTSAT, and LOWERBOUND with and without core boosting. Especially on the set covering and spot5 families, all algorithms benefit greatly' from core boosting. The core-boosted configurations often solve more than twice as many instances as the variant without core boosting. With core boosting, all solvers solve more set-cover-ep benchmarks in less cumulative runtime than without core boosting.

Figure 4 shows a per-instance runtime comparison of P-MINIMAL, BIOPT-SAT, and LOWERBOUND, respectively, with and without core boosting. It can be seen that many of the spot5 instances that could not be solver without core boosting within the 1.5-hour time limit become trivial to solve after core boosting: five spot5 instances that P-MINIMAL does not solve without core boosting are solved in under 5 s after core boosting. On the other benchmark families, core boosting both allows for solving more instances and also drastically reduces solving times.

The time spent in core boosting is for a great majority of the benchmark instances negligible compared to time spent in the MO-MaxSAT solvers: only for

Table 1. Number of solved instances (#) and cumulative runtime over solved instances in 10^3 s for each algorithm with and without core boosting (CB).

Algorithm	CB	set-cover-sc		set-cover-ep		packup		lidr		ftp		spot5	
		#	$\sum t$	#	$\sum t$	#	$\sum t$	#	$\sum t$	#	$\sum t$	#	$\sum t$
P-MINIMAL	no	14	22.74	34	26.82	13	10.10	**7**	7.97	**6**	3.53	2	0.33
	yes	**34**	37.62	**39**	20.53	13	**7.14**	6	4.01	5	1.35	**13**	1.27
BIOPTSAT	no	8	4.84	18	8.09	8	0.65	6	4.61	**7**	8.56	2	0.35
	yes	**16**	8.37	**19**	3.76	8	**0.24**	6	**2.44**	5	1.32	**13**	1.07
LOWERBOUND	no	5	1.97	15	15.01	8	0.44	5	6.61	**6**	3.16	2	0.69
	yes	**13**	3.55	**18**	6.86	8	**0.22**	5	**5.04**	5	1.39	**13**	6.79

10 (resp. 9) instances more than 5% of the runtime was spent in core boosting in conjunction with *P*-MINIMAL (resp., BIOPTSAT or LOWERBOUND). Core boosting timed out on only a single benchmark instance that was solved without core boosting.

As core boosting can be viewed as preprocessing, we also compare its impact on solver runtimes to the impact of the recently-proposed MO-MaxSAT preprocessor MaxPre 2.1 [27] implementing liftings of SAT and MaxSAT preprocessing techniques to MO-MaxSAT. Table 2 shows the effect of core boosting and MaxPre on the number of instances solved by *P*-MINIMAL, BIOPTSAT, and LOWERBOUND in terms of the change in number of solved instances. Overall, the positive impact of core boosting is more significant than that of MaxPre. However, as MaxPre has a somewhat more positive impact on ftp and lidr families, an interesting direction for further work would be to study how to interleave core boosting and the various MaxPre preprocessing techniques for maximal positive overall impact on runtimes.

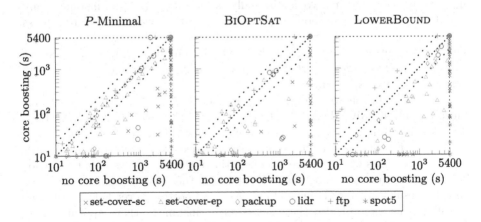

Fig. 4. Comparison of the per-instance CPU time of the *P*-MINIMAL (left), BIOPTSAT (middle), and LOWERBOUND (right) algorithms with and without core boosting.

Table 2. Change in number of solved instances ($\Delta\#$) through core boosting (CB) and preprocessing with MaxPre.

Algorithm	Prepro.	set-cover-sc	set-cover-ep	packup	lidr	ftp	spot5
		$\Delta\#$	$\Delta\#$	$\Delta\#$	$\Delta\#$	$\Delta\#$	$\Delta\#$
P-MINIMAL	CB	**+20**	**+5**	±0	−1	−1	**+11**
	MaxPre	+1	−1	−1	±0	+3	+1
BIOPTSAT	CB	**+8**	**+1**	±0	±0	−2	**+11**
	MaxPre	±0	±0	±0	±0	+2	+1
LOWERBOUND	CB	**+16**	**+6**	+1	±0	−1	**+11**
	MaxPre	+1	±0	−1	+1	+1	±0

4.2   Impact of Core Boosting on Search Space and Instance Size

Finally, we analyze the effects of core boosting on the search space and constraint encodings during search, and how these relate to changes in solving time. Due to space constraints, we focus on presenting results for P-MINIMAL; the data for BIOPTSAT and LOWERBOUND shows the same trends.

Figure 5 (left) relates the impact of core boosting on solver performance with reduction of search space achieved by core boosting. The change in search space (on the x-axis) is measured as the (hyper)volume of the search space after core boosting relative to the original volume, i.e., a value of 50% represents that the search space volume was halved with 100% representing that core boosting has no effect. In detail, the measure is $\frac{V(\mathcal{O}^{\mathrm{cb}})}{V(\mathcal{O})} \cdot 100$, where $V(\mathcal{O}) = \prod_{O \in \mathcal{O}} \sum_{l_i \in \mathrm{LITS}(O)} c_i$ is the search space volume of a given set of objectives, i.e., the product of the objective coefficient sums. The impact of core boosting on solver performance is measured as $\frac{t_{\mathrm{no\ cb}} - t_{\mathrm{cb}}}{t_{\mathrm{no\ cb}} + t_{\mathrm{cb}}}$, with $t_{\mathrm{(no)\ cb}}$ denoting solving time of P-MINIMAL with and without core boosting. We additionally assign value 1 (-1) for instances only solved with (without) core boosting. Positive (negative) values therefore express a positive (negative) impact of core boosting in terms of decreased solving time, with 0 representing no impact. We observe that core boosting has the strongest positive impact on solver performance on those instances that it significantly reduces the search space of.

Figure 5 (right) shows the combined number of clauses in the PB constraint encodings in the MO-MaxSAT solver with and without core boosting. Here clauses were counted at beginning of search based on the same initial solution to eliminate differences due to diverging search trajectories. For most benchmark families—esp. ones with unit coefficients in the objectives—the size of the encodings decreases due to core boosting. However, on the set covering instances core boosting results in larger PB encodings. As core boosting nevertheless decreases overall solving time of also these set covering instances, there appears to be no clear correlation between the changes in encoding sizes and solving times in general.

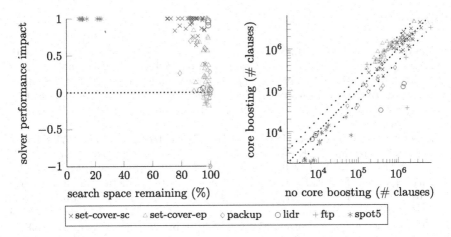

Fig. 5. Left: relating the impact of core boosting on solver performance with reduction of search space achieved (P-Minimal). Right: number of clauses in all objective encodings with and without core boosting.

As a further remark, we also experimented with resetting the internal SAT solver between core boosting and invocation of P-Minimal to check whether keeping the SAT solver state (learned clauses, variable activities, and polarities) can have an effect on overall runtimes. We observed no meaningful difference between resetting the SAT solver and keeping it alive throughout.

5   Conclusions

We proposed core boosting as an MO-MaxSAT reformulation technique that maintains all Pareto-optimal solutions. Core boosting increases the objective offsets of an MO-MaxSAT instance to match the ideal point without removing any Pareto-optimal solutions through reformulating the MO-MaxSAT instance. This results in a more restricted search space for a subsequently called MO-MaxSAT solver. Through tight integration into SAT-based MO-MaxSAT solvers, our empirical evaluation suggests that core boosting often has a significant positive impact on the runtimes of recently-proposed MO-MaxSAT algorithms, allows for solving more instances, and is more impactful than present MaxSAT-based MO-MaxSAT preprocessing techniques. The adaptation of core boosting to multi-objective generalizations of core-guided optimization algorithms proposed beyond MaxSAT, including CP and pseudo-Boolean optimization, is an interesting direction for further work.

Acknowledgments. Work financially supported by Research Council of Finland (grants 342145, 356046). The authors thank the Finnish Computing Competence Infrastructure for computational and data storage resources.

Disclosure of Interests. The authors have no competing interests to declare.

References

1. Andres, B., Kaufmann, B., Matheis, O., Schaub, T.: Unsatisfiability-based optimization in clasp. In: Dovier, A., Costa, V.S. (eds.) Technical Communications of the 28th International Conference on Logic Programming, ICLP 2012, 4–8 September 2012, Budapest, Hungary. LIPIcs, vol. 17, pp. 211–221. Schloss Dagstuhl—Leibniz-Zentrum für Informatik (2012). https://doi.org/10.4230/LIPICS.ICLP.2012.211
2. Ansótegui, C., Bonet, M.L., Levy, J.: Solving (weighted) partial MaxSAT through satisfiability testing. In: Kullmann, O. (ed.) SAT 2009. LNCS, vol. 5584, pp. 427–440. Springer, Heidelberg (2009). https://doi.org/10.1007/978-3-642-02777-2_39
3. Ansótegui, C., Didier, F., Gabàs, J.: Exploiting the structure of unsatisfiable cores in MaxSAT. In: Yang, Q., Wooldridge, M.J. (eds.) Proceedings of the Twenty-Fourth International Joint Conference on Artificial Intelligence, IJCAI 2015, Buenos Aires, Argentina, 25–31 July 2015, pp. 283–289. AAAI Press (2015). http://ijcai.org/Abstract/15/046
4. Ansótegui, C., Gabàs, J.: WPM3: an (in)complete algorithm for weighted partial MaxSAT. Artif. Intell. **250**, 37–57 (2017). https://doi.org/10.1016/J.ARTINT.2017.05.003
5. Bacchus, F., Järvisalo, M., Martins, R.: Maximum satisfiability. In: Biere, A., Heule, M., van Maaren, H., Walsh, T. (eds.) Handbook of Satisfiability—Second Edition, Frontiers in Artificial Intelligence and Applications, vol. 336, pp. 929–991. IOS Press (2021). https://doi.org/10.3233/FAIA201008
6. Bailleux, O., Boufkhad, Y.: Efficient CNF encoding of Boolean cardinality constraints. In: Rossi, F. (ed.) CP 2003. LNCS, vol. 2833, pp. 108–122. Springer, Heidelberg (2003). https://doi.org/10.1007/978-3-540-45193-8_8
7. Belov, A., Morgado, A., Marques-Silva, J.: SAT-based preprocessing for MaxSAT. In: McMillan, K., Middeldorp, A., Voronkov, A. (eds.) LPAR 2013. LNCS, vol. 8312, pp. 96–111. Springer, Heidelberg (2013). https://doi.org/10.1007/978-3-642-45221-5_7
8. Berg, J., Demirović, E., Stuckey, P.J.: Core-boosted linear search for incomplete MaxSAT. In: Rousseau, L.-M., Stergiou, K. (eds.) CPAIOR 2019. LNCS, vol. 11494, pp. 39–56. Springer, Cham (2019). https://doi.org/10.1007/978-3-030-19212-9_3
9. Berg, J., Järvisalo, M.: Weight-aware core extraction in SAT-based MaxSAT solving. In: Beck, J.C. (ed.) CP 2017. LNCS, vol. 10416, pp. 652–670. Springer, Cham (2017). https://doi.org/10.1007/978-3-319-66158-2_42
10. Berg, J., Saikko, P., Järvisalo, M.: Subsumed label elimination for maximum satisfiability. In: Kaminka, G.A., et al. (eds.) ECAI 2016—22nd European Conference on Artificial Intelligence, 29 August–2 September 2016, The Hague, The Netherlands—Including Prestigious Applications of Artificial Intelligence (PAIS 2016). Frontiers in Artificial Intelligence and Applications, vol. 285, pp. 630–638. IOS Press (2016). https://doi.org/10.3233/978-1-61499-672-9-630
11. Le Berre, D., Parrain, A.: The Sat4j library, release 2.2. J. Satisf. Boolean Model. Comput. **7**(2–3), 59–64 (2010). https://doi.org/10.3233/SAT190075
12. Biere, A., Heule, M., van Maaren, H., Walsh, T. (eds.): Handbook of Satisfiability—Second Edition. Frontiers in Artificial Intelligence and Applications, vol. 336. IOS Press (2021). https://doi.org/10.3233/FAIA336
13. Biere, A., Järvisalo, M., Kiesl, B.: Preprocessing in SAT solving. In: Biere, A., Heule, M., van Maaren, H., Walsh, T. (eds.) Handbook of Satisfiability—Second

Edition. Frontiers in Artificial Intelligence and Applications, vol. 336, pp. 391–435. IOS Press (2021). https://doi.org/10.3233/FAIA200992

14. Cabral, M., Janota, M., Manquinho, V.M.: SAT-based leximax optimisation algorithms. In: Meel, K.S., Strichman, O. (eds.) 25th International Conference on Theory and Applications of Satisfiability Testing, SAT 2022, 2–5 August 2022, Haifa, Israel. LIPIcs, vol. 236, pp. 29:1–29:19. Schloss Dagstuhl—Leibniz-Zentrum für Informatik (2022). https://doi.org/10.4230/LIPICS.SAT.2022.29

15. Cortes, J., Lynce, I., Manquinho, V.M.: New core-guided and hitting set algorithms for multi-objective combinatorial optimization. In: Sankaranarayanan, S., Sharygina, N. (eds.) TACAS 2023, Part II. LNCS, vol. 13994, pp. 55–73. Springer, Cham (2023). https://doi.org/10.1007/978-3-031-30820-8_7

16. Devriendt, J., Gocht, S., Demirovic, E., Nordström, J., Stuckey, P.J.: Cutting to the core of pseudo-Boolean optimization: combining core-guided search with cutting planes reasoning. In: Thirty-Fifth AAAI Conference on Artificial Intelligence, AAAI 2021, Thirty-Third Conference on Innovative Applications of Artificial Intelligence, IAAI 2021, The Eleventh Symposium on Educational Advances in Artificial Intelligence, EAAI 2021, Virtual Event, 2–9 February 2021, pp. 3750–3758. AAAI Press (2021). https://doi.org/10.1609/AAAI.V35I5.16492

17. Eén, N., Sörensson, N.: Temporal induction by incremental SAT solving. Electron. Notes Theor. Comput. Sci. 89(4), 543–560 (2003). https://doi.org/10.1016/S1571-0661(05)82542-3

18. Eén, N., Sörensson, N.: Translating pseudo-Boolean constraints into SAT. J. Satisf. Boolean Model. Comput. 2(1–4), 1–26 (2006). https://doi.org/10.3233/SAT190014

19. Ehrgott, M.: Multicriteria Optimization, 2nd edn. Springer, Cham (2005). https://doi.org/10.1007/3-540-27659-9

20. Ehrgott, M., Gandibleux, X.: Bound sets for biobjective combinatorial optimization problems. Comput. Oper. Res. 34(9), 2674–2694 (2007). https://doi.org/10.1016/J.COR.2005.10.003

21. Fu, Z., Malik, S.: On solving the partial MAX-SAT problem. In: Biere, A., Gomes, C.P. (eds.) SAT 2006. LNCS, vol. 4121, pp. 252–265. Springer, Heidelberg (2006). https://doi.org/10.1007/11814948_25

22. Gange, G., Berg, J., Demirović, E., Stuckey, P.J.: Core-guided and core-boosted search for CP. In: Hebrard, E., Musliu, N. (eds.) CPAIOR 2020. LNCS, vol. 12296, pp. 205–221. Springer, Cham (2020). https://doi.org/10.1007/978-3-030-58942-4_14

23. Guerreiro, A.P., et al.: Exact and approximate determination of the Pareto front using minimal correction subsets. Comput. Oper. Res. 153, 106153 (2023). https://doi.org/10.1016/J.COR.2023.106153

24. Heras, F., Larrosa, J., de Givry, S., Schiex, T.: 2006 and 2007 Max-SAT Evaluations: Contributed instances. J. Satisf. Boolean Model. Comput. 4(2–4), 239–250 (2008). https://doi.org/10.3233/SAT190046

25. Ignatiev, A., Morgado, A., Marques-Silva, J.: RC2: an efficient MaxSAT solver. J. Satisf. Boolean Model. Comput. 11(1), 53–64 (2019). https://doi.org/10.3233/SAT190116

26. Ihalainen, H., Berg, J., Järvisalo, M.: Refined core relaxation for core-guided MaxSAT solving. In: Michel, L.D. (ed.) 27th International Conference on Principles and Practice of Constraint Programming, CP 2021, Montpellier, France (Virtual Conference), 25–29 October 2021. LIPIcs, vol. 210, pp. 28:1–28:19. Schloss Dagstuhl—Leibniz-Zentrum für Informatik (2021). https://doi.org/10.4230/LIPICS.CP.2021.28

27. Jabs, C., Berg, J., Ihalainen, H., Järvisalo, M.: Preprocessing in SAT-based multi-objective combinatorial optimization. In: Yap, R.H.C. (ed.) 29th International Conference on Principles and Practice of Constraint Programming, CP 2023, 27–31 August 2023, Toronto, Canada. LIPIcs, vol. 280, pp. 18:1–18:20. Schloss Dagstuhl—Leibniz-Zentrum für Informatik (2023). https://doi.org/10.4230/LIPICS.CP.2023.18

28. Jabs, C., Berg, J., Niskanen, A., Järvisalo, M.: MaxSAT-based bi-objective Boolean optimization. In: Meel, K.S., Strichman, O. (eds.) 25th International Conference on Theory and Applications of Satisfiability Testing, SAT 2022, 2–5 August 2022, Haifa, Israel. LIPIcs, vol. 236, pp. 12:1–12:23. Schloss Dagstuhl—Leibniz-Zentrum für Informatik (2022). https://doi.org/10.4230/LIPICS.SAT.2022.12

29. Janota, M., Lynce, I., Manquinho, V.M., Marques-Silva, J.: PackUp: tools for package upgradability solving. J. Satisf. Boolean Model. Comput. 8(1/2), 89–94 (2012). https://doi.org/10.3233/SAT190090

30. Joshi, S., Martins, R., Manquinho, V.: Generalized totalizer encoding for pseudo-Boolean constraints. In: Pesant, G. (ed.) CP 2015. LNCS, vol. 9255, pp. 200–209. Springer, Cham (2015). https://doi.org/10.1007/978-3-319-23219-5_15

31. Koshimura, M., Nabeshima, H., Fujita, H., Hasegawa, R.: Minimal model generation with respect to an atom set. In: Peltier, N., Sofronie-Stokkermans, V. (eds.) Proceedings of the 7th International Workshop on First-Order Theorem Proving, FTP 2009, Oslo, Norway, 6–7 July 2009. CEUR Workshop Proceedings, vol. 556. CEUR-WS.org (2009). https://ceur-ws.org/Vol-556/paper06.pdf

32. Koshimura, M., Zhang, T., Fujita, H., Hasegawa, R.: QMaxSAT: a partial MaxSAT solver. J. Satisf. Boolean Model. Comput. 8(1/2), 95–100 (2012). https://doi.org/10.3233/SAT190091

33. Maliotov, D., Meel, K.S.: MLIC: a MaxSAT-based framework for learning interpretable classification rules. In: Hooker, J. (ed.) CP 2018. LNCS, vol. 11008, pp. 312–327. Springer, Cham (2018). https://doi.org/10.1007/978-3-319-98334-9_21

34. Marler, R., Arora, J.: Survey of multi-objective optimization methods for engineering. Struct. Multidisc. Optim. 26, 369–395 (2004). https://doi.org/10.1007/s00158-003-0368-6

35. Marques, R., Russo, L.M.S., Roma, N.: Flying tourist problem: flight time and cost minimization in complex routes. Expert Syst. Appl. 130, 172–187 (2019). https://doi.org/10.1016/J.ESWA.2019.04.024

36. Marques-Silva, J., Lynce, I., Malik, S.: Conflict-driven clause learning SAT solvers. In: Biere, A., Heule, M., van Maaren, H., Walsh, T. (eds.) Handbook of Satisfiability—Second Edition. Frontiers in Artificial Intelligence and Applications, vol. 336, pp. 133–182. IOS Press (2021). https://doi.org/10.3233/FAIA200987

37. Marques-Silva, J., Planes, J.: On using unsatisfiability for solving maximum satisfiability. Computing Research Repository abs/0712.1097 (2007). http://arxiv.org/abs/0712.1097

38. Martins, R.: ASP to MaxSAT: metro, ShiftDesign, TimeTabling and BioRepair. In: Ansotegui, C., Bacchus, F., Järvislo, M., Martins, R. (eds.) MaxSAT Evaluation 2017: Solver and Benchmark Descriptions. Department of Computer Science Series of Publications B, vol. B-2017-2, p. 27. University of Helsinki (2017). http://hdl.handle.net/10138/228949

39. Martins, R., Joshi, S., Manquinho, V., Lynce, I.: Incremental cardinality constraints for MaxSAT. In: O'Sullivan, B. (ed.) CP 2014. LNCS, vol. 8656, pp. 531–548. Springer, Cham (2014). https://doi.org/10.1007/978-3-319-10428-7_39

40. Morgado, A., Dodaro, C., Marques-Silva, J.: Core-guided MaxSAT with soft cardinality constraints. In: O'Sullivan, B. (ed.) CP 2014. LNCS, vol. 8656, pp. 564–573. Springer, Cham (2014). https://doi.org/10.1007/978-3-319-10428-7_41

41. Narodytska, N., Bacchus, F.: Maximum satisfiability using core-guided MaxSAT resolution. In: Brodley, C.E., Stone, P. (eds.) Proceedings of the Twenty-Eighth AAAI Conference on Artificial Intelligence, 27–31 July 2014, Québec City, Québec, Canada, pp. 2717–2723. AAAI Press (2014). https://doi.org/10.1609/AAAI.V28I1.9124

42. Paxian, T., Raiola, P., Becker, B.: On preprocessing for weighted MaxSAT. In: Henglein, F., Shoham, S., Vizel, Y. (eds.) VMCAI 2021. LNCS, vol. 12597, pp. 556–577. Springer, Cham (2021). https://doi.org/10.1007/978-3-030-67067-2_25

43. Piotrów, M.: UWrMaxSat: Efficient solver for MaxSAT and pseudo-Boolean problems. In: 32nd IEEE International Conference on Tools with Artificial Intelligence, ICTAI 2020, Baltimore, MD, USA, 9–11 November 2020, pp. 132–136. IEEE (2020). https://doi.org/10.1109/ICTAI50040.2020.00031

44. Soh, T., Banbara, M., Tamura, N., Le Berre, D.: Solving multiobjective discrete optimization problems with propositional minimal model generation. In: Beck, J.C. (ed.) CP 2017. LNCS, vol. 10416, pp. 596–614. Springer, Cham (2017). https://doi.org/10.1007/978-3-319-66158-2_38

45. Terra-Neves, M., Lynce, I., Manquinho, V.M.: Multi-objective optimization through Pareto minimal correction subsets. In: Lang, J. (ed.) Proceedings of the Twenty-Seventh International Joint Conference on Artificial Intelligence, IJCAI 2018, 13–19 July 2018, Stockholm, Sweden, pp. 5379–5383. ijcai.org (2018). https://doi.org/10.24963/IJCAI.2018/757

46. Paxian, T., Reimer, S., Becker, B.: Dynamic polynomial watchdog encoding for solving weighted MaxSAT. In: Beyersdorff, O., Wintersteiger, C.M. (eds.) SAT 2018. LNCS, vol. 10929, pp. 37–53. Springer, Cham (2018). https://doi.org/10.1007/978-3-319-94144-8_3

Fair Minimum Representation Clustering

Connor Lawless[✉] and Oktay Günlük

Operations Research and Information Engineering, Cornell University,
Ithaca, NY, USA
{cal379,ong5}@cornell.edu

Abstract. Clustering is an unsupervised learning task that aims to partition data into a set of clusters. In many applications, these clusters correspond to real-world constructs (e.g., electoral districts, playlists, TV channels) whose benefit can only be attained by groups when they reach a minimum level of representation (e.g., 50% to elect their desired candidate). In this paper, we study the k-means clustering problem with the additional constraint that each group (e.g., demographic group) must have a minumum level of representation in at least a given number of clusters. We formulate the problem through a mixed-integer optimization framework and present an alternating minimization algorithm, called MiniReL, that directly incorporates the fairness constraints. While incorporating the fairness criteria leads to an NP-Hard assignment problem within the algorithm, we present computational approaches that make the algorithm practical even for large datasets. Numerical results show that this approach can produce fair clusters with practically no increase in the clustering cost across standard benchmark datasets.

Keywords: Fair Clustering · Lloyds Algorithm · Integer Programming · K-means Clustering

1 Introduction

Clustering is an unsupervised learning task that aims to partition data points into sets of similar data points called clusters [39]. Clustering is widely used due to its broad applicability in domains such as customer segmentation [27], grouping content together for entertainment platforms [14], and identifying subgroups within a clinical study [38] amongst others. However the wide-spread application of clustering, and machine learning broadly, to human-centric applications has raised concerns about its disparate impact on minority groups and other vulnerable demographics. Motivated by a flurry of recent results highlighting bias in many automated decision making tasks such as facial recognition [10] and criminal justice [33], researchers have begun focusing on mechanisms to ensure machine learning algorithms are *fair* to all those affected. One of the challenges of fairness in an unsupervised learning context, compared to the supervised setting, is the lack of ground truth labels. Consequently, instead of enforcing approximately equal error rates across groups, fair clustering generally aims to ensure that composition of the clusters or their centers (for settings like k-means and k-medians clustering) represent all groups fairly.

A common approach to fair clustering is to require each cluster to have a fair proportion of itself represented by each group (i.e., via balance [12] or bounded representation [3]). This approach aims to balance the presence of each group in each cluster and therefore tries to spread each group uniformly across the clusters. Notice that this approach might not be desirable in settings where a group only gains a significant benefit from the cluster when they reach a minimum level of representation in that cluster. Consider the problem of clustering a set of media (e.g., songs, tv shows) into cohesive segments (e.g., playlists, channels). A natural fairness consideration in designing these segments would be to ensure that there is sufficient representation for different demographic groups. In these settings the benefit of the representation is only realized when a large percentage of the segment is associated with a demographic group (i.e., so that listeners can consistently watch or hear programming that speaks to them). This is even legislated in some countries, for example Canadian television channels are required to have at least 50% Canadian programming [9]. Note that, in this setting, spreading a minority demographic group across all clusters ensures that the demographic group will never have majority representation in any segment.

As another example, consider a simple voting system for a committee where the goal is to first cluster voters (e.g., employees, faculty) into different constituencies that can then elect a committee representative. Here, a proportionally fair clustering would assign a minority group that represents 30% of the vote equally among each cluster. However, the minority group only gets a benefit (i.e., the ability to elect a candidate of their choice) if they have at least 50% representation in the cluster. In this paper we introduce a new notion of fairness in clustering that addresses this issue. Specifically, we introduce *minimum representation fairness* (MR-fairness) which requires each group to have a certain number of clusters where they cross a given minimum representation threshold (i.e., 50% in the voting example).

Arguably the most popular algorithm for clustering is Lloyd's algorithm for k-means clustering [24]. It is an iterative algorithm that alternates between fixing cluster centers and assigning each point to the closest cluster center, and is guaranteed to return a locally optimal solution (i.e., no perturbation of the cluster centers around the solution leads to a better clustering cost). However, using this algorithm can lead to clusters that violate MR-fairness. As an example, consider the Adult dataset which contains census data for 48842 individuals in 1994 [15]. Suppose we wanted to cluster these individuals into groups that represent different stakeholder groups for a local committee and geographic contiguity was not a concern. A natural fairness criteria in this setting would be to ensure that there are a sufficient number of groups where minority groups (i.e., non-white in this dataset) have majority voting power. Despite the fact that only approximately 85% of the dataset is white, every cluster produced by Lloyd's algorithm is dominated by white members even when the number of clusters is as high as twenty. This highlights the need for a new approach to address fair minority representation.

In this paper we introduce a modified version of Lloyd's algorithm for k-means clustering that ensures MR-fairness, henceforth referred to as MINImum REpresentation fair Lloyd's algorithm (MiniReL for short). The key modification behind our approach is to replace the original greedy assignment step with a new optimization problem that finds the minimum cost assignment while ensuring fairness. In contrast to the standard clustering setting, we show that finding a minimum cost clustering that respects MR-fairness is NP-Hard even when the cluster centers are already fixed. We show that this optimization problem can be solved via integer programming (IP) in practice and introduce two computational approaches to improve the run-time including warm-starting our algorithm with the output of the standard Lloyd's algorithm and pre-assigning groups to specific clusters to help break symmetry and reduce the size of the IP that needs to be solved. We empirically show that our approach is able to construct fair clusters which have nearly the same clustering cost as those produced by Lloyd's algorithm.

1.1   Minimum Representation Fair Clustering Problem

The input to the standard clustering problem is a set of n m-dimensional data points $\mathcal{X} = \{x^i \in \mathbb{R}^m\}_{i=1}^n$. Note that assuming the data points to have real-valued features is not a restrictive assumption as categorical features can be converted to real-valued features through the one-hot encoding scheme. The goal of the clustering problem is to partition the data points into a set of K clusters $\mathcal{C} = \{C_1, \ldots, C_K\}$ in such a way that some measure of clustering cost is minimized. Let $\mathcal{K} = \{1, \ldots, K\}$ be the set of clusters. We focus on the popular k-means clustering metric which aims to find both a clustering and a set of centers $c_k \in \mathbb{R}^m$ for each cluster so as to minimize the sum of the squared distance between each point and its cluster center. Formally:

$$\min_{\mathcal{C}, c_1, \ldots, c_K} \sum_{k \in \mathcal{K}} \sum_{x^i \in C_k} \|x^i - c_k\|^2$$

where c_k denotes the center of each cluster C_k. In the absence of any additional constraints, for a given set of cluster centers the optimal cluster assignment is to simply assign each point to the nearest cluster center. Thus the problem can be viewed as an optimization over the choice of cluster centers.

In the fair clustering setting, each data point belongs to one or more groups $g \in \mathcal{G}$ (i.e., gender, race). Let X_g be the set of data points that belong to group g. Note that unlike other fair machine learning work, we do not assume that the groups form a partition of the data points. For instance, if the groups correspond to race and gender then a data point can belong to more than one group. The key intuition behind MR-fair clustering is that individuals belonging to a group gain material benefit only when they have a minimum level of representation in their cluster. We denote this minimum representation threshold $\alpha \in (0, 1]$, and define the associated notion of an α-represention as follows:

Definition 1 (α-representation). *A group $g \in \mathcal{G}$ is said to be α-represented in a cluster C_k if*

$$|C_k \cap X_g| \geq \alpha |C_k|$$

Note that α represents the minimum threshold needed for a given group to receive benefit from a cluster and thus depends on the application. For instance, most voting systems require majority representation (i.e., $\alpha = 0.5$). Our framework also allows for α to be group-dependent (i.e., α_g for each group g), however in most applications of interest α is a fixed threshold regardless of group. For a given clustering $\mathcal{C}$, group g, and α, let $\Lambda(\mathcal{C}, X_g, \alpha)$ be the number of clusters where group g has α-representation. In MR-fairness, each group g has a parameter β_g that specifies a minimum number of clusters where that group should have α-representation.

Definition 2 (Minimum representation fairness). *A given clustering $\mathcal{C} = \{C_1, \ldots, C_K\}$ is said to be an $(\alpha, \boldsymbol{\beta})$-minimum representation fair clustering if for every group $g \in \mathcal{G}$:*

$$\Lambda(\mathcal{C}, X_g, \alpha) \geq \beta_g$$

for a given $\boldsymbol{\beta} = \{\beta_g \in \mathbb{Z}^+\}_{g \in \mathcal{G}}$.

The definition of MR-fairness is flexible enough that the choice of $\boldsymbol{\beta}$ can (and should) be specialized to each application as well as the choice of α. Also note that up to $\lfloor 1/\alpha \rfloor$ groups can be α-represented in a cluster. In the remainder of the paper we explore two different natural choices for $\boldsymbol{\beta}$ that mirror fairness definitions in the fair classification literature. The first sets β_g to be equal for all groups:

$$\beta_g = \left\lfloor \frac{1}{|\mathcal{G}|} \lfloor \alpha^{-1} \rfloor K \right\rfloor \quad \forall g \in \mathcal{G}$$

which we denote **cluster statistical parity**. The second set choice sets β_g to be proportional to the size of the group:

$$\beta_g = \left\lfloor \frac{|X_g|}{n} \lfloor \alpha^{-1} \rfloor K \right\rfloor \quad \forall g \in \mathcal{G}$$

which we denote **cluster equality of opportunity**. Note that in both cases, β_g should be at most K (i.e., if α is very small, set $\beta_g = K$).

Combining the standard k-means clustering problem with MR-fairness criteria gives the following formal optimization problem:

Definition 3 (Minimum representation fair k-means problem). *For a given $\alpha \in (0, 1]$ and $\boldsymbol{\beta} = \{\beta_g \in \mathbb{Z}^+\}_{g \in \mathcal{G}}$, the minimum representation fair k-means problem is:*

$$\min_{\mathcal{C}, c_1, \ldots, c_K} \sum_{k \in \mathcal{K}} \sum_{x^i \in C_k} \|x^i - c_k\|^2 \quad \textbf{s.t.} \quad \Lambda(\mathcal{C}, X_g, \alpha) \geq \beta_g \ \forall g \in \mathcal{G}$$

The main difference between the fair and the standard versions of the k-means clustering problem is that greedily assigning data points to their closest cluster center may no longer be feasible for the fair version (i.e., assigning a data point to a farther cluster center may be necessary to meet the fairness criteria). Thus the problem can no longer be viewed simply as an optimization problem over cluster centers.

1.2 Related Work

A recent flurry of work in fair clustering has given rise to a number of different notions of fairness. One broad line of research, started by the seminal work of [12], puts constraints on the *proportion* of each cluster that comes from different groups. This can be in the form of balance [2,4,6–8,11,12,29–31,36,40] which ensures each group has relatively equal representation, or a group specific proportion such as the bounded representation criteria [2,3,5,6,17,22,23,25,36] or maximum fairness cost [11]. MR-fairness bares a resemblance to this line of work as it puts a constraint on the proportion of a group in a cluster, however instead of constraining a fixed proportion across all clusters it looks holistically across all clusters and ensures that threshold is met in a baseline number of clusters. Another line of work tries to minimize the *worst case average clustering cost* (i.e., k-means cost) over all the groups, called social fairness [1,19,20,32]. Most similar to our algorithmic approach is the Fair Lloyd algorithm introduced in [19] which studies social fairness. They also present a modified version of Lloyd's algorithm that converges to a local optimum. As a consequence of the social fairness criterion their approach requires a modified center computation step that can be done in polynomial time. MR-fairness, however, requires a modified cluster assignment step that is NP-hard which we solve via integer programming.

Most similar to MR-fairness is diversity-aware fairness introduced in [37] and the related notion of fair summarization [13,26,28]. These notions of fairness require that amongst all the cluster centers selected, a minimum number comes from each group. MR-fairness differs in that our criteria is not tied to the group membership of the cluster center selected but the proportion of each group in a given cluster. Our notion of fairness is more relevant in settings where the center cannot be prescribed directly, but is only a function of its composition (i.e., in voting where members of a 'cluster' elect an official).

1.3 Main Contributions

We summarize our main contributions as follows:

- We introduce a novel definition of fairness for clustering called MR-fairness, which requires that a specified number of clusters should have at least α percent members from a given group.
- We formulate the problem of finding a MR-fair k-means clustering in a mixed integer optimization framework, and introduce a new heuristic algorithm MiniReL, based on Lloyd's algorithm for clustering, to find a local optimum.

- We show that unlike other notions of proportional fairness, this problem can not be approximated by adjusting unfair cluster centers.
- We show that incorporating MR-fairness into Lloyd's algorithm leads to a NP-Hard sub-problem. To tackle this issue, we develop computational techniques that make the our approach tractable even for large datasets.
- We present numerical results to demonstrate that MiniReL is able to construct MR-fair clusterings with only a modest increase in run-time and with little to no increase in clustering cost compared to the standard k-means clustering algorithm.

The remainder of the paper is organized as follows. In Sect. 2 we present a mixed integer optimization formulation for the MR-fair clustering problem and introduce MiniReL. In Sect. 3 we introduce computational approaches to help our algorithm scale to large datasets. Finally Sect. 4 presents a numerical study of MiniReL compared to the standard k-means algorithm.

2   Mixed Integer Optimization Framework

We start by formulating the MR-fair k-means clustering problem as a mixed-integer program with a non-linear objective. We use binary variable z_{ik} to denote if data point x^i is assigned to cluster k, and variable $c_k \in \mathbb{R}^d$ to denote the center of cluster k. The binary variable y_{gk} indicates whether group g is α-represented in cluster k. We can now formulate the MR-fair clustering problem as follows:

$$\min \quad \sum_{x^i \in \mathcal{X}} \sum_{k \in \mathcal{K}} \|x^i - c_k\|_2^2 z_{ik} \tag{1}$$

$$\text{s.t.} \quad \sum_{k \in \mathcal{K}} z_{ik} = 1 \qquad \forall x^i \in \mathcal{X} \tag{2}$$

$$\sum_{x^i \in X_g} z_{ik} + M(1 - y_{gk}) \geq \alpha \sum_{x^i \in \mathcal{X}} z_{ik} \qquad \forall g \in \mathcal{G}, k \in \mathcal{K} \tag{3}$$

$$\sum_{k \in \mathcal{K}} y_{gk} \geq \beta_g \qquad \forall g \in \mathcal{G} \tag{4}$$

$$u \geq \sum_{x^i \in \mathcal{X}} z_{ik} \geq l \qquad \forall k \in \mathcal{K} \tag{5}$$

$$z_{ik}, y_{gk} \in \{0, 1\} \qquad \forall x^i \in \mathcal{X}, g \in \mathcal{G}, k \in \mathcal{K} \tag{6}$$

$$c_k \in \mathbb{R}^m \qquad \forall k \in \mathcal{K} \tag{7}$$

The objective (1) is to minimize the sum of the squared distances between the data points and the centers of the clusters they are assigned to. Constraint (2) ensures that each data point is assigned to exactly one cluster. Constraint (3) tracks whether a group g is α-represented in cluster k, and includes a big-M which can be set to αn. Finally, constraint (4) enforces that each group g is α-represented in at least β_g clusters. In many applications of interest, it might also be necessary to add a constraint on the size of the clusters. Constraint (5) captures this notion where l and u represent the lower and upper bounds on the cardinality of each cluster respectively. Note that in cases where the cardinality constraint is used, the big-M in constraint (3) can be reduced to αu. For all our experiments we set $l = 1$ to ensure that exactly k clusters are returned by the algorithm. Adding a lower bound also ensures that each group is α-represented in non-trivial clusters. Note that without this lower bound every group would be trivially α-represented in an empty cluster according to Definition 1.

To solve problem (1)–(7) in practice, we introduce a modified version of Lloyd's algorithm called MiniReL in Sect. 2.1. Incorporating MR-fairness into Lloyd's algorithm requires solving a fair assignment problem, which we discuss in Sect. 2.2. Finally in Sect. 2.3 we show that, unlike other notions of fairness in clustering, first solving the clustering problem without fairness constraints and then finding a fair assignment of the data points to these centers can lead to arbitrarily poor results under MR-fairness which further justifies the use of an iterative approach such as MiniReL.

2.1   MiniReL Algorithm for Fair Clustering

Solving the optimization problem outlined in the preceding section to optimality is computationally challenging as it is an integer optimization problem with a non-linear objective. To solve the problem in practice, we introduce a modified version of Lloyd's algorithm which we call *the Minimum Representation Fair Lloyd's Algorithm* (MiniReL) that alternates between adjusting cluster centers and fairly assigning data points to clusters to converge to a local optimum, see Algorithm 1.

Given a fixed set of cluster assignments (i.e., when variables z are fixed in (1)–(6)) the optimal choice of c_k is simply the mean value of data points assigned to C_k. For a given (fixed) set of cluster centers c_k we denote the problem (1)–(6) the *fair minimum representation assignment (FMRA) problem* which is a linear integer program. Note that if the FMRA problem is infeasible, it provides a certificate that no MR-fair clustering with the given α, β exists. We analyze the computational complexity of this problem in Sect. 2.2. While integer programs do not always scale well to large datasets, in our computational experiments we observed that FMRA can be solved to optimality in a reasonable amount of time even for datasets with tens of thousands of data points. In Sect. 3 we describe the computational techniques that help scale our algorithm.

A natural question is whether MiniReL converges to a locally optimal solution as Lloyd's algorithm does. When discussing local optimality, it is important to formally define the local neighborhood of a solution. In the absence of fairness constraints, data points must be assigned to the closest centers to minimize cost. Consequently, a clustering is locally optimal if perturbing the centers does not improve the clustering cost. In our setting, we define a local change as any perturbation to a cluster center, or an individual change to cluster assignment (i.e., moving a data point from one cluster to another). With this notion of a local neighborhood, the following result shows that the MiniReL converges to a local optimum in finite time. Note that while MiniReL converges to a local

Algorithm 1. Minimum Representation Fair Lloyd's Algorithm (MiniReL)

Input: Data $\mathcal{X}$, Number of clusters K, Fairness parameters β, α
Output: Cluster assignments $\{C_k\}_{k \in \mathcal{K}}$

1: Initialize $\{c_k\}_{k \in \mathcal{K}}$ (for example by uniformly at random sampling K data points).
2: **repeat**
3:   Solve fair assignment problem (1)–(6) for fixed cluster centers $\{c_k\}_{k \in \mathcal{K}}$ to get $\mathbf{z}^*$
4:   Compute the cost of solution *current_cost*
5:   **for** $k = 1, 2, \ldots, K$ **do**
6:     Set $N_k = \sum_{x^i \in \mathcal{X}} z_{ik}$, $c_k = \frac{1}{N_k} \sum_{x^i \in \mathcal{X}} z_{ik} x^i$.
7:   **end for**
8:   Compute the cost of solution *improved_cost*
9: **until** *improved_cost=current_cost*
10: Set $C_k = \{x^i \in \mathcal{X} : z_{ik} = 1\}$ $\forall k \in \mathcal{K}$
11: **return** $\{C_k\}_{k=1}^{K}$

optimum, the solution may be arbitrarily worse than the global optimum as is the case for Lloyd's algorithm.

Theorem 1. *MiniReL converges to a local optimum in finite time.*

Proof. We start by noting that given an assignment of data points to clusters, the optimal cluster centers for this assignment are precisely the ones computed as in Step 6 and any deviation from these centers increases the cost. Consequently, if *improved_cost = current_cost* in Step 9, then the cluster centers used in Step 3 and the ones computed in Step 6 (in the last iteration) must be identical. Therefore, if Algorithm 1 terminates, then the current assignment of the data points to the current cluster centers is optimal and the cost cannot improve by changing their assignment due to Step 3. In addition, Step 6 guarantees that perturbing cluster centers cannot not improve the cost for the current assignment.

It remains to show that the algorithm will terminate after a finite number of iterations. Similar to the proof for Lloyd's algorithm, we leverage the fact that there are only a finite number of partitions of the data points. By construction, the objective value decreases in each iteration of the algorithm, and thus can never cycle through any partition multiple times as for a given partition we use the optimal cluster centers when computing *improved_cost* in Step 8. Thus the algorithm can visit each partition at most once and thus must terminate in finite time. □

2.2 Fair Minimum Representation Assignment Problem

While the optimal assignment step in Lloyd's algorithm can be done in polynomial time, the following result shows that the FMRA problem is NP-Hard.

Theorem 2. *The fair minimum representation assignment problem is NP-Hard.*

Proof. The reduction is from the exact cover by 3 sets (X3C) problem, with $3q$ elements $\mathcal{U} = \{u_1, \ldots, u_{3q}\}$ and $t = q + r$ subsets $\mathcal{W} = \{W_1, \ldots, W_t\}$ where each subset $W_i \subseteq \mathcal{U}$ and $|W_i| = 3$. Recall that X3C is one of Karps 21 NP-Complete problems and its goal is to select a collection of subsets $\mathcal{W}^* \subseteq \mathcal{W}$ such that all elements of $\mathcal{U}$ are covered exactly once.

Given an instance of the X3C problem, we first construct an undirected bipartite graph $G = (\mathcal{U} \cup \mathcal{W}, E)$ which we then convert into a FMRA instance following a similar construction as [18]. Graph G has a node for each $u_i \in \mathcal{U}$ and a node w_j for each $W_j \in \mathcal{W}$. There is an edge $\{u_i, w_j\}$ between nodes u_i and w_j provided that $u_i \in W_j$. Note that with slight abuse of notation, we use $\mathcal{W}$ both for the subsets in the X3C instance and the nodes corresponding to them in the graph G.

Using graph G, we construct an instance of the FMRA problem with two groups by interpreting each vertex in the graph as a data point where points corresponding to vertices in $\mathcal{U}$ and $\mathcal{W}$ belong to separate groups. We set $K = t$ (i.e., one cluster for each subset), $\alpha = 0.75$, $\beta_W = t - q$ and $\beta_U = q$ (minimum number of α-represented clusters for each group). We place the (fixed) K centers at the data points associated with $\mathcal{W}$. The distances between pairs of data points are set to be the length of the shortest path distances between the corresponding vertices in the graph.

We now argue that the optimal solution to the FMRA instance has a cost of $3q$ if and only if there exists a feasible solution to the X3C instance. First note that any feasible solution to the FMRA instance has a cost of at least $3q$ as each point $u_i \in \mathcal{U}$ incurs a cost of at least 1 and there are $3q$ such points. Next, note that given a feasible solution $\mathcal{W}^* \subseteq \mathcal{W}$ to the X3C instance, we can construct a solution to the corresponding FMRA instance with cost $3q$ by creating one cluster for each $W_j \in \mathcal{W}^*$ in the solution. These q clusters contain w_j together with the 3 elements $u_i \in W_j$. The remaining $t - q$ clusters contain a single point w_j, one for each $W_j \in \mathcal{W} \setminus \mathcal{W}^*$. Note that these clusters satisfy the fairness constraints. Furthermore, letting the center of each cluster to be the w_j that it contains gives a cost of $3q$.

Finally we argue that if the FMRA instance has a cost of $3q$, then the X3C instance is feasible. If the solution to the FMRA has cost $3q$, then (i) each point $u_i \in \mathcal{U}$ must be assigned to a center $w_j \in \mathcal{W}$ such that $u_i \in W_j$, and, (ii) each point $w_j \in \mathcal{W}$ must be assigned to the center w_j and consequently, each cluster must contain exactly one $w_j \in \mathcal{W}$. In addition, the fairness constraint for the $\mathcal{W}$ group requires that at least $t - q$ clusters must have 75% of its points belonging to the $\mathcal{W}$ group. As there is exactly one $w_j \in \mathcal{W}$ in each cluster, this can only happen if at least $t - q$ clusters have no points from $\mathcal{U}$ group in them. Consequently all $3q$ points form the $\mathcal{U}$ group are assigned to at most q clusters. As all $u_i \in \mathcal{U}$ must be assigned to a center $w_j \in \mathcal{W}$ such that $u_i \in W_j$, and $|W_j| = 3$, the solution to the FMRA instance must have exactly $t - q$ clusters with no points from $\mathcal{U}$ group and q clusters with one w_j and three $u_i \in W_j$, which gives a feasible solution to the X3C instance. $\qquad\square$

2.3   A Natural Solution Approach and an Inapproximability Result

One natural approach [6,16,18] used for other notions of fairness in clustering is to first obtain cluster centers by solving the clustering problem without fairness constraints and then find a fair assignment of the data points to these centers. This has been shown to provide an overall approximation guarantee for the full optimization problem in some settings even when the centers are only approximately optimal for the clustering problem without fairness constraints [6,16,18]. Unfortunately, we show that in our MR-fairness setting this approach, which we call *one-shot fair adjustment*, can lead to arbitrarily bad solutions. Let z^*, c^* be the optimal solution to the MR-fair clustering problem. Let c^*_{UF} be the optimal centers for the clustering problem without fairness constraints and z^*_{FA} be the optimal (fair) assignment of data points to centers c^*_{UF}. Note both z^* and z^*_{FA} are feasible fair assignments. Let $COST(z,c) = \sum_{x^i \in \mathcal{X}} \sum_{k \in \mathcal{K}} \|x^i - c_k\|_2^2 z_{ik}$ be the objective for the clustering problem.

Theorem 3. *There does not exist a constant $M > 0$ such that:*

$$COST(z^*_{FA}, c^*_{UF}) \leq M \cdot COST(z^*, c^*)$$

In other words, fairly assigning data points to (approximately) optimal unfair centers can lead to arbitrarily bad performance relative to the optimal solution of the MR-fair clustering problem.

Proof. To prove the claim we will construct an instance in two dimensions with four data points and three groups (Red, Yellow, and Blue). The first two points are located at $(0,0)$ and belong to the red and yellow groups, respectively. The next two data points belong to the blue group and are located at $(\gamma, 0)$ and (γ, ϵ) for some $\gamma, \epsilon > 0$. For the fair clustering problem, we set $K = 3$, $\alpha > 0.5$, and $\beta_R = \beta_Y = \beta_B = 1$.

Note that any optimal solution to the clustering problem without fairness constraints will place the cluster centers c^*_{UF} at $(0,0), (\gamma, 0), (\gamma, \epsilon)$ with cost 0. It is straightforward to see that fairly assigning points to these fixed centers requires assigning either the red or yellow data point to the center at $(\gamma, 0)$ and assigning the blue data point at $(\gamma, 0)$ to the center at (γ, ϵ) yielding a cost of $\gamma^2 + \epsilon^2$. Now consider the optimal solution to the fair clustering problem which selects centers at $(0,0), (0,0), (\gamma, \epsilon/2)$. It assigns the red data point to the first center at $(0,0)$, the yellow data point to the second center at $(0,0)$, and both blue data points to the center at $(\gamma, \epsilon/2)$ which satisfies the fairness constraints and gives a total cost of $\epsilon^2/2$. The ratio of the costs of the two solutions is $\frac{\gamma^2 + \epsilon^2}{\epsilon^2/2}$ which can be made arbitrarily large by increasing γ, completing the proof. □

Note that using the simple example constructed above, one can argue that the cost of the optimal fair clustering can be arbitrarily larger than the cost of the optimal clustering without fairness. Also note that the above proof also shows that one can get an arbitrarily bad solution when the (unfair) cluster centers are chosen using a constant factor approximation algorithm.

3   Scaling MiniReL

In MiniReL, the computational bottleneck is solving the fair assignment problem which is an IP. To improve the run-time of MiniReL, we introduce two key computational approaches that reduce the number of fair assignment problems that need to be solved, and improve the speed at which they can be solved.

3.1   Warm-Starting MiniReL

To reduce the number of iterations needed to converge in MiniReL, we warm-start the initial cluster centers with the final centers of the unfair variants of Lloyd's algorithm. The key intuition behind this approach is that it allows us to leverage the polynomial time assignment problem for the majority of iterations, and only requires solving the fair assignment problem to adjust the locally optimal unfair solution to a fair one. To incorporate warm-starting into MiniReL, we replace step 1 in Algorithm 1 with centers generated from running Lloyd's algorithm. In practice we found this warm-starting procedure led to a 75% reduction on average in the number of MiniReL iterations compared to randomly initiated centers across benchmark datasets used in our numerical studies.

3.2   Pre-fixing Group Assignment

One computational shortcoming of the fair assignment IP model is the use of big-M constraints which are well known to lead to weak continuous relaxations and by extension longer computation times. These constraints are needed to track which groups are α-represented in which clusters. One approach to avoid these constraints is to simply pre-fix which groups need to have α-representation in which clusters in advance (i.e., akin to fixing the y_{gk} variables). This removes the need for the y variables and the associated big-M constraints and breaks symmetry in the IP (i.e., removes permutations of feasible cluster assignments), dramatically improving the computation time. In problems where only a single group can have α-represention in a cluster (i.e., a data point can only be part of one group and $\alpha > 0.5$), pre-fixing preserves an optimal solution to the full problem. However, in more complicated settings pre-fixing may remove all optimal solutions and becomes a heuristic for improving run-time. It is worth noting that the MiniReL algorithm is itself a heuristic, and thus the pre-fixing scheme has ambiguous effects on the cost of the solution as it may cause the algorithm to converge to a better local optimum.

To find a good pre-fixing of the y variables, we use the clusters and centers obtained from Lloyd's algorithm and formulate a small integer program. Note that this solution most likely does not satisfy the fairness constraints. For a given cluster k and group g, let $q_{kg} \geq 0$ be the additional number of points from group g needed to make this group α-represented in cluster k (i.e., smallest $q_{kg} \geq 0$ that satisfies $q_{kg} + |C_k \cap X_g| \geq \alpha(q_{kg} + |C_k|)$). Let $c_{(x)} = \text{argmin}_{c \in \{c_1, \dots, c_K\}} \{\|x - c\|_2^2\}$ denote the closest center for point $x \in \mathcal{X}$. To find a good pre-fixing, we estimate the (myopic) increase in cost m_{gk} to make group g α-represented in cluster k as follows:

$$m_{gk} = \min_{X \subset X_g \backslash C_K : |X| = q_{kg}} \sum_{x \in X} \left(\|x - c_k\|_2^2 - \|x - c_{(x)}\|_2^2 \right)$$

We can now formulate an IP to perform the pre-fix assignment as follows:

$$\min \quad \sum_{(g,k) \in \mathcal{W}} m_{gk} y_{gk} \tag{8}$$

$$\text{s.t.} \quad \sum_{k \in \mathcal{K} : (g,k) \in \mathcal{W}} y_{gk} \geq \beta_g \qquad \forall g \in \mathcal{G} \tag{9}$$

$$\sum_{g \in \mathcal{G}} y_{gk} \leq \left\lfloor \frac{1}{\alpha} \right\rfloor \qquad \forall k \in \mathcal{K} \tag{10}$$

$$y_{gk} \in \{0, 1\} \qquad \forall g \in \mathcal{G}, k \in \mathcal{K} \tag{11}$$

where $y_{g,k}$ is a binary variable indicating whether group g will be α-represented in cluster k. The objective (8) is a proxy for the cost of pre-fixing. Constraint (9) ensures enough clusters are allocated to each group to meet the MR-fairness constraint. Finally constraint (10) ensures no cluster is assigned more groups than can simultaneously have α-represention.

Note that the size of this formulation does not depend on the size of the dataset and it has only $|\mathcal{K}| \times |\mathcal{G}|$ variables. As mentioned earlier, when $\alpha > 0.5$ and the groups are disjoint (i.e., no group can be in multiple groups) then this pre-fixing scheme simply removes symmetry. In other cases, this operation is a heuristic that leads to substantial speedups to the overall solution time. In practice we found that solving this IP took under 1 s for all instances tested in this paper. The resulting speedup is very significant when compared to running MiniReL with the full IP model for fair assignment, at times leading to over a 400x speedup in some instances, as shown in Fig. 1 which compares the runtimes of the two approaches under cluster equality of opportunity. This speedup also comes at practically no increase in terms of clustering cost as shown in Fig. 2. Results for cluster statistical parity were similar and are excluded for brevity.

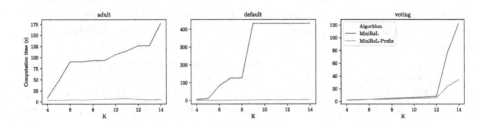

Fig. 1. Impact of pre-fixing on runtime of the MiniReL Algorithm.

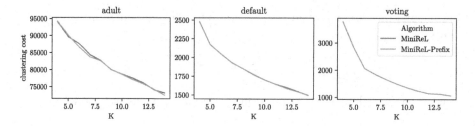

Fig. 2. Impact of pre-fixing on the clustering cost of MiniReL solution.

4   Numerical Results

To benchmark our approach, we evaluate it on three datasets that have been used in recent work in fair clustering: adult ($n = 48842$, $m = 14$) [15], default ($n = 30000$, $m = 24$) [15], and Brunswick County voting data ($n = 49190$, $m = 2$) [34] where n denotes the number of data points and m denotes the initial number of features. We pre-process the Brunswick County voting data by geocoding the raw addresses to latitude and longitude using the US census bureau geocoding tool and retain all data with a successful geocoding that belong to white and Black voters. For each dataset we use one sensitive feature to represent group membership - namely gender for both adult (67% Male, 33% Female) and default (60% Male, 40% Female), and race for voting (92% white, 8% Black). For all datasets we normalize all real-valued features to be between $[0, 1]$ and convert all categorical features to be real-valued via the one-hot encoding scheme. For datasets that were originally used for supervised learning, we remove the target variable and do not use the sensitive attribute as a feature for the clustering itself.

We compare MiniReL with the standard Lloyd's algorithm for k-means using the implementation available in scikit-learn [35] with a k-means++ initialization. We evaluate both algorithms with respect to cluster statistical parity (i.e., every group must be α-represented in the same number of clusters), and cluster equality of opportunity (i.e., every every group must be α-represented in a number of clusters proportional to the size of the group). For all k-means results we run Lloyd's algorithm with 100 different random seeds. We report the clustering with the lowest clustering cost (k-means), the fairest clustering with respect to cluster statistical parity (k-means-SP), and the fairest clustering with respect to cluster equality of opportunity (k-means-EqOp). We implemented MiniReL in Python with Gurobi 10.0 [21] for solving all IPs. All experiments were run on a personal computer with 16 GB of RAM and 2.7 GHz Quad-Core Intel Core i7 processor. To initialize the cluster centers, we use the warm start scheme outlined in Sect. 3.1. We ran MiniReL with β set for both cluster statistical parity (*MiniReL-SP*) and cluster equality of opportunity (*MiniReL-EqOp*). For the following experiments we set $\alpha = 0.51$ to represent majority representation in a cluster.

Figure 3 shows the maximum deviation from the fairness parameters β_g for the different algorithms (i.e., $\max_g \max(\beta_g - \Lambda(C, X_g, \alpha), 0)$). Across all three

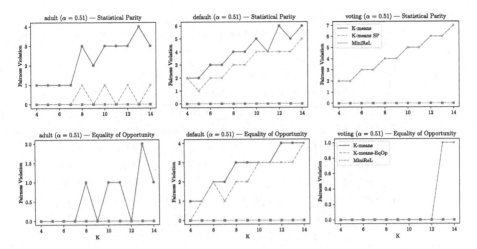

Fig. 3. Maximum deviation from β_g for Lloyd's algorithm and MiniReL under both notions of fairness.

datasets we can see that k-means can lead to outcomes that violate MR-fairness constraints significantly, and that selecting the fairest clustering has only marginal improvement. This is most stark in the default dataset where there is as much as an 11 cluster gap (6 cluster fairness violation) between the two groups despite having similar proportions in the dataset. In contrast, the MiniReL algorithm is able to generate fair clusters under both notions of fairness and for all three datasets. Table 1 shows the average computation time in seconds for Lloyd's algorithm and MiniReL under cluster statistical parity. We display results for the cluster statistical parity as it was the more computationally demanding setting requiring more MiniReL iterations than cluster equality of opportunity. As expected, the harder assignment problem in MiniReL leads to higher overall computation times when compared to the standard Lloyd's algorithm. However, MiniReL is still able to solve large problems in under 40 s demonstrating that the approach is of practical use.

One remaining question is whether fairness comes at the expense of the cost of the clustering. Figure 4 shows the k-means clustering cost for Lloyd's algorithm

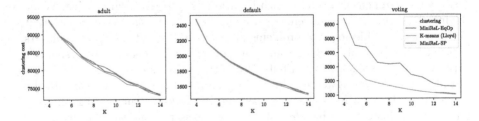

Fig. 4. k-means clustering cost of Lloyd's algorithm and MiniReL.

Table 1. Average computation time (standard deviation) in seconds over 10 random seeds. Final row includes average computation time over all random seeds and setting of K for each dataset.

K	adult		default		voting	
	Lloyd	MiniReL	Lloyd	MiniReL	Lloyd	MiniReL
4	0.6 (0.03)	13.4 (0.25)	0.3 (0.02)	10.8 (0.18)	0.8 (0.04)	2.1 (0.03)
5	0.7 (0.04)	14.3 (0.37)	0.4 (0.03)	11.2 (0.14)	0.7 (0.02)	2.4 (0.04)
6	0.8 (0.04)	15.8 (1.08)	0.5 (0.05)	12.0 (0.33)	0.7 (0.03)	3.4 (0.04)
7	0.9 (0.04)	15.8 (1.03)	0.6 (0.05)	12.3 (0.35)	1.1 (0.07)	3.5 (0.08)
8	1.0 (0.05)	17.2 (0.54)	0.8 (0.09)	13.3 (0.29)	1.0 (0.09)	3.8 (0.06)
9	1.1 (0.06)	17.8 (0.91)	0.8 (0.08)	13.5 (0.37)	1.1 (0.08)	4.5 (0.18)
10	1.3 (0.04)	19.1 (0.9)	0.9 (0.07)	14.8 (0.29)	1.4 (0.07)	5.0 (0.12)
11	1.4 (0.04)	19.6 (1.16)	1.0 (0.06)	14.4 (0.3)	1.5 (0.12)	5.4 (0.21)
12	1.4 (0.04)	21.4 (3.77)	1.1 (0.11)	15.6 (0.55)	1.7 (0.18)	6.1 (0.3)
13	1.5 (0.05)	20.9 (3.01)	1.3 (0.18)	16.0 (0.36)	2.2 (0.19)	24.3 (0.32)
14	1.7 (0.07)	22.8 (3.02)	1.5 (0.21)	16.7 (0.35)	2.4 (0.18)	34.7 (0.3)
15	1.8 (0.08)	21.4 (1.7)	1.5 (0.11)	17.5 (1.68)	2.5 (0.20)	38.1 (0.41)
mean	**1.2**	**18.3**	**0.9**	**14.0**	**1.3**	**8.7**

and MiniReL under both definitions of fairness. Although there is a small increase in the cost when using MiniReL the overall cost closely matches that of the standard k-means algorithm in 5 out of 6 instances, with the sole exception being the voting dataset under statistical parity, showing that we can gain fairness at practically no additional increase to clustering cost.

5   Conclusion

In this paper we introduce a novel definition of group fairness for clustering that requires each group to have a minimum level of representation in a specified number of clusters. This definition is a natural fit for a number of real world examples, such as voting and entertainment segmentation. To create fair clusters we introduce a modified version of Lloyd's algorithm called MiniReL that solves an assignment problem in each iteration via integer programming. While solving the integer program remains a computational bottleneck, we note that our approach is able to solve problems of practical interest, including datasets with tens of thousands of data points, and provides a mechanism to design fair clusters when Lloyd's algorithm fails. While this paper focused on the k-means setting, we also note that our alternating minimization approach can also be extended to the k-medians clustering setting. In the k-medians setting, the centers for each cluster must be selected from the dataset $\mathcal{X}$ and the clustering cost can be any distance metric. Extending our approach to the k-medians setting would require a revised center computation step that selects the best center from the input dataset. Implementing such an extension and evaluating its empirical performance remains a promising direction for future work.

References

1. Abbasi, M., Bhaskara, A., Venkatasubramanian, S.: Fair clustering via equitable group representations. In: Proceedings of the 2021 ACM Conference on Fairness, Accountability, and Transparency, pp. 504–514. ACM, New York (2021)
2. Ahmadian, S., et al.: Fair hierarchical clustering. In: Advances in Neural Information Processing Systems, vol. 33, pp. 21050–21060 (2020)
3. Ahmadian, S., Epasto, A., Kumar, R., Mahdian, M.: Clustering without over-representation. In: Proceedings of the 25th ACM SIGKDD International Conference on Knowledge Discovery & Data Mining, pp. 267–275. ACM, New York (2019)
4. Backurs, A., Indyk, P., Onak, K., Schieber, B., Vakilian, A., Wagner, T.: Scalable fair clustering. In: International Conference on Machine Learning, pp. 405–413. PMLR, Long Beach (2019)
5. Bandyapadhyay, S., Fomin, F.V., Simonov, K.: On coresets for fair clustering in metric and Euclidean spaces and their applications (2020). arXiv:2007.10137
6. Bera, S., Chakrabarty, D., Flores, N., Negahbani, M.: Fair algorithms for clustering. In: Wallach, H., Larochelle, H., Beygelzimer, A., d' Alché-Buc, F., Fox, E., Garnett, R. (eds.) Advances in Neural Information Processing Systems, vol. 32. Curran Associates, Inc., Vancouver (2019). https://proceedings.neurips.cc/paper/2019/file/fc192b0c0d270dbf41870a63a8c76c2f-Paper.pdf
7. Bercea, I.O., et al.: On the cost of essentially fair clusterings (2018). arXiv:1811.10319
8. Böhm, M., Fazzone, A., Leonardi, S., Schwiegelshohn, C.: Fair clustering with multiple colors (2020). arXiv:2002.07892
9. Brownwell, C.: CRTC relaxes quotas on Canadian content for tv broadcasters (2015). https://financialpost.com/technology/crtc-relaxes-quotas-on-canadian-content-for-tv-broadcasters
10. Buolamwini, J., Gebru, T.: Gender shades: intersectional accuracy disparities in commercial gender classification. In: Conference on Fairness, Accountability and Transparency, pp. 77–91. PMLR, New York City (2018)
11. Chhabra, A., Vashishth, V., Mohapatra, P.: Fair algorithms for hierarchical agglomerative clustering (2020). arXiv:2005.03197
12. Chierichetti, F., Kumar, R., Lattanzi, S., Vassilvitskii, S.: Fair clustering through fairlets. In: Guyon, I., et al. (eds.) Advances in Neural Information Processing Systems, vol. 30. Curran Associates, Inc., Long Beach (2017). https://proceedings.neurips.cc/paper/2017/file/978fce5bcc4eccc88ad48ce3914124a2-Paper.pdf
13. Chiplunkar, A., Kale, S., Ramamoorthy, S.N.: How to solve fair k-center in massive data models. In: International Conference on Machine Learning, pp. 1877–1886. PMLR, Remote (2020)
14. Daudpota, S.M., Muhammad, A., Baber, J.: Video genre identification using clustering-based shot detection algorithm. Signal Image Video Process. **13**(7), 1413–1420 (2019)
15. Dua, D., Graff, C., et al.: UCI machine learning repository (2017)
16. Esmaeili, S., Brubach, B., Srinivasan, A., Dickerson, J.: Fair clustering under a bounded cost. In: Advances in Neural Information Processing Systems, vol. 34, pp. 14345–14357 (2021)

17. Esmaeili, S., Brubach, B., Tsepenekas, L., Dickerson, J.: Probabilistic fair clustering. In: Advances in Neural Information Processing Systems, vol. 33, pp. 12743–12755 (2020)
18. Esmaeili, S.A., Duppala, S., Dickerson, J.P., Brubach, B.: Fair labeled clustering. In: Proceedings of the 28th ACM SIGKDD Conference on Knowledge Discovery and Data Mining, pp. 327–335 (2022)
19. Ghadiri, M., Samadi, S., Vempala, S.: Socially fair k-means clustering. In: Proceedings of the 2021 ACM Conference on Fairness, Accountability, and Transparency, pp. 438–448. ACM, Online (2021)
20. Goyal, D., Jaiswal, R.: Tight FPT approximation for socially fair clustering (2021). arXiv:2106.06755
21. Gurobi Optimization, LLC: Gurobi Optimizer Reference Manual (2022). https://www.gurobi.com
22. Harb, E., Lam, H.S.: Kfc: A scalable approximation algorithm for k-center fair clustering. In: Advances in Neural Information Processing Systems, vol. 33, pp. 14509–14519 (2020)
23. Huang, L., Jiang, S., Vishnoi, N.: Coresets for clustering with fairness constraints. In: Advances in Neural Information Processing Systems, vol. 32 (2019)
24. Jain, A.K., Murty, M.N., Flynn, P.J.: Data clustering: a review. ACM Comput. Surv. (CSUR) 31(3), 264–323 (1999)
25. Jia, X., Sheth, K., Svensson, O.: Fair colorful k-center clustering. In: Bienstock, D., Zambelli, G. (eds.) IPCO 2020. LNCS, vol. 12125, pp. 209–222. Springer, Cham (2020). https://doi.org/10.1007/978-3-030-45771-6_17
26. Jones, M., Nguyen, H., Nguyen, T.: Fair k-centers via maximum matching. In: International Conference on Machine Learning, pp. 4940–4949. PMLR, Virtual (2020)
27. Kansal, T., Bahuguna, S., Singh, V., Choudhury, T.: Customer segmentation using k-means clustering. In: 2018 International Conference on Computational Techniques, Electronics and Mechanical Systems (CTEMS), pp. 135–139. IEEE, Belgaum (2018)
28. Kleindessner, M., Awasthi, P., Morgenstern, J.: Fair k-center clustering for data summarization. In: International Conference on Machine Learning, pp. 3448–3457. PMLR, Long Beach (2019)
29. Kleindessner, M., Samadi, S., Awasthi, P., Morgenstern, J.: Guarantees for spectral clustering with fairness constraints. In: International Conference on Machine Learning, pp. 3458–3467. PMLR, Long Beach (2019)
30. Le Quy, T., Roy, A., Friege, G., Ntoutsi, E.: Fair-capacitated clustering. In: EDM. EDM, Paris (2021)
31. Liu, S., Vicente, L.N.: A stochastic alternating balance k-means algorithm for fair clustering (2021). arXiv:2105.14172
32. Makarychev, Y., Vakilian, A.: Approximation algorithms for socially fair clustering. In: Conference on Learning Theory, pp. 3246–3264. PMLR, Boulder (2021)
33. Mehrabi, N., Morstatter, F., Saxena, N., Lerman, K., Galstyan, A.: A survey on bias and fairness in machine learning (2019). arXiv:1908.09635
34. NCSBE: North Carolina voter registration data (2020). https://www.ncsbe.gov/results-data/voter-registration-data
35. Pedregosa, F., et al.: Scikit-learn: machine learning in Python. J. Mach. Learn. Res. 12, 2825–2830 (2011)
36. Schmidt, M., Schwiegelshohn, C., Sohler, C.: Fair coresets and streaming algorithms for fair k-means clustering (2018). arXiv:1812.10854

37. Thejaswi, S., Ordozgoiti, B., Gionis, A.: Diversity-aware k-median: clustering with fair center representation (2021). arXiv:2106.11696
38. Wang, Y., et al.: Unsupervised machine learning for the discovery of latent disease clusters and patient subgroups using electronic health records. J. Biomed. Inform. **102**, 103364 (2020)
39. Xu, R., Wunsch, D.: Survey of clustering algorithms. IEEE Trans. Neural Netw. **16**(3), 645–678 (2005)
40. Ziko, I.M., Yuan, J., Granger, E., Ayed, I.B.: Variational fair clustering. In: Proceedings of the AAAI Conference on Artificial Intelligence, vol. 35, pp. 11202–11209. AAAI, Virtual (2021)

Proof Logging for the Circuit Constraint

Matthew J. McIlree[1]([✉]) [iD], Ciaran McCreesh[1] [iD], and Jakob Nordström[2,3] [iD]

[1] University of Glasgow, Glasgow, Scotland
m.mcilree.1@research.gla.ac.uk
[2] University of Copenhagen, Copenhagen, Denmark
[3] Lund University, Lund, Sweden

Abstract. Proof logging in constraint programming is an approach to certifying a conclusion reached by a solver. To allow for this, different propagators must be augmented to produce justifications for any inferences they make, so that an independent proof checker can certify correctness. The Circuit constraint is used to enforce a Hamiltonian cycle on a set of vertices, e.g. for vehicle routing. Maintaining consistency for the Circuit constraint is hard, so various ad-hoc propagation techniques have been devised and implemented in solvers. We show that standard Circuit constraint inference rules can be efficiently justified within a pseudo-Boolean proof system, either by using a simple sequence of cutting planes steps or through a conditional counting argument.

Keywords: Proof logging · Circuit · Constraint propagation

1 Introduction

A constraint programming (CP) solver that implements *proof logging* is able to provide a strong correctness guarantee for every result it produces. Alongside any answer, it outputs a formal proof that rigorously demonstrates that the answer is correct. This is already standard practice in the field of Boolean satisfiability (SAT) solving [8,32], and will, we believe, be crucial for the acceptance of CP in safety-critical applications. Proof logging is achieved by having a solver output justifications for all its reasoning steps in the language of a sound and complete proof system. Previous work [10,16,25] has shown that it is possible to do this efficiently for many important constraint propagation algorithms, by using a *pseudo-Boolean* (PB) proof system that is based on *cutting planes* [3,7] along with further strengthening rules. These proofs are written in a machine-readable format that can be independently verified using the *VeriPB* proof checker [2].

A common feature of all propagation algorithms that have been considered before is that they enforce a strong level of local consistency among the variables in scope. For example, AllDifferent [10], SmartTable and Regular [25] enforce domain consistency (DC), while LinearEquality [16] enforces bounds consistency (BC). To show a proof logging procedure for these constraints is comprehensive, it is sufficient to establish that any DC or BC inference can be justified,

© The Author(s), under exclusive license to Springer Nature Switzerland AG 2024
B. Dilkina (Ed.): CPAIOR 2024, LNCS 14743, pp. 38–55, 2024.
https://doi.org/10.1007/978-3-031-60599-4_3

(a) Valid assignment.   (b) Invalid assignment.   (c) Partial assignment.

Fig. 1. Interpretation of assignments for six variables constrained by Circuit.

since this demonstrates that certification is possible regardless of how that inference is actually computed. The same approach is no longer viable when dealing with more complex propagators that have less clearly defined notions of consistency, as it is harder to capture clearly in advance what propagations are to be expected. If pseudo-Boolean proof logging is to be applicable to CP in general, it is important to show that these types of propagators do not present any fundamental barriers to its adoption.

In this paper we present proof logging for propagation of the (Hamiltonian) Circuit constraint. Enforcing domain consistency for Circuit is known to be NP-hard [19] and so it is generally propagated via ad-hoc propagation rules [4,12,30]. We therefore work from the simplest checking and basic lookahead inferences, up to more advanced propagation techniques based on depth-first search and identification of strongly connected components. In each case we briefly outline the situation in which the inference applies, and find that it can be justified either by a simple sequence of cutting planes steps, or through a conditional counting argument. The latter consists of identifying a vertex which cannot reach every other vertex under some conditions, and deriving PB constraints over auxiliary variables that establish the set of reachable vertices is too small. We have implemented and tested these techniques, building a complete certifying Circuit propagator comparable in propagation strength to well known open-source CP solver implementations [6,14,21], and have been able to produce and verify proofs using this for a variety of instance sizes.

2   Preliminaries

The Circuit constraint uses a successor representation to treat a set of variables $X_0 \ldots X_{n-1}$, each with domain $\{0, \ldots, n-1\}$ as the vertices of a directed graph. At any stage in the solving process, an edge (i,j) is viewed as being present in the graph if and only if j is still in the domain of the variable X_i. Circuit requires that any assignment represents a *Hamiltonian cycle*, with the value of X_i representing the successor of i in a tour that visits all vertices; see Fig. 1. This is useful for modelling problems such as vehicle routing [22,23], activity scheduling [9] and other graph problems [4,13]. In CP solvers, propagation for a global Circuit constraint is generally achieved by first at least partially propagating an AllDifferent and then attempting further propagation based on the fact that there can be no sub-cycles. At a minimum the algorithm should check whether any sub-cycles are encoded by the current partial assignment and back-track if so [4], but further lookahead and ad-hoc propagation rules are possible [12,30].

2.1   Requirements for Proof Logging

To follow the CP proof logging methodology of previous work [16], we are required to compile any CP problem to a *pseudo-Boolean* (PB) format. This is a separate model of the problem that is only used for certification and should be kept independent of the solving process. In a PB model we are only allowed 0–1 variables (PB variables), and all constraints must be integer linear inequalities (PB constraints) over *literals*, where a literal ℓ is a PB variable or its negation $\bar{x} = 1 - x$. We also allow *reified* PB constraints of the form $\bigwedge r \implies \sum_{i=1}^{k} a_i \ell_i \geq A$, where $\bigwedge r$ is a conjunction of literals: these are syntactic sugar for $\sum K r + \sum_{i=1}^{k} a_i \ell_i \geq A$ for K chosen to be sufficiently large. We can also reify the negation of a PB constraint on the negation of each of the literals in r, which allows us to define $\bigwedge r \iff \sum_{i=1}^{k} a_i \ell_i \geq A$.

We can create a PB model from a CP problem by associating each integer variable X with a set of bit variables $x_{b0}, x_{b1}, x_{b2}, \ldots$ sufficient to represent every value in the variable's domain using a two's complement representation. We then encode restrictions on X imposed by CP constraints by adding pseudo-Boolean constraints over these bit variables. Since this process is not verified in itself, we should choose simple encodings that establish a clear correspondence between satisfying assignments of the CP and PB models. To aid this, we can employ auxiliary PB variables $x[i]_{=j}$ which are defined through reification to be true precisely when the bit representation of the variable X_i evaluates to the value j. For example if we have k bits and want to define $x[4]_{=4}$ we would have PB constraints equivalent to $x[4]_{=3} \iff \sum_{i=0}^{k-1} 2^i x[4]_{bi} = 3$. This gives us flexibility for PB encodings since we can make use of either bitwise or direct representations for variables depending on what gives us the most straightforward encoding of each CP constraint. We might also define other auxiliary variables that are not directly tied to the values of CP variables for use as flags, selectors, or counters.

For a Circuit constraint on n variables $X_0, \ldots, X_{n-1}$, we already know how to achieve proof-logging for the AllDifferent component [10]. A simple PB encoding of AllDifferent consists of constraints on each distinct pair of variables (X_l, X_r) that enforce either $X_l < X_r$ or $X_r < X_l$ depending on a selector bit f_{lr}.

We are then left with the task of defining PB constraints that encode the elimination of subcycles. For this we can take inspiration from known SAT encodings, since a logical clause such as $x \lor \bar{y} \lor \bar{z}$ is always equivalent to a PB constraint $(x + y + \bar{z} \geq 1)$, and so PB formulas can be viewed as a superset of conjunctive normal form. There are many possible options for such encodings [17], with different trade-offs, but since our chosen encoding will only be used for certification and not solving, compactness and obvious correctness is much more important than strong propagation properties. We make use of a PB encoding that is a simplified version of the SAT encoding given by Zhou [33, Sec. 4.2], and first

Algorithm 1 Procedure for constructing a PB encoding of a subcycle elimination constraint on variable $X_0, \ldots, X_{n-1}$

1: **define** $P_0 = 0$;
2: **for all** $i \in \{0, \ldots, n-1\}$ and $j \in \{1, \ldots, n-1\}$
3:     **define** $x[i]_{=j} \implies P_j - P_i \geq 1$
4:     **define** $x[i]_{=j} \implies -P_j + P_i \geq -1$
5: **for all** $i \in \{1, \ldots, n-1\}$
6:     **define** $x[i]_{=0} \implies P_i \geq n-1$
7:     **define** $x[i]_{=0} \implies -P_i \geq -n+1$

---

define an additional set of auxiliary bit variables $\{p[i]_{b0}, p[i]_{b1}, \ldots\}$ for each variable X_i, sufficient to represent the range of integers $0 \ldots n - 1$. We will use P_i as a shorthand for the sum $\sum_j 2^j p[i]_{bj}$, and conceptually treat P_i as a variable in itself, encoded with a sequence of PB bit variables. These bits are then constrained to represent the "position" of X_i in the circuit relative to X_0, which is arbitrarily designated as the start vertex. We do this as shown in Algorithm 1: define $P_0 = 0$, and then require $P_j = P_i + 1$ whenever $x[i]_{=j}$ is true, unless $j = 0$, in which case require $P_i = n - 1$. A satisfying assignment to these PB constraints is only possible when the cycle obtained by following the successors starting from X_0 visits every vertex. The condition $P_i = j$ can then be interpreted as encoding the fact that "the vertex represented by X_i is the j_{th} vertex visited after vertex 0 in the Hamiltonian cycle".

Once a valid encoding has been produced, a proof logging CP solver can justify its reasoning steps by deriving further PB constraints, and recording them in a proof log file that can eventually be used certify whatever result it arrives at. Further explanation of how this can be achieved in general for a backtracking-based CP solver is given by Gocht et al. [16], but for our purposes the key idea is that whenever the solver backtracks it should be possible to derive a PB constraint that encodes the negated conjunction of the currently guessed assignments. For example, if the solver guesses $(X_0 = 2, X_1 = 3, X_2 = 1)$ and then discovers a contradiction before assigning the remaining variables it should derive the PB constraint

$$\overline{x[0]}_{=2} + \overline{x[1]}_{=3} + \overline{x[2]}_{=3} \geq 1. \tag{1}$$

As a shorthand we will sometimes denote the solver's guessed assignments by $\mathcal{G}$, and use $\bigwedge \mathcal{G}$ to denote the conjunction of pseudo-Boolean literals encoding them. So in general a backtracking justification is of the form $\bigwedge \mathcal{G} \implies 0 \geq 1$.

This is somewhat similar to the way *lazy clause generation* solvers work [27], except the "explaining" constraints do not inform the solving process, and they have to be formally derived from the model and previously derived constraints via *VeriPB's* sound and complete proof rules, rather than simply asserted. As mentioned, these rules are based on the cutting planes proof system: so we can derive linear combinations of PB constraints, divide with rounding, *saturate* constraints to minimise coefficients, and also use the axiom that any single literal is at least 0; see Buss and Nordström [3] for more details. Additionally, if any auxiliary PB variables required are not already defined in the original PB formula, *VeriPB* allows them to be introduced dynamically whenever needed as extension variables during the proof. This is an application of the *VeriPB*'s *redundance-based strengthening* rule, and for our purposes is only needed to introduce fresh variables reified on arbitrary constraints [2]. The backtracking justification (1) itself should be derived via a further rule: the *reverse unit propagation* (RUP) rule. This allows derivation of a PB constraint D if the verifier can obtain a contradiction by iteratively enforcing bounds consistency on the negation of D along with constraints in the original formula and any previously derived constraints. The iterative consistency process is the pseudo-Boolean generalisation of unit propagation from SAT, and can be performed efficiently by the verifier [15], allowing "obvious" facts to be made available when checking RUP derivations. In the example above, since (1) is a clause, the negation asserts that all the literals are false, i.e. $\overline{x[0]}_{=2} = \overline{x[1]}_{=3} = \overline{x[2]}_{=3} = 0$, and so for the RUP check to succeed we would need these assignments to trigger propagations that eventually lead the verifier to a contradiction when propagating over the constraints in the PB model along with those already derived in the proof log.

When the solver performs sophisticated reasoning via bespoke propagation algorithms, we are able to guarantee that deriving the backtracking clause in this way will be possible providing any inferences made by a CP propagator at the given level of search are also be available to the verifier via unit propagation when performing the RUP check. We can do this by ensuring that $\bigwedge \mathcal{G} \implies y \geq 1$ is in the proof log, where y is a PB literal that encodes the propagator's inference. So if, under the sequence of guesses used in (1), and prior to backtracking, a Circuit propagator is able to infer say, $X_3 = 0$, we should somehow derive

$$x[0]_{=2} \wedge x[1]_{=3} \wedge x[2]_{=3} \implies x[3]_{=0} \geq 1; \tag{2}$$

$$\text{i.e.} \quad \overline{x[0]}_{=2} + \overline{x[1]}_{=3} + \overline{x[2]}_{=3} + x[3]_{=0} \geq 1. \tag{3}$$

These justifications are then interleaved with the backtracking clauses, resulting in the complete proof being essentially a description of the solver's backtracking search tree, expressed using RUP steps. What we show in this paper is that a range of standard Circuit propagation inferences can indeed be efficiently justified by deriving these intermediate pseudo-Boolean constraints.

3   Proof Logging for Simple Circuit Propagators

The minimum requirement for a Circuit sub-cycle elimination algorithm is that it is *checking*: it should return contradiction if a total assignment of the CP variables contains a small cycle. This requires no justification under our PB formula (as produced by Algorithm 1), since repeated bounds consistency (unit propagation) will immediately establish a contradiction. In particular, $P_0 = 0$ together with $X_0 = v_1$ will fix $P_{v_1} = 1$, and then this together with $X_{v_1} = v_2$ will fix $P_{v_2} = 2$, and so on. Since there must be a small cycle, say of length

$m < n$, passing through X_0 (as we are assuming AllDifferent has been correctly enforced), at some point unit propagation will attempt to fix the value of P_{v_m} when it has already been fixed to a smaller value, arriving at a contradiction.

A better checking propagator can also return contradiction on *partial* assignments, when they encode a small cycle. This is what Francis and Stuckey [12] call *check*, and is a key component of the NoSubtour propagator of Pesant et al. [29] and the similar NoCycle propagator of Caseau and Laburthe [4]. Such solver reasoning does require some justification in the proof, since a small cycle encoded by the partial assignment might not set X_0, and the corresponding PB variables for this are required to set off the chain reaction of unit propagation and achieve the inconsistent setting of p variables. To create such a justification we can use the cutting planes *addition rule* to add together all the corresponding constraints for the position variables in the cycle, allowing us to unit propagate a contradiction under the solver's guesses.

In particular, if a sequence of guesses $\mathcal{G}$ includes a small cycle of length $m < n$ not passing through 0 and consisting of vertices $(v_1, \ldots, v_m)$, we would add together each of the constraints of the form

$$x[v_i]_{=v_{i+1}} \implies P_{v_{i+1}} - P_{v_i} \geq 1 \qquad (4)$$

(from line 3 in Algorithm 1) for each guessed assignment $(X_{v_i} = v_{i+1}) \in \mathcal{G}$ identified by the propagator as being part of the cycle. Recall from Sect. 2.1 that these reified PB constraints are actually represented as

$$K \cdot \overline{x[v_i]}_{=v_{i+1}} + P_{v_{i+1}} - P_{v_i} \geq 1 \qquad (5)$$

for some sufficiently large K, and hence each successive addition cancels the previous P_{v_i} value. This results in the constraint

$$K \cdot \overline{x[v_1]}_{=v_2} + \cdots + K \cdot \overline{x[v_m]}_{=v_1} - P_{v_1} + P_{v_2} - \cdots$$
$$\cdots - P_{v_{m-1}} + P_{v_m} - P_{v_m} + P_{v_1} \geq m, \qquad (6)$$

which telescopes to

$$K \cdot \overline{x[v_1]}_{=v_2} + \cdots + K \cdot \overline{x[v_m]}_{=v_1} \geq m. \qquad (7)$$

With this constraint present in the proof log, unit propagation of the small cycle in $\mathcal{G}$ will obviously lead to $0 \geq m$, a contradiction, and hence (7) is adequate to allow justification within the proof framework for the solver backtracking.

The above idea can be easily be extended to produce justifications for a basic lookahead version of the *check* propagator, called *prevent* by Francis and Stuckey [12], which is described in the literature [4,29,30]. This filters domains by disallowing any further assignments that would immediately complete a sub-cycle. So if a sequence of guesses $\mathcal{G}$ includes the encoding of a *chain* of vertices $(v_1, \ldots, v_m)$, *prevent* would remove v_1 from the domain of X_{v_m}, and this can be justified by first deriving (7) exactly as above, which then allows us to derive $\bigwedge \mathcal{G} \implies \overline{x[v_m]}_{=v_1}$ by RUP, as required.

4   Proof Logging for Stronger Propagation

There are several possibilities for stronger propagation for the Circuit constraint, although there is no general consensus between solvers on which forms are worthwhile in practice. This paper does not argue for one propagation strategy over any other; rather, our focus is to show that whatever propagator is chosen, it should be feasible to implement a proof logging version of it. We will demonstrate that it is possible to provide pseudo-Boolean proof logging for Circuit propagators that make use of more complex reasoning by considering a further propagator and set of associated possible inferences. This algorithm is based on analysis of the depth-first spanning tree obtained during a search of the domain graph for *strongly connected components* (SCCs). Stuckey and Francis call it the SCC algorithm [12] and versions of it are implemented in the solvers *Gecode* [14], *Chuffed* [6], *JaCoP* [21], and *CP-SAT* [28] among others.

Let $G = (V, E)$ be a graph, and let $\mathcal{R}$ be the (directed) *reachability* relation on G—for $v, w \in V, (v, w) \in \mathcal{R}$ if and only if there exists a path from v to w. We will denote by $\text{REACH}(v)$ the set $\{w : (v, w) \in \mathcal{R}\}$, i.e. the set of all vertices in G reachable from v. The core observation used by the SCC algorithm for Circuit propagation is that if the graph contains a Hamiltonian circuit, then it can only have a single strongly connected component, which means every vertex must be reachable from every other vertex. Thus, if we identify more than one strongly connected component in the graph induced by the current domains of variables in scope we can backtrack early, as no satisfying Circuit assignment is possible.

At any given point in the process of solving a constraint satisfaction problem involving a Circuit constraint on variables $X_0, \ldots, X_{n-1}$, let G be the graph that has a directed edge (v, w) whenever w is still in the domain of X_v. To simplify the discussion of proof logging for the SCC algorithm, we will assume in what follows that if we can identify a vertex v in this graph such that $|\text{REACH}(v)| < |G|$ then we can run a proving procedure `ReachTooSmall`(v) that derives in the proof log a contradiction subject to the current sequence of guesses i.e. $\bigwedge \mathcal{G} \Rightarrow 0 \geq 1$. Furthermore, we will assume that if we have an additional "assumption" PB literal ℓ that encodes a further restriction on the graph so that $|\text{REACH}(v)| < |G|$ we can similarly run `ReachTooSmall`(v) and derive $\bigwedge \mathcal{G} \wedge \ell \Rightarrow 0 \geq 1$. We will later outline in Sect. 4.2 and Sect. 4.3 how `ReachTooSmall` can construct this argument using proof steps recognised by *VeriPB*.

The SCC propagator is based on *Tarjan's algorithm* [31], which uses the fact that strongly connected components always form subtrees of a depth-first spanning forest of the graph. It initiates a depth-first search (DFS) from a chosen arbitrary vertex v_0, and immediately returns contradiction if any of its descendants are identified as the root of an SCC. To justify backtrack in this case we can run `ReachTooSmall`(w), where w is the root of the identified SCC. This will always prove contradiction as v_0 cannot possibly be reachable from w, otherwise v_0 would also be part of the SCC and hence w would not be the SCC root.

A vertex can only be identified as the root of an SCC once all of its descendants have been visited during the DFS. So if none of v_0's descendants are identified as SCC roots, it must be that all the vertices reachable from v_0 comprise a single SCC. In this case, either DFS has visited every vertex, in which case there is no contradiction for Circuit, or else there is some vertex not reachable from v_0 and the propagator returns a contradiction. The latter can clearly be justified by ReachTooSmall(v_0).

Backtracking when the domain graph is disconnected or contains more than one SCC seems to be the most commonly implemented technique for SCC propagation, based on our examination of source code for open source solvers. Several solvers such as *Gecode* and *Chuffed* also implement further ad-hoc propagation opportunities when multiple distinct subtrees are explored below v_0. In each of the following cases we state a propagation rule applicable as part of the SCC algorithm and briefly indicate how ReachTooSmall can be used to justify these too. Figure 2 gives an illustration for each.

1. Prune any edge (w, v_0) where v_0 is the starting vertex and w is not in the *earliest* visited subtree [12]. To justify this we use $x[w]_{=v_0}$ as an assumption literal and run ReachTooSmall(r), where r is the root of a subtree visited earlier than the one containing w. Unit propagation of $x[w]_{=v_0}$ will force $x[w']_{=v_0} = 0$ for all $w' \neq w$ due to the encoding of AllDifferent, so the assumption excludes any edges from descendants of r leading to v_0. Since vertices in this earlier subtree cannot have any edges leading to vertices in w's subtree or later, otherwise they would have been traversed as part of the same subtree by DFS, it follows that r cannot reach w. Hence, ReachTooSmall(r) can be used to establish $\bigwedge \mathcal{G} \wedge x[w]_{=v_0} \Rightarrow 0 \geq 1$. See Fig. 2a.
2. Prune any edge (v_0, w) where v_0 is the starting vertex and w is not in the latest visited subtree [30]. Similarly, we use $x[v_0]_{=w}$ as an assumption, and this time run ReachTooSmall(w) to obtain a contradiction under the assumption. Since w can only reach vertices in its own subtree or earlier, and v_0 no longer has edges to the later subtrees, it is clear than not everything can be reached from w. See Fig. 2b.
3. Prune any edge (v, w), where w is v's first child, and no edges from vertices in the subtree rooted at w lead to vertices visited earlier in the DFS than v [12]. Here we can run ReachTooSmall(w) under the assumption $x[v]_{=w}$, since fixing the successor of v to be w eliminates any possibility of reaching any nodes visited earlier than v from w. See Fig. 2c.
4. Prune any edge (v, w) that skips a subtree, that is, where v is in the i_{th} visited subtree and w visited earlier than the root of the $(i-1)th$ subtree [30]. Intuitively this rule is sound because if the edge (v, w) were used in the circuit we would have to visit the initial node v_0 between visiting w and visiting the root r of the $(i-1)th$ subtree, but also visit v_0 between visiting r and visiting v, and both of these cannot be simultaneously true. A single assumption and ReachTooSmall argument is not always sufficient to justify this pruning, but our intuition *can* be encoded using two ReachTooSmall arguments and more complex assumptions. If r is the root of the $(i-1)th$ subtree, we assume first that r must be seen at some point *between* w and v_0 (denoted as $w \prec r \prec v_0$) and then run ReachTooSmall(v). This will derive a contradiction as every path from the subtree containing w to the

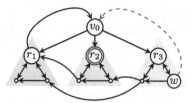

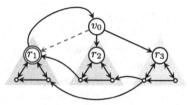

(a) Justifying the "prune skip to root" inference. If the dotted edge (w, v_0) is used, (r_1, v_0) is eliminated and so there is no way to reach v_0 from r_2.

(b) Justifying the "prune root" inference. If the dotted edge (v_0, r_1) is used, (v_0, r_2) and (v_0, r_3) are eliminated and so there is no way to reach e.g. r_2 from r_1.

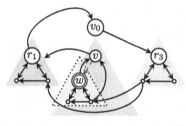

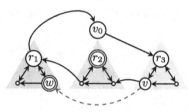

(c) Justifying the "prune within" inference. If the dotted edge (v, w) is used, (v, r_1) is eliminated and so there is no way to reach e.g. v_0 from w.

(d) Justifying the "prune skip" inference. We can disprove the ordering assumption $w \prec r_2 \prec v_0$ with ReachTooSmall(w) and disprove the ordering assumption $r_2 \prec v \prec v_0$ with ReachTooSmall(r_2). These together imply that (v, w) cannot be used.

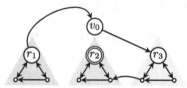

(e) Justifying "no backedges" contradiction. There are no backedges from the subtree rooted at r_2, and since "prune skip" inferences have already been made, there is then no way to reach any nodes earlier than r_2 from r_2.

Fig. 2. Illustrations of how each SCC inference can be justified. Three distinct subtrees explored by a DFS of the domain graph starting at v_0 are indicated with triangles. The dashed edge is the one assumed to be used as part of the circuit (via the corresponding variable assignment), and the double ringed node is the one passed to the ReachTooSmall procedure.

subtree rooted at r must pass through v_0 and so there will be no way to reach v_0 without violating the assumption. We can similarly assume that v must be seen between r and v_0 ($r \prec v \prec v_0$) and establish a contradiction using ReachTooSmall(v). Note that this establishes the negation of our two assumptions, namely that v_0 must be seen both between w and r ($v \prec v_0 \prec r$) and between r and v ($r \prec v_0 \prec w$) which is impossible if w is the immediate successor of v. So altogether, if we can encode these ordering assumptions in pseudo-Boolean form, and run ReachTooSmall subject to them, we should also be able to justify this pruning inference. See Fig. 2d.

5. Return a contradiction if there are no *backedges* identified after exploring any subtree later than the first [30]. Backedges are edges from a node in the i_{th} subtree to a node in the $(i-1)_{th}$. To justify contradiction in a case where a subtree rooted at w has no backedges we can run ReachTooSmall(w). Since any edges that skip subtrees have been removed at this point, by rule 3., and the only edges left leading to the initial node v_0 come from the earliest subtree, by rule 1., the only way to escape the subtree rooted at w would be through a backedge, and so ReachTooSmall(w) will establish a contradiction. Similarly, if there is only a single backedge (v, w) we can justify the fixing of $X_v = w$, by first assuming that it is not taken, i.e. $\overline{x[v]}_{=w}$, and running ReachTooSmall(v). See Fig. 2e.

These are all the inference rules we implemented in our prototype certifying Circuit propagator, as discussed in Sect. 5. We observe, however, that similar strategies may be used to introduce proof logging for other ad-hoc techniques. For example, if the algorithm is based on identifying *strong bridges* [18] and requiring them to be part of the solution, clearly a ReachTooSmall argument must be applicable if the bridge is assumed to be excluded. Another set of inferences can be applied if the Circuit is first relaxed to a path constraint [11], and the structural filtering of the reduced graph used here is essentially a generalisation of rule 4 ("prune skip") and so should be amenable to justification using ReachTooSmall and ordering assumptions.

4.1   Proving a Set Reachable from Vertex 0 is Too Small

We have shown in the previous section that all the inferences performed by a typical SCC propagator can be justified within a proof log if we are able to construct a sequence of PB steps ReachTooSmall(v) that establishes a contradiction for any vertex v in the graph G induced by the current domains of variables where $|\text{REACH}(v)| \leq |G|$. It needs to be possible to construct these steps subject to three kinds of assumption, namely, assuming an edge is required, assuming an edge is disallowed, and an "ordering assumption": assuming that a particular vertex must be seen between two other vertices. We now give a sketch for how such an argument can be constructed. First we will show, by way of example, how to construct it when running from the 0 index vertex, ReachTooSmall(0), without assumptions, as this is the simplest case. We later show how this can be modified to work for an arbitrary vertex v, and then finally show how the assumptions can be taken into account.

The idea is to collect possible position values (as defined in Algorithm 1) in a breadth-first search from the starting node. We create auxiliary variables $p[i]_{=k}$

defined through reification to be true if and only if the bit sum P_i is equal to k, and we aim to derive sets of PB constraints enforcing AtLeast1 and AtMost1 requirements over all of the possible i values for each $k \in \{0, \ldots, |\text{REACH}(0)|\}$. As an example, suppose the domain graph under a sequence of guesses $\mathcal{G}$ is as represented in Fig. 1c. Clearly $\text{REACH}(0) = \{0, 1, 5\}$, which has fewer that 6 elements, so we should be able to run ReachTooSmall(0). In this particular case the procedure would derive constraints (8) to (11) which show that for each $k \in \{0, 1, 2, 3\}$ at least one of the vertices 0, 1, 5 must have position value k. It would then derive corresponding constraints (12) to (14) which express the fact that each vertex can have at most one position value. Note that these are all reified on the sequence of solver guesses, but the $\bigwedge \mathcal{G} \Rightarrow$ is omitted for compactness.

AtLeast1 constraints:

$$p[0]_{=0} \geq 1 \quad (8)$$
$$p[1]_{=1} + p[5]_{=1} \geq 1 \quad (9)$$
$$p[0]_{=2} + p[1]_{=2} + p[5]_{=2} \geq 1 \quad (10)$$
$$p[0]_{=3} + p[1]_{=3} + p[5]_{=3} \geq 1 \quad (11)$$

AtMost1 constraints:

$$-p[0]_{=0} \quad -p[0]_{=2} - p[0]_{=3} \geq -1 \quad (12)$$
$$-p[1]_{=1} - p[1]_{=2} - p[1]_{=3} \geq -1 \quad (13)$$
$$-p[5]_{=1} - p[5]_{=2} - p[5]_{=3} \geq -1 \quad (14)$$

Using the addition rule the procedure can then derive the sum of all these constraints, and by construction everything on the left-hand side will cancel out, leaving $\mathcal{G} \Rightarrow 0 \geq 1$, as required. This process similar to how Hall violators for AllDifferent are derived by Elffers et al. [10].

It remains to show how the AtLeast1 and AtMost1 constraints can be derived using *VeriPB* proof rules. The first AtLeast1 (8) can be introduced by RUP, since $p[0]_{=0}$ propagates directly from the encoding. Then, each subsequent constraint can be derived from the previous constraint by first deriving some intermediate reified constraints by RUP, adding them together, and applying the *saturation* rule [3], which reduces any unnecessarily large coefficients. For example to derive (10) from (9) we would use the following proof steps:

$$x[1]_{=0} + x[1]_{=5} \geq 1 \qquad \text{(RUP)} \qquad (15)$$
$$\overline{p[1]}_{=1} + \overline{x[1]}_{=0} + p[0]_{=2} \geq 1 \qquad \text{(RUP)} \qquad (16)$$
$$\overline{p[1]}_{=1} + \overline{x[1]}_{=5} + p[5]_{=2} \geq 1 \qquad \text{(RUP)} \qquad (17)$$
$$\overline{p[1]}_{=1} + p[0]_{=2} + p[5]_{=2} \geq 1 \ ((15) + (16) + (17), \text{sat.}) \qquad (18)$$
$$\overline{p[5]}_{=1} + p[0]_{=2} + p[1]_{=2} \geq 1 \qquad \text{(similarly)} \qquad (19)$$
$$p[0]_{=2} + p[1]_{=2} + p[2]_{=2} \geq 1 \quad (9) + (18) + (19), \text{sat.}) \qquad (20)$$

To derive each of the AtMost1 constraints, we first introduce constraints $\overline{p[i]}_{=k} + \overline{p[i]}_{=l} \geq 1$ by RUP for each distinct pair of values (l, k) values possible for P_i. We then add these together but divide by j after adding the j_{th} constraint to recover the required constraint.

4.2   Proving a Set Reachable from an Arbitrary Vertex is Too Small

The above example establishes the general structure of the ReachTooSmall procedure: we collect AtLeast1 constraints over auxiliary position variables until we have more values than variables, and then add recovered AtMost1 constraints to these to obtain contradiction. However, the specifics of deriving the AtLeast1 constraints depended on us starting from the 0 vertex, as this is required in the encoding to be 0. There is nothing particularly special about the 0 vertex, but without requiring some position label $P_i = 0$ there would be n isomorphic solutions to the PB model for each arbitrary choice of starting vertex in a corresponding solution to the CP model. For our justifications from Sect. 4 to work, we need to be able to run ReachTooSmall(v) from an arbitrary vertex, and so we need a way to start the breadth-first search for possible positions without necessarily knowing with the position of the first node might be.

The idea is to dynamically introduce a new set of position labels $\{q[r, i] : 1 \leq i \leq n\}$ for a given starting vertex r, that are tied to the value of the P_i variables but represent what would be obtained if the value of each P_i was *shifted back* modulo n so that $P_r = 0$. Specifically we should have $q[r, i] = P_i - P_r \bmod n$. This preserves the useful property that if $X_i = j$ then $q[r, j] = q[r, i] + 1 \bmod n$, as is true for the p variables. By construction we must have $q[r, r] = 0$, and so we should be able to collect sets of possible $q[r, i]$ variables for each subsequent value and use this to construct our ReachTooSmall argument as before.

As with the other auxiliary variables, flags for these q variables can be introduced in the proof as needed using *VeriPB's* redundance-based strengthening rule. We do require some additional $d[r, i]$ flags to encode the definitions in pseudo-Boolean form, to correct for when the difference $P_i - P_r$ is less than 0. Specifically, whenever we require a variable $q[r, i]_{=k}$ we introduce the following.

$$d[r, i] \implies \qquad\qquad P_r - P_i \geq 1 \qquad\qquad (21)$$

$$\overline{d[r, i]} \implies \qquad\qquad P_i - P_r \geq 1 \qquad\qquad (22)$$

$$q[r, i]_{\geq k} \iff \qquad P_i - P_r + nd[r, i] \geq k \qquad\qquad (23)$$

$$q[r, i]_{\geq k} \iff q[r, i]_{\geq k} + \overline{q[r, i]}_{\geq k+1} \geq 2 \qquad\qquad (24)$$

These can each be introduced by redundance, and only need to be defined once for each combination of r, i, and k. One technicality is that for the constraint (22) we do require a subproof that establishes $P_i \neq P_r$ in order to apply redundance, but this is straightforward since we can pay a one-time cost to recover an AllDifferent constraint (AtLeast1 and AtMost1 constraints) over the p variables at the very start of the proof.

With these in place, we can outline the general procedure for constructing a ReachTooSmall argument from an arbitrary vertex, which is shown in Algorithm 2. All of the statements marked with **derive** can be derived from the PB model and the previous statements either with a single RUP step, or by a sequence of cutting planes steps followed by a RUP step. For lines 10, 12, and 23 this is just adding up the defining constraints for the auxiliary variables involved so that any p and d variables cancel out—we omit the details for brevity.

Algorithm 2 Procedure for constructing the proof, $\texttt{ReachToSmall}(r)$, showing that a sequence of guesses $\bigwedge \mathcal{G}$ imply contradiction.

1: **derive** $q[r,r]_{=0} \geq 0$
2: $\texttt{reached} \leftarrow \{r\}$; $\texttt{lastReached} \leftarrow \{r\}$; $\texttt{valuesSeen}[r] \leftarrow \{0\}$; $k \leftarrow 1$
3: **while** $k \leq |\texttt{reached}|$
4:     $\texttt{newReached} \leftarrow \emptyset$
5:     **for** $i \in \texttt{lastReached}$
6:         **for** $j \in \text{domain}(x[i])$
7:             $\texttt{newReached} \leftarrow \texttt{newReached} \cup \{j\}$
8:             $\texttt{valuesSeen}[j] \leftarrow \texttt{valuesSeen}[j] \cup \{k\}$
9:             **if** $j \neq r$
10:                 **derive** $q[r,i]_{=k-1} \wedge x[i]_{=j} \implies q[r,j]_{=k} \geq 1$ ▷ *Add defs. and RUP*
11:             **else**
12:                 **derive** $q[r,i]_{=k-1} \wedge x[i]_{=j} \implies 0 \geq 1$ ▷ *Add defs. and RUP*

13:         **derive** $\bigwedge \mathcal{G} \implies \sum_{j \in \text{domain}(x[i])} x[i]_{=j} \geq 1$ ▷ *RUP*

14:         ▷ *Add up line 13 with all constraints last derived at lines 10 and 12*   ◁
15:         **derive** $\bigwedge \mathcal{G} \wedge q[r,i]_{=k-1} \implies \sum_{j \in \text{domain}(x[i])} q[r,j]_{=k} \geq 1$

16:     ▷ *AL1 constraint: Add up last AL1 constraint and all lines last derived at 15* ◁
17:     **derive** $\bigwedge \mathcal{G} \implies \sum_{j \in \texttt{newReached}} q[r,j]_{=k} \geq 1$

18:     $\texttt{lastReached} \leftarrow \texttt{newReached}$; $\texttt{reached} \leftarrow \texttt{reached} \cup \texttt{newReached}$
19:     $k \leftarrow k + 1$

20: **for** $j \leftarrow 0$ **to** n
21:     **if** $\texttt{valuesSeen}[i] \neq \emptyset$
22:         **for** $(l,k) \in \texttt{valuesSeen}[j] \times \texttt{valuesSeen}[j]$ **where** $l < k$
23:             **derive** $\overline{q[r,j]_{=l}} + \overline{q[r,j]_{=k}} \geq 1$ ▷ *NotBoth constraint: Add up definitions*

24:         ▷ *AM1 constraint: Add up NotBoth constraints, dividing by i after each*
              step adding the i_{th} NothBoth constraint   ◁
25:         **derive** $\sum_{k \in \texttt{valuesSeen}[j]} -q[r,j]_{=k} \geq -1$

26: **derive** $\bigwedge \mathcal{G} \implies 0 \geq 1$ ▷ *Add up AL1 and AM1 constraints.*

4.3   Proving Reach is Too Small with Assumptions

The procedure outlined in Algorithm 2 requires minimal modification to work with assignment assumptions. Assuming $X_i = j$ or $X_i \neq j$ means including a PB variable encoding the assumption as an additional guess in $\bigwedge \mathcal{G}$, and skipping any domain values excluded by this assumption (either trivially or by AllDifferent) when iterating through the domains on line 5. This will allow the procedure to derive $\bigwedge \mathcal{G} \wedge x[i]_{=j} \implies 0 \geq 1$ or $\bigwedge \mathcal{G} \wedge \overline{x[i]}_{=j} \implies 0 \geq 1$, as required.

More care is required to encode and use ordering assumptions as discussed in Sect. 4. If we want to force the `ReachTooSmall`(r) to assume that $r \prec a \prec b$, that is, a vertex a must be visited before b when following a path from r, we first have to encode this assumption and reify it with its own flag. We can use auxiliary variables $d[i, j]$, defined as in (21) and (22) to do this:

$$a_{r \prec a \prec b} \iff \overline{d[r, a]} + \overline{d[a, b]} + \overline{d[b, r]} \geq 2 \qquad (25)$$

We can then include $a_{r \prec a \prec b}$ as an additional "guess" and use it to exclude $q[r, b]_{=k}$ from any `AtLeast1` constraint where, for $k' < k$, $q[r, a]_{=k'}$ was not part of a previous `AtLeast1`. This is achieved by deriving $q[r, a]_{\geq 1}$ after, line 1, and subsequently $q[r, a]_{\geq k}$ after line 15, and using these to derive $a_{r \prec a \prec b} \wedge q[r, i]_{=k+1} \wedge x[i]_{=b} \implies 0 \geq 1$ instead of line 10 whenever $j = b$. Once again these each amount to steps adding up definition constraints, followed by a RUP step, and we omit the details for brevity.

We note that the encoding of ordering assumptions (25) allows the justification of the prune skip inference with two conditional `ReachTooSmall` arguments, as discussed in Sect. 4. When `ReachTooSmall` arrives at a contradiction under an ordering assumption, the negation is established, and we can add up definitions for e.g. $\bar{a}_{j \prec r \prec v_0}$, $\bar{a}_{r \prec i \prec v_0}$, $x[i]_{=j}$ to cancel out p and d variables and arrive at a final contradiction for this inference.

5   Implementation and Evaluation

We have implemented proof logging versions of the *check*, *prevent* and *SCC* propagators (with all inference rules discussed in Sect. 5) using the techniques described in this paper as part of the auditable *Glasgow Constraint Solver* project [24], and we included in our implementation all the inference methods available in the Circuit propagators of *Gecode* [14]. We tested our implementation[1] by solving randomly generated travelling salesperson problems (TSPs), with graphs ranging in size from 3 to 40 vertices. The potential of proof logging

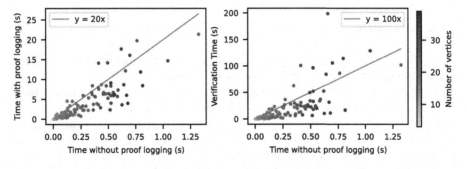

Fig. 3. Scatter plot of results of solving randomly generated TSP instances.

---

[1] https://zenodo.org/records/10848992.

as a powerful debugging and development tool was immediately apparent from this, as initial proof failures immediately indicated bugs in our implementation such as *prevent* trying to disallow full circuits in certain situations, or *SCC* trying to apply an incorrect inference based on the structure of the graph. This aligns with the results of previous research projects, where the implementation of proof logging uncovered hard-to-find bugs in well-tested combinatorial solvers [1,5,20]. Once the bugs were addressed, all proofs were verified as correct using *VeriPB*. The performance data from our evaluation is shown in Fig. 3.

There is clearly a cost in terms of overhead from enabling proof logging, although the exact slowdown is very dependent on hardware since we are writing to disk and using a non-optimised text-based proof format. What is clear is that the overhead is not unreasonable, with time to produce the proofs scaling roughly in proportion the time taken to solve without proof logging. This is what we would expect: our proof procedures for *check* and *prevent* output exactly one sequence of cutting planes steps for each subcycle prevented or disallowed, and so clearly are not doing significantly more work than the propagators themselves. Similarly, the `ReachTooSmall` procedure is called once for every inference (except the "prune skip" inference where it is called twice) and can at worst generate a number of proof steps proportional to $n \cdot E$ where E is the number of edges currently encoded by the domains of variables. Since Tarjan's algorithm itself runs in $O(n + E)$, we are satisfied that we are roughly within a linear factor of the amount of work done by the propagator and that our proof logging procedure is practical and free from any exponential blow up.

Additionally, we tested the TSP instance from the *MiniCP* benchmark suite of Michel et al. [26]. This was created for testing CP solver speed, and took the *Glasgow Constraint Solver* 44.9407 s to solve without proof logging (using full Circuit and AllDifferent propagation, not the simple propagation used by MiniCP). With proof logging, it took 3603.84 s (∼1 h) to solve, and *VeriPB* needed 585893.41 s (∼ 1 week) to verify the produced proof. This, together with previously implemented constraints [16], brings the *Glasgow Constraint Solver* in line with *MiniCP* in terms of propagators implemented and instance modelling capabilities.

6   Conclusion

We have exhibited the first certifying Circuit propagator using *VeriPB* proof logging, showing that ad-hoc inference rules with complicated notions of consistency can be included in an auditable constraint solver. In particular, we found that a range of standard inference types could make use of similar proof procedures, taking advantage of concepts such as connectedness and vertex ordering despite the proof system having no native representantions of these notions, or even of a graph. We expect that the core concepts exemplified here: such as counting reachable vertices under implications; creating shifted auxiliary labels; and running proof procedures under ordering assumptions will be useful for other constraints, and for proof logging combinatorial solving more generally.

Acknowledgements. Ciaran McCreesh was supported by a Royal Academy of Engineering research fellowship, and by the Engineering and Physical Sciences Research Council [grant number EP/X030032/1]. Jakob Nordström was supported by the Swedish Research Council grant 2016-00782 and the Independent Research Fund Denmark grant 9040-00389B. For the purpose of open access, the authors have applied a creative commons attribution (CC BY) licence to any author accepted manuscript version arising from this work.

References

1. Berg, J., Bogaerts, B., Nordström, J., Oertel, A., Vandesande, D.: Certified core-guided MaxSAT solving. In: Pientka, B., Tinelli, C. (eds.) CADE 2023. LNCS, vol. 14132, pp. 1–22. Springer, Cham (2023). https://doi.org/10.1007/978-3-031-38499-8_1
2. Bogaerts, B., Gocht, S., McCreesh, C., Nordström, J.: Certified symmetry and dominance breaking for combinatorial optimisation. J. Artif. Intell. Res. **77**, 1539–1589 (2023). Preliminary version in AAAI 2022
3. Buss, S.R., Nordström, J.: Proof complexity and SAT solving. In: Biere, A., Heule, M.J.H., van Maaren, H., Walsh, T. (eds.) Handbook of Satisfiability, Frontiers in Artificial Intelligence and Applications, vol. 336, chap. 7, 2nd edn., pp. 233–350. IOS Press (2021)
4. Caseau, Y., Laburthe, F.: Solving small TSPs with constraints. In: Naish, L. (ed.) Logic Programming, Proceedings of the Fourteenth International Conference on Logic Programming, Leuven, Belgium, 8–11 July 1997, pp. 316–330. MIT Press (1997)
5. Cheung, K.K.H., Gleixner, A., Steffy, D.E.: Verifying integer programming results. In: Eisenbrand, F., Koenemann, J. (eds.) IPCO 2017. LNCS, vol. 10328, pp. 148–160. Springer, Cham (2017). https://doi.org/10.1007/978-3-319-59250-3_13
6. Chu, G., Stuckey, P.J., Schutt, A., Ehlers, T., Gange, G., Francis, K.: Chuffed, a lazy clause generation solver (2023). https://github.com/chuffed/chuffed
7. Cook, W., Coullard, C.R., Turán, Gy.: On the complexity of cutting-plane proofs. Discrete Appl. Math. **18**(1), 25–38 (1987). https://doi.org/10.1016/0166-218X(87)90039-4
8. Cruz-Filipe, L., Heule, M.J.H., Hunt, W.A., Kaufmann, M., Schneider-Kamp, P.: Efficient certified RAT verification. In: de Moura, L. (ed.) CADE 2017. LNCS (LNAI), vol. 10395, pp. 220–236. Springer, Cham (2017). https://doi.org/10.1007/978-3-319-63046-5_14
9. Di Gaspero, L., Urli, T.: A CP/LNS approach for multi-day homecare scheduling problems. In: Blesa, M.J., Blum, C., Voß, S. (eds.) HM 2014. LNCS, vol. 8457, pp. 1–15. Springer, Cham (2014). https://doi.org/10.1007/978-3-319-07644-7_1
10. Elffers, J., Gocht, S., McCreesh, C., Nordström, J.: Justifying all differences using pseudo-boolean reasoning. In: The Thirty-Fourth AAAI Conference on Artificial Intelligence, AAAI 2020, The Thirty-Second Innovative Applications of Artificial Intelligence Conference, IAAI 2020, The Tenth AAAI Symposium on Educational Advances in Artificial Intelligence, EAAI 2020, New York, NY, USA, 7–12 February 2020, pp. 1486–1494. AAAI Press (2020)
11. Fages, J.G., Lorca, X.: Improving the asymmetric TSP by considering graph structure (2012). https://doi.org/10.48550/arXiv.1206.3437
12. Francis, K.G., Stuckey, P.J.: Explaining circuit propagation. Constraints **19**(1), 1–29 (2014). https://doi.org/10.1007/s10601-013-9148-0

13. Gaspero, L.D., Rendl, A., Urli, T.: Balancing bike sharing systems with constraint programming. Constraints **21**(2), 318–348 (2016). https://doi.org/10.1007/s10601-015-9182-1
14. Gecode Team: Gecode: generic constraint development environment (2023). http://www.gecode.org
15. Gocht, S.: Certifying correctness for combinatorial algorithms: by using pseudo-Boolean reasoning. Ph.D. thesis, Lund University, Sweden (2022)
16. Gocht, S., McCreesh, C., Nordström, J.: An auditable constraint programming solver. In: Solnon, C. (ed.) Proceeding of the 28th International Conference on Principles and Practice of Constraint Programming. Leibniz International Proceedings in Informatics (LIPIcs), vol. 235, pp. 25:1–25:18. Schloss Dagstuhl – Leibniz-Zentrum für Informatik, Dagstuhl (2022). https://doi.org/10.4230/LIPIcs.CP.2022.25
17. Heule, M.J.H.: Chinese remainder encoding for Hamiltonian cycles. In: Li, C.-M., Manyà, F. (eds.) SAT 2021. LNCS, vol. 12831, pp. 216–224. Springer, Cham (2021). https://doi.org/10.1007/978-3-030-80223-3_15
18. Italiano, G.F., Laura, L., Santaroni, F.: Finding strong bridges and strong articulation points in linear time. Theoret. Comput. Sci. **447**, 74–84 (2012). https://doi.org/10.1016/j.tcs.2011.11.011
19. Karp, R.M.: Reducibility among combinatorial problems. In: Miller, R.E., Thatcher, J.W., Bohlinger, J.D. (eds.) Complexity of Computer Computations. The IBM Research Symposia Series, pp. 85–103. Springer, Boston (1972). https://doi.org/10.1007/978-1-4684-2001-2_9
20. Kraiczy, S., McCreesh, C.: Solving graph homomorphism and subgraph isomorphism problems faster through clique neighbourhood constraints. In: Zhou, Z. (ed.) Proceedings of the Thirtieth International Joint Conference on Artificial Intelligence, IJCAI 2021, Virtual Event/Montreal, Canada, 19–27 August 2021, pp. 1396–1402. ijcai.org (2021). https://doi.org/10.24963/IJCAI.2021/193
21. Kuchcinski, K., Szymanek, R.: JaCoP - Java constraint programming solver. In: CP Solvers: Modeling, Applications, Integration, and Standardization, Co-located with the 19th International Conference on Principles and Practice of Constraint Programming (2013)
22. Lam, E., Van Hentenryck, P.: A branch-and-price-and-check model for the vehicle routing problem with location congestion. Constraints **21**(3), 394–412 (2016). https://doi.org/10.1007/s10601-016-9241-2
23. Lam, E., Van Hentenryck, P., Kilby, P.: Joint vehicle and crew routing and scheduling. In: Pesant, G. (ed.) CP 2015. LNCS, vol. 9255, pp. 654–670. Springer, Cham (2015). https://doi.org/10.1007/978-3-319-23219-5_45
24. McCreesh, C., McIlree, M.: The Glasgow constraint solver. GitHub repository (2023). https://github.com/ciaranm/glasgow-constraint-solver
25. McIlree, M.J., McCreesh, C.: Proof logging for smart extensional constraints. In: Yap, R.H.C. (ed.) 29th International Conference on Principles and Practice of Constraint Programming (CP 2023). Leibniz International Proceedings in Informatics (LIPIcs), vol. 280, pp. 26:1–26:17. Schloss Dagstuhl – Leibniz-Zentrum für Informatik, Dagstuhl (2023). https://doi.org/10.4230/LIPIcs.CP.2023.26
26. Michel, L.D., Schaus, P., Van Hentenryck, P.: MiniCP: a lightweight solver for constraint programming. Math. Program. Comput. **13**(1), 133–184 (2021). https://doi.org/10.1007/s12532-020-00190-7
27. Ohrimenko, O., Stuckey, P.J., Codish, M.: Propagation = lazy clause generation. In: Bessière, C. (ed.) CP 2007. LNCS, vol. 4741, pp. 544–558. Springer, Heidelberg (2007). https://doi.org/10.1007/978-3-540-74970-7_39

28. Perron, L., Didier, F.: CP-SAT. https://developers.google.com/optimization/cp/cp_solver/
29. Pesant, G., Gendreau, M., Potvin, J.Y., Rousseau, J.M.: An exact constraint logic programming algorithm for the traveling salesman problem with time windows. Transp. Sci. **32**(1), 12–29 (1998). https://doi.org/10.1287/trsc.32.1.12
30. Schulte, C., Tack, G.: Weakly monotonic propagators. In: Gent, I.P. (ed.) CP 2009. LNCS, vol. 5732, pp. 723–730. Springer, Heidelberg (2009). https://doi.org/10.1007/978-3-642-04244-7_56
31. Tarjan, R.: Depth-first search and linear graph algorithms. SIAM J. Comput. **1**(2), 146–160 (1972). https://doi.org/10.1137/0201010
32. Wetzler, N., Heule, M.J.H., Hunt, W.A.: DRAT-trim: efficient checking and trimming using expressive clausal proofs. In: Sinz, C., Egly, U. (eds.) SAT 2014. LNCS, vol. 8561, pp. 422–429. Springer, Cham (2014). https://doi.org/10.1007/978-3-319-09284-3_31
33. Zhou, N.-F.: In pursuit of an efficient SAT encoding for the Hamiltonian cycle problem. In: Simonis, H. (ed.) CP 2020. LNCS, vol. 12333, pp. 585–602. Springer, Cham (2020). https://doi.org/10.1007/978-3-030-58475-7_34

Probabilistic Lookahead Strong Branching via a Stochastic Abstract Branching Model

Gioni Mexi[1]([✉]) [ID], Somayeh Shamsi[2] [ID], Mathieu Besançon[1,3] [ID],
and Pierre Le Bodic[2] [ID]

[1] Zuse Institute Berlin, Berlin, Germany
`mexi@zib.de`
[2] Monash University, Clayton, Australia
{`somayeh.shamsi,pierre.lebodic`}`@monash.edu`
[3] Université Grenoble Alpes, Inria, LIG, Grenoble, France
`mathieu.besancon@inria.fr`

Abstract. Strong Branching (SB) is a cornerstone of all modern branching rules used in the Branch-and-Bound (BnB) algorithm, which is at the center of Mixed-Integer Programming solvers. In its full form, SB evaluates all variables to branch on and then selects the one producing the best relaxation, leading to small trees, but high runtimes. State-of-the-art branching rules therefore use SB with working limits to achieve both small enough trees and short run times. So far, these working limits have been established empirically. In this paper, we introduce a theoretical approach to guide how much SB to use at each node within the BnB. We first define an abstract stochastic tree model of the BnB algorithm where the geometric mean dual gains of all variables follow a given probability distribution. This model allows us to relate expected dual gains to tree sizes and explicitly compare the cost of sampling an additional SB candidate with the reward in expected tree size reduction. We then leverage the insight from the abstract model to design a new stopping criterion for SB, which fits a distribution to the dual gains and, at each node, dynamically continues or interrupts SB. This algorithm, which we refer to as Probabilistic Lookahead Strong Branching, improves both the tree size and runtime over MIPLIB instances, providing evidence that the method not only changes the amount of SB, but allocates it better.

Keywords: Mixed-Integer Programming · Branch-and-Bound · Strong Branching · Abstract Branching Tree · Stochastic Model

1 Introduction

Mixed-integer programming (MIP) is a powerful framework to model and solve optimization problems that include combinatorial structures. It consists in

---

G. Mexi and S. Shamsi—The first two authors contributed equally to the work presented in this paper.

© The Author(s), under exclusive license to Springer Nature Switzerland AG 2024
B. Dilkina (Ed.): CPAIOR 2024, LNCS 14743, pp. 56–73, 2024.
https://doi.org/10.1007/978-3-031-60599-4_4

minimizing a linear objective function over a polyhedron described by linear (in)equality constraints, where a subset of the decision variables must take integer values.

Despite the NP-hardness of general MIP solving, intensive research on the development of efficient algorithms grounded in polyhedral theory [11,18] and carefully designed combinations of these algorithms made generic solvers capable of optimizing to proven optimality problems that were previously considered out of reach.

The core of most MIP solvers is the Branch-and-Bound algorithm (BnB) [1]. It recursively partitions the solution space of the MIP into subproblems and computes their continuous Linear Programming (LP) relaxations. If the LP solution to a subproblem is integral, i.e., it is a solution to the MIP, then the best MIP solution is updated. If the optimal value of the LP relaxation of a subproblem is not better than the current best integer solution known for the MIP, this subproblem can be pruned. Otherwise, the subproblem is further divided. We refer interested readers to Conforti et al. [11] for a recent overview on MIP and to Achterberg [1] for the algorithmic aspects of MIP solving.

Modern solvers obtained outstanding progress in speed and scale over the last decades from improvements in presolving, cutting planes, primal heuristics, and the algorithmic choices present at multiple phases of the solving process. One crucial component is *branching*, or variable selection, which consists, given a relaxation solution $\hat{\mathbf{x}}$, in finding a variable x_j that violates its integrality constraint to partition the space with two subproblems with constraints $x_j \leq \lfloor \hat{x}_j \rfloor$ and $x_j \geq \lceil \hat{x}_j \rceil$ respectively. The set of integer variables that have a fractional value in the relaxation solution $\hat{\mathbf{x}}$ of a node are referred to as (branching) candidates. Although branching on any fractional candidate ensures that BnB attains the MIP optimum, the choice of the candidate has a significant impact on how implicit the enumeration is, i.e. how fast the search is. Indeed, [1] reports that branching on an arbitrary candidate leads to trees 5.43 times larger on average than the best branching rule at the time, which leads to a slowdown factor of 2.26 compared to the state-of-the-art at that time.

One branching technique is *strong branching* (SB), originally proposed in [5]. SB assesses each branching candidate by solving the two LPs resulting from the partition mentioned above. The differences in optimal value of the two LPs compared to the current node are referred to as *downgain, upgain* when adding the bound constraint forcing the variable "down" or "up" with respect to its continuous relaxation value, respectively. These gains in relaxation value or *dual gains* are a key metric to evaluate the progress of the tree towards a proof of optimality.

Despite looking "only" one level of depth down the tree, SB is a robust predictor for candidates that lead to small tree sizes, relative to other tractable branching rules. SB can in particular, for some classes of instances, offer theoretical guarantees compared to the optimal tree size [12]. In addition to its predictive power for variable selection, strong branching offers additional benefits in modern MIP solvers, including node pruning, primal solutions, and domain propagation.

The current default branching algorithm in SCIP is an extension of reliability [3] and hybrid branching [2]. It uses in particular strong branching calls for variables that are considered *unreliable* (i.e., which have not been branched on a certain number of times) and then relies on a weighted score dominated by pseudocosts for reliable variables. Even when all candidates are still unreliable – e.g., at the root node – evaluating all of them would be prohibitively expensive, solving twice as many LPs as there are candidates. If the branching rule calls strong branching to evaluate candidates, a *lookahead* strategy is used: if a certain number of successive variables are evaluated without any improvement over the best candidate, the procedure stops and returns the current best. This maximum lookahead value is essentially proportional to a fixed parameter L. Conceptually, this problem can be viewed as a generalized secretary problem with interview cost [6,15] or a variant of Pandora's box problem [9,24], in which indistinct candidates with hidden rewards have to be evaluated sequentially to maximize a net difference between the selected candidate's reward and the total interview costs. We will refer to the number of unsuccessful successive candidates as the lookahead value throughout this paper and will propose a novel stopping criterion equipped with a dynamic lookahead replacing the static L value.

A probabilistic model of variable dual gains was proposed in Hendel [17] in order to determine whether a given variable is unreliable and to choose which one to select for branching. More precisely, the dual gains of any variable are modeled as random variables following a normal distribution (i.i.d. across all nodes), and a parametric statistical test is used to determine a probability that a variable can yield a greater gain than the incumbent. This per-variable distribution can only provide guidance deep in the tree when a variable has been branched on several times to offer statistically significant inference. In contrast, our probabilistic model represents the dual gains of all variables at the current node as drawn from a distribution, making our approach applicable even when some variables have not been sampled yet, and therefore from the very first nodes of the tree. Furthermore, our model can be adapted to any distribution for which the cumulative function can be evaluated, alleviating the need for restricting assumptions such as normality.

Contributions. We define an abstract model of the Branch-and-Bound algorithm, based on the Multi-Variable Branching (MVB) model of Le Bodic and Nemhauser [19], incorporating unknown variable gains akin to Pandora's box problem and a particular form of tree restart. We coin the new model as Pandora's MVB, or PVB. We determine additional assumptions under which strong branching should continue or stop and branch on the best candidate found so far, defining a stopping criterion for our Probabilistic Lookahead Strong Branching (PL-SB) algorithm.

We study the dual gain distributions at nodes of the SCIP BnB algorithm where SB is performed on MIPLIB instances and construct a mixed distribution which captures these observed samples well. Beyond our work, we anticipate that this finding will help future research MIP research. We perform simulation runs on dual gains of MIP instances, comparing the total number of nodes obtained with SCIP current maximum lookahead criterion and PL-SB. These first simulations show that PL-SB is capable of reducing the expected tree size in Pandora's

MVB, even when the distribution is not perfectly known but fitted to the dual gains observed so far.

Finally, we design a practical strong branching algorithm incorporating working limits within PL-SB, and implement it within the SCIP default reliability branching rule. We conduct extensive experiments to assess its reliable performance on established MIP benchmarks, yielding a reduction in both time and number of nodes, two results that could not be achieved by simply changing some parameters of the current SB strategy.

2   Parameter Tuning on Strong Branching Yields No Improvements

The reliability pseudocost branching technique used in SCIP has been engineered to perform well across an array of instances of various sizes, difficulties, and structures, designed with careful tradeoffs between predictive power and However, with the tremendous progress achieved in MIP solving in the last decades [18] and the successive algorithmic changes, one could assume that the default parameters of the branching algorithm do not achieve optimal performance anymore and that simply tuning the available parameters already achieves significant performance improvements.

In the default configuration of SCIP, strong branching is restricted by a maximum lookahead L^{max} defined as:

$$L^{\mathrm{max}} = (1 + c^{\mathrm{uninit}}/c^{\mathrm{all}}) \cdot L,$$

where $c^{\mathrm{uninit}}/c^{\mathrm{all}}$ is the fraction of uninitialized branching candidates and L is the lookahead parameter which by default is fixed to 9. Moreover, the solver restricts the number of strong branching simplex iterations to:

$$\gamma^{\mathrm{max}} = \gamma^{\mathrm{node}} + K,$$

where γ^{node} is the total number of regular node LP simplex iterations and K is the number of additional iterations in SB, which by default is fixed to 10^6.

The choice of values for L and K is based on rigorous empirical studies on diverse instance sets, and as we show in Table 1 it is not straightforward to find alternative values that improve the overall performance of the solver. Table 1 shows the results of preliminary experiments on the MIPLIB 2017 benchmark [16] with three different random seeds for different values for L and K. To compare the running time and number of nodes of different settings we use the shifted geometric mean. The first row is SCIP with default parameter values and is compared with each of the three shown settings. We report on the quotient of running time for "affected" instances, i.e., instances on which the changed parameters affect the solver default behavior, and the quotient of running time and number of nodes for "affected-solved" instances, i.e., affected instances solved by both default SCIP and the setting that we compare it with.

Decreasing the default lookahead parameter L from 9 to 7 leads to worse performance on all affected instances and on affected instances solved by both settings. This behavior can be explained by the fact that worse branching decisions are made because of the smaller maximum lookahead value. Similarly, increasing the lookahead parameter L to 11 leads to worse overall performance, an effect that is not as obvious, since intuitively, better branching decisions are made. However, SCIP limits the budget of simplex iterations allowed in strong branching. Therefore, it can occur that calling strong branching more frequently at the start of the solving process, e.g., by increasing the parameter L, may restrict the calls of strong branching later in the search. To assess the effect of the two parameters, we conduct a final experiment. Here, we increase the budget of LP simplex iterations by setting $K = 1.2 \cdot 10^6$. However, overall, the final setting also leads to performance deterioration compared to the default settings.

Table 1. Evaluating SCIP performance for different values of the strong branching parameters L and K.

Setting	affected		affected-solved	
	#	time(%)	time(%)	nodes(%)
$L = 9, K = 10^6$ (default)	–	100.0	100.0	100.0
$L = 7, K = 10^6$	278	104.3	104.4	104.8
$L = 11, K = 10^6$	269	102.1	104.0	105.0
$L = 11, K = 1.2 \cdot 10^6$	271	102.4	105.2	106.5

Finally, we mention that we also experimented with other parameters which also did not lead to conclusive trends or performance improvements. Even though on average all settings from Table 1 are worse than the default setting, the virtual best uses 16% fewer nodes. This indicates that a dynamic decision of whether or not the solver should continue applying SB until the maximum lookahead parameter is reached could potentially improve the overall performance of the solver. This virtual best is obtained by applying a single rule throughout the tree, that potential gain could even become higher by applying a rule at each node where a SB decision is made.

3   Pandora's MVB: A Stochastic Abstract Model for Branch-and-Bound

The preliminary experiment of Sect. 2 leads to the conclusion that the parameters controlling strong branching in SCIP are already tuned, but the virtual best configuration shows that adjusting the parameters on a per-instance basis could yield significant improvements. This implies that a new algorithm that more adaptively drives strong branching in SCIP is necessary to better allocate computational resources.

Given the complexity of MIP solvers and the myriad of "adaptive" techniques that could be tapped into, there are virtually infinitely many algorithms that could be designed for this purpose. Undoubtedly, many would work, in the sense that they would improve performance on reference benchmarks, but at the cost of increased complexity and degrees of freedom in the solver configuration space, without an understanding of why the method works. We instead focus on building a theoretical model for the BnB process that is well-adapted to represent strong branching and use it as a theoretical foundation for an algorithm so that the improvements it incurs are well understood and can be transferred and generalized to other solvers, methods, and problem types.

Therefore, we start by constructing an abstract BnB model extending the approach of Le Bodic and Nemhauser [19], which has proven to capture important aspects of BnB well enough to improve performance in SCIP. However, all the models proposed in Le Bodic and Nemhauser [19] suppose that the dual gains are fixed throughout the tree and known in advance, which renders an algorithm like SB completely superfluous in this setting.

For this reason, we design in this section an extended abstract model in which the dual gains are fixed, but not known until the corresponding variable is branched on. This is not true in general. However, MIP solvers internally estimated the dual gains of variables by a pair of values, the *pseudocosts*. This assumption is useful to analyze the resulting abstract models. Based on this model, we design in Sect. 4 a stopping criterion for SB based on the likelihood that additional strong branching iterations provide a better branching candidate, as measured by the tree size.

3.1 Background on BnB Abstract Modeling

The original paper on abstract tree models for BnB [19] sets a foundation for studying the properties of BnB algorithms. The approach focuses on proving a given bound (i.e. closing a given dual gap) by branching on variables with fixed and known dual gains. It provides a theoretical framework to better understand BnB and has proven useful for improving solver performance, providing a new rule for variable selection in branching, now a component of SCIP's BnB implementation since in SCIP 7 [13].

In the abstract models of [19], a variable x_j is represented by a pair of values (l_j, r_j). The value l_j (resp. r_j) models the left (resp. right) dual gain of x_j, i.e. the change in objective value in the LP relaxation, of the left (resp. right) child when x_j is branched on. This means that, when x_j is branched on at node i, the bound on the left child of i is improved by l_j, and the bound on the right child of i is improved by r_j. The gains (l_j, r_j) assigned to x_j remain constant throughout the tree in the abstract model, regardless of the node depth or the dual bound value at that node. Given a dual gap G, a tree closes the gap G (i.e., proves the bound G) if each leaf has a value of at least G.

Definition 1 (Tree dual gap). *The dual gap G_i closed at node i is given by:*

$$G_i = \begin{cases} 0 & \text{if } i \text{ is the root node,} \\ G_h + l_j & \text{if } i \text{ is the left child of node } h, \\ G_h + r_j & \text{if } i \text{ is the right child of node } h, \end{cases}$$

where node h is the parent of i, if there is one, G_h is the gap at h, and (l_j, r_j) is the pair of gains of the variable branched on at h.

One of the problems defined using this abstract model is the Multiple Variable Branching (MVB) problem. The MVB model is a simplified version of the branch and bound tree, where n variables are given in input, and where one variable x_j with gains (l_j, r_j) must be branched on at each node to build a tree that closes a gap G. In MVB, one variable can be branched on multiple times on a path from the root to a leaf. However, because of the simplicity of MVB, some aspects of a modern implementation of a BnB, such as SB, cannot be modeled. Indeed, in a BnB in which the dual gains of the variables would be fixed and known in advance as they are in MVB, information discovery from candidate sampling as performed by SB would not be relevant. Therefore, in Sect. 3.2, we introduce an abstract model in which the dual gains of the variables are unknown.

3.2   An Abstract BnB Model with Stochastic Variable Gains

We extend MVB to incorporate variable gains that are not known in advance. They are still fixed throughout the tree, so the gains of a variable are fully revealed once the variable is branched on, including if the variable is evaluated by strong branching, even if it is not selected as a branching candidate.

We will represent SB in an abstract BnB model using *restarts*, a common technique from constraint and mixed-integer programming. We define a restart as the removal of all nodes from the tree, with two important properties: (a) the nodes evaluated before a restart are still accounted for in the total cost of the tree, and (b) the observed dual gains remain valid after a restart.

Performing SB therefore consists in branching on one variable, recording its dual gain, restarting the tree, and iterating. We can now define the PVB problem on our abstract BnB model:

Definition 2 (Pandora's Multi-Variable Branching Problem). *Given $n \in \mathbb{Z}_+$ variables with random dual gains in $\mathbb{R}^2_+$ that follow their own distribution, a target gap $G \in \mathbb{R}_+$, and $k \in \mathbb{Z}_+$, can a BnB tree that closes the gap G be built using at most k nodes, using each variable as many times as needed, where the random dual gains of a variable are revealed when it is first branched on, and where restarts are allowed?*

Note that one can effectively simulate SB in PVB by restarting after branching on one variable, and finally build the final tree without restarting.

Note that PVB does not make any assumption on whether and how the dual gains of the variables are distributed and related. In Sect. 4.2, we will assume that the dual gains are i.i.d. and follow a distribution that has a defined cumulative distribution function. Section 5 shows that this assumption is realistic by assessing several distribution families.

4 Probabilistic Lookahead: A Stopping Criterion for Strong Branching on Pandora's MVB

Section 2 shows the potential of more selectively using SB in MIP solvers. In this section, we design a SB strategy for PVB, and we compare it to SCIP's default rule on PVB instances. Simulation experiments on instances of the new abstract model show that the new criterion is substantially better in total LP evaluations than the fixed maximum lookahead stopping criterion used in SCIP.

4.1 Necessary Assumptions

The smallest tree that closes a gap G by repeatedly branching on a single variable is referred to as *SVB tree*, in line with [19]. In order to establish a stopping criterion that is grounded in theory, we first make a number of explicit assumptions:

A0 The minimum size MVB tree is the SVB tree of the best variable.

A1 The size of an SVB tree that closes G by only branching on variable with gains (g_j, g_j) is equal to the size of the tree that only branches on a variable with gains (l_j, r_j), where $g_j = \sqrt{l_j r_j}$ is the geometric mean of dual gains of variable x_j.

A2 The geometric mean of the dual gain of all variables follows the same distribution.

Assumption A0 does not hold in general, and simple counterexamples can be constructed [4,19], but a key result from [19, Theorem 7] is that, for large gaps, the size of an optimal MVB tree grows at the same rate as the size of the best SVB tree. Therefore, in this paper, we use the SVB tree size as a measure of the quality of a variable. Assumption A1 is perhaps the furthest away from actual BnB trees, based on early experiments. However, the numerical improvements shown in Sect. 4.3 and 7 show that this approximation is close enough to be representative of the quality of the variables. We show that Assumption A2 is close to what can be observed on dual gains of real instances separately in Sect. 5, as these results are interesting in and of themselves.

We focus on the geometric mean of a variable because its square, the *product*, is a state-of-the-art scoring function for variables [1]. Hence, it has been shown that it is a good practical heuristic for ranking branching candidates. We now use the assumptions above to design a stopping criterion for SB.

4.2 Probabilistic Lookahead: A Stopping Criterion for Strong Branching on Pandora's MVB

We now use Pandora's MVB to design a new stopping criterion for SB, which we call *Probabilistic Lookahead*. At each SB iteration i, the tree is restarted, and an arbitrary variable x_j is branched on, which reveals its dual gains (l_j, r_j). We denote S_i the set of variables that have known dual gains at iteration i. The stopping criterion then decides whether to further iterate and expand S_{i+1} or stop and use a variable in S_i. Using Assumption A1, we estimate the SVB tree size as $2^{d_j+1} - 1$, where $d_j = \lceil \frac{G}{g_j} \rceil$, which is the exact SVB tree size of variable (g_j, g_j). Therefore, according to this estimate, the best variables found so far are those that minimize $d_i^* = \min_{x_j \in S_i} d_j$ or equivalently maximize the mean dual gain.

In the following statement and throughout, we use the term "size" when referring to the nodes of a tree, and we write "nodes used" to refer to all nodes created so far, including those that have been discarded by a restart.

Proposition 1. *If we stop SB at the end of iteration $i > 0$ with a depth d_i^*, the minimum number of nodes used to construct the MVB tree is:*

$$t_i(G) = 2^{d_i^*+1} - 1 + 2 \cdot i.$$

Proof. First, according to Assumption A1, the total size of the best MVB tree is the size of the best SVB variable, therefore we can suppose without loss of optimality that the same variable is branched on throughout the tree. Second, we can assume that the variable branched on is in S_i, as otherwise fewer nodes are required by stopping SB at $i = 0$, i.e., branching on a variable on which we have no information. Therefore, the minimum-size tree is a perfect binary tree, where all leaves have depth d_i^*, and whose size is thus $2^{d_i^*+1} - 1$. Accounting for the i SB iterations, the total number of nodes used is $2^{d_i^*+1} - 1 + 2 \cdot i$. □

In order to decide whether to stop, we need to estimate the size of the tree if we continue iterating SB. Following Assumption A2, we suppose that we can compute the probability that the next variable will decrease the depth required, i.e. that $d_{i+1}^* < d_i^*$. For $d = \{1, \ldots, d_i^*\}$, we denote p_d the probability that $d_{i+1}^* = d$.

Theorem 1. *The expected number of nodes used if we stop SB at the end of iteration $i + 1$ is:*

$$E[t_{i+1}(G)] = \sum_{d=1}^{d_i^*} ((2^{d+1} - 1) \cdot p_d) + 2 \cdot (i + 1). \tag{1}$$

Proof. At iteration $i + 1$, the minimum depth of the tree is $d = 1$, and the maximum depth is d_i^*. Hence $\sum_{d=1}^{d_i^*} p_d = 1$. Therefore, the expected value of the size of the final tree, if SB stops at iteration $i + 1$, is $\sum_{d=1}^{d_i^*} (2^{d+1} - 1) \cdot p_d$. Additionally counting the $i + 1$ strong branching calls, each of them producing two nodes, we obtain (1). □

The Probabilistic Lookahead stopping criterion then consists in comparing $t_i(G)$ and $E[t_{i+1}(G)]$ at the end of every iteration i, to decide whether to continue SB or to stop it and branch on the best candidate found so far.

4.3 Evaluating the Probabilistic Lookahead on Pandora's MVB

In this section, we investigate the effectiveness of the Probabilistic Lookahead SB stopping criterion on instances of Pandora's MVB. In order to create realistic instances, we extract dual gains observed during the solution of MIP instances. Specifically, we use the *full* SB rule of SCIP, with all other parameters at default values, i.e. we perform SB on all branching candidates, and exit after the root node. Together with a gap G, all dual gains collected for one MIP instance make a PVB instance.

The experiment compares different stopping criteria when applied to Pandora's MVB in terms of total number of nodes. The baseline is the state-of-the-art lookahead strategy (see Sect. 2), which is the default in SCIP and which we refer to as "Fixed Lookahead". We compare this to the Probabilistic Lookahead SB, with different dynamically-fitted distributions: mixed exponential and mixed Pareto, as they are the best two found in Sect. 5, and the exponential to assess the impact of representing the mass point at 0.

Because we observed that the performance of these stopping criteria is similar relative to the others on all instances, here we only present results on the Trento1 MIPLIB instance. For each gap value G, we conduct 1000 experiments. The average overall number of nodes and number of nodes used for SB are presented in Table 2.

The outcome demonstrates that the abstract model performs significantly better than the SCIP strategy in terms of number of nodes required to close the gap, across all examined cases.

Table 2. Comparison of the number of nodes with different stopping criteria for Pandora's MVB problem based on dual gains from the Trento1 instance. The number of nodes produced by SB is given in parentheses.

Gap	Fixed lookahead	Fitted distribution			Full SB
		Exp.	Mixed Exp.	Mixed Pareto	
1000	112 (64)	91 (42)	**90** (44)	**90** (44)	1009 (966)
2000	161 (64)	137 (42)	**136** (44)	145 (52)	1051 (966)
3000	210 (66)	189 (44)	**185** (46)	194 (62)	1093 (966)
4000	338 (66)	**235** (44)	236 (46)	259 (68)	1133 (966)
5000	1267 (64)	306 (48)	**302** (50)	348 (96)	1175 (966)
6000	1810 (64)	370 (52)	**365** (52)	387 (110)	1217 (966)
6500	26258 (64)	462 (56)	415 (54)	**409** (116)	1237 (966)

5   The Geometric Mean of Dual Gains Follows a Mixed Distribution

Before presenting the algorithm we design for strong branching, we assess the accuracy of Assumption A2, namely, that there is a probability distribution that captures well the geometric mean of dual gains. In practice, the dual bound changes when branching on a given variable depend heavily on the particular structure of the problem. Fixing decision variables that impact the rest of the problem structure – e.g., design variables in a network design problem – will result in structural changes that are likely to change the relaxation bound a lot more than variables that have a more "local" influence on – e.g., routing variables in the same network design example. We compare the distribution of mean dual gains for the MIPLIB 2017 benchmark instances at the root node with different distributions fitted with their maximum likelihood estimators. Three requirements for the distribution we fit are having a zero probability on negative gains, allowing for a mass point at zero, and having a support that is a proper superset of the interval $[0, g_{\max}]$, since we want to estimate the probability of a new sample having a value greater than the current maximum mean dual gain $g_{\max}$, which would be zero if the support is equal to the interval, e.g., with the uniform or empirical cumulative distributions. The reason for the mass point at zero is the dual gain often being zero for many variables, i.e., the LP optimum after branching remains on the original optimal face. To accommodate this characteristic, we model dual gains with a mixed continuous-discrete distribution:

$$P[G \leq g] = p_0 + (1 - p_0)F_D(g; \theta),$$

with p_0 the probability of a zero dual gain and $F_D(\cdot\,; \theta)$ the cumulative distribution function of a continuous probability distribution D with parameters θ. We tested several continuous distribution candidates including the uniform, log-normal, exponential, Pareto, and normal distributions. Note that the normal distribution was used mostly as a "control", since its nonempty support on negative numbers makes it a poor modeling choice. We fit the aforementioned distributions only on nonzero gains and evaluate their fitness with the Kolmogorov-Smirnov non-parametric test implemented in `Distributions.jl` [7]. This test evaluates the maximum distance between the empirical cumulative distribution function of the data and the cumulative function of the tested distribution (with parameters estimated from the data). The results hint that the exponential and Pareto distributions appear as good candidates, based on a proportion of dual gains series on which the null hypothesis fails to be rejected (20% of the dual gains series for Pareto, 33% for exponential, when evaluating series with at least 10 dual gains, and rejecting the null hypothesis for a p value of 5%), although we notice a difference between the fitness of the distribution and its performance on Pandora's MVB and within the actual solving process of a solver in the next section. Future work will investigate this relationship and gap between the predictive capabilities of a distribution and its performance when used within the probabilistic lookahead algorithm.

6   Probabilistic Lookahead Within SCIP's Strong Branching

Based on the abstract model for strong branching and a probability distribution fitted on observed dual gains, we can compute a probability that sampling a new strong branching candidate attains a higher mean dual gain than the current one, and correspondingly, reduces the total number of LPs to evaluate. Equipped with this estimation, we can dynamically decide whether to sample a new candidate or stop strong branching, and then branch on the candidate offering the highest predicted score, either with the dual gain sampled by strong branching or with the pseudocosts for the other reliable candidates. We summarize our method in Algorithm 1. One central difference of the algorithm integrated within reliability pseudocost branching compared to the abstract model presented in Sect. 3.2 is that it uses a weighed sum as a score function, aggregating not only the dual gains (or pseudocosts for reliable candidates), but also a range of other criteria used for branching, such as inference score and conflict analysis [2], the fraction of the times a variable led to node cutoff, the score of cuts corresponding to the variable [21], or dual degeneracy [14]. The weights of these scores however remain orders of magnitude lower than pseudocosts or dual gains, effectively making them act as tie-breakers, we thus consider that our model – although ignoring these additional aspects – captures the essential information.   The small perturbation ϵ is used to produce shifted geometric mean values g_j that can be compared, even when one of the down- or upgain is close to zero. We note that when applying strong branching, there is a limit on the number of LP iterations per candidate, mitigating pathological cases where branching yields a high number of simplex iterations.

7   Computational Experiments

We implement our algorithm as a branching rule in SCIP 8 [8] generalizing the hybrid branching algorithm [2]. The modified branching rule used for the computational experiments is available and open-source[1] and will be integrated in the next major release of SCIP.

In our computational experiments, in order to lessen the effects of performance variability [20] and obtain accurate comparisons, we use a large test set of diverse problem instances. In particular, our test set consists of the official benchmark set MIPLIB 2017 [16], with random seeds 0 to 4, amounting to 1200 instance-seed pairs. For the remainder of this paper, we will refer to each instance-seed pair as an *instance*.

All experiments are run in exclusive mode (one job per machine) on a cluster equipped with Intel Xeon Gold 5122 CPUs running at 3.60GHz, where each run is restricted to a 4-hour time limit and 48GB of memory. After preliminary

---

[1] The branching rule is integrated in github.com/matbesancon/scip on the strongbranching_dynamiclookahead_2 branch.

Algorithm 1: Probabilistic Lookahead Strong Branching

Input   : Set of fractional candidates, $\phi = 0.6$
Output: Updated pseudocosts, best k among unreliable candidates
$L^{\max} \leftarrow (1 + c^{\text{uninit}}/c^{\text{all}}) \cdot L$;
$\gamma^{\max} = \gamma^{\text{node}} + K$;
for x_j **in** at most C fractional unreliable candidates **do**

    Apply SB on x_j to compute the down- and upgain Δ^-, Δ^+;
    Update the pseudocosts with the gains Δ^-, Δ^+;
    $g_j \leftarrow \sqrt{(\Delta^- + \epsilon) \cdot (\Delta^+ + \epsilon)} - \epsilon$;
    Update the distribution with the new geometric mean gain g_j;
    $s_j \leftarrow \text{score}(g_j)$;
    $k \leftarrow \text{argmax}_{k \in 1..j} s_k$;
    if s_k has not changed for $L^{\max}$ iterations **then**
        | break;
    end
    if strong branching LP iterations exceed the limit $\gamma^{\max}$ **then**
        | break;
    end
    if s_k not changed for $\phi \cdot L^{\max}$ iterations **and** enough nonzero samples **then**
        Test the expected tree size now and with one more SB iteration;
        if no expected improvement **then**
            | break;
        end
    end
end

---

experiments, we observed that the mixed Pareto distribution not only fits the data as observed in Sect. 5, but also works better than the exponential distribution, which is the other one we assessed our method with. Our method can only stop strong branching when enough non-zero samples are collected, and we reached 60% of the maximum lookahead.

Table 3 compares default SCIP ("default") to SCIP using our PL-SB strategy ("dynamic"). Besides the number of instances solved to optimality, the main measure we are interested in is the shifted geometric mean of running time and number of BnB nodes. The shift for the running time is set to 1 s and for the nodes to 100. We compare the two settings on instances where the solver behavior is affected by our dynamic strong branching method ("affected"), and also on the whole test set ("all"). Moreover, we show results for the subcategories with solved instances by both settings ("*-solved"), and also harder instances solved in more than 1000 s by at least one of the two settings ("*-solved-1000+"). In the last two columns, we show for each category the relative difference of the two settings as a percentage. We highlight relative differences exceeding 1% by presenting them in bold in the table.

On the entire test set ("all"), we solve ten more instances with the dynamic SB strategy and observe a substantial improvement in both runtime and number of nodes on instances affected by our method. On "affected", there is a decrease in both running time and nodes. In particular, for harder instances, we observe a 8% decrease in running time and a 9% decrease in the number of nodes. On the entire test set, we observe a slight improvement of 1–2% in terms of both running time and number of nodes. This marginal impact is due to the limited number of instances affected by the PL-SB strategy. Our method stops strong branching earlier than the lookahead value in 20% of the calls.

Table 3. Performance comparison of SCIP (default) and SCIP using the PL-SB strategy (dynamic)

Subset	instances	default			dynamic			relative	
		solved	time	nodes	solved	time	nodes	time	nodes
affected	235	223	944.6	11019	**233**	907.1	10423	**0.96**	**0.94**
affected-solved	221	221	795.0	9596	221	771.7	9105	**0.97**	**0.95**
affected-solved-1000+	110	110	4062.7	42394	110	3756.5	38712	**0.92**	**0.91**
all	1200	685	1591.9	6317	**695**	1578.9	6225	0.99	**0.98**
all-solved	683	683	302.2	3254	683	299.0	3198	0.99	**0.98**
all-solved-1000+	254	254	3225.4	30179	254	3117.2	29013	**0.97**	**0.96**

The results highlight that the working limits on SB (LP iterations and maximum lookahead) significantly impact both tree size and runtime. We therefore perform an additional experiment with a 2-hour time limit, 20% more SB simplex iterations (amounting to 120000 iterations) and a lookeahead of 11 instead of 9; we then compare the static lookahead against our dynamic algorithm with this new setting in Table 4. The time limit was reduced compared to the first experiment to mitigate the computational burden of the overall experiments.

In this setting, we solve four more instances with the PL-SB algorithm and observe a substantial improvement in both runtime and number of nodes on affected instances solved by both methods. In particular, we obtain 11% fewer

Table 4. Performance comparison of SCIP (default) and SCIP using the PL-SB strategy (dynamic) after increasing the maximum lookahead L to 11 and the maximum number of simplex LP iterations K in SB to $1.2 \cdot 10^6$.

Subset	instances	default			dynamic			relative	
		solved	time	nodes	solved	time	nodes	time	nodes
affected	188	179	577.4	5588	**183**	546.9	5281	**0.95**	**0.94**
affected-solved	174	174	479.8	4675	174	450.5	4368	**0.94**	**0.93**
affected-solved-1000+	74	74	2644.3	19718	74	2452.1	17619	**0.93**	**0.89**
all	1200	632	1168.3	4570	**636**	1164.3	4541	1.00	0.99
all-solved	627	627	225.4	2605	627	223.6	2566	0.99	**0.98**
all-solved-1000+	201	201	2266.3	23660	201	2239.8	22993	0.99	**0.97**

nodes on hard affected instances, and a 7% smaller runtime. Overall, we observe a 5% improvement on affected instances. We however note that this changed setting is not performing as well as the PL-SB algorithm with default SB parameters presented in Table 3, when compared on the basis of the 2-hour time limit. The key conclusion we draw is that the dynamic lookahead mechanism improves over its static counterpart for different values of the SB parameters.

SCIP is also an exact global solver for Mixed-Integer Nonlinear Programs (MINLP), which remain much more challenging than their linear counterparts. A lot of algorithmic questions considered answered for MIPs still need further investigation for MINLPs, see [8,23] for an overview of the formulation, methods, and implementation involved. In particular, some techniques that are essential to the performance of MIP solving may be insignificant or detrimental on MINLPs; motivating us to assess the proposed Probabilistic Lookahead Strong Branching on MINLPs.

We test our method against the default lookahead on the `minlpdev-solvable`[2] subset of the MINLPLIB [10,22] which is the standard to assess the performance of SCIP on MINLPs. We run the 169 instances with 5 seeds, resulting in 845 instance-seed pairs. We exclude any instance-seed pair leading to numerical errors on any of the settings, leaving 817 instance-seed pairs. The time limit for this experiment is one hour since this captures the bulk of the instances already (808 out of 817 instances). On affected instances, our method produces modest improvements, 2% in time and 4% in the number of nodes.

The results are presented in Table 5, highlighting a performance improvement consistent with that observed on MIPs. Unlike on MIP instances, using the exponential instead of the Pareto distribution produces slightly better results, with a 3% reduction in time and 4% reduction in the number of nodes. This experiment highlights that the new algorithm improves the performance of SCIP on a diverse set of MINLP instances with the same effect of fewer nodes and reduced time that we observed on MIPs. It also suggests that a single distribution might not be the best to capture dual gains in all situations, opening the question of adaptively selecting a distribution to fit based on instance attributes.

Table 5. Performance comparison of SCIP (default) and SCIP using the PL-SB strategy (dynamic) on MINLP instances

Subset	instances	default			dynamic			relative	
		solved	time	nodes	solved	time	nodes	time	nodes
affected	140	140	47.2	1604	140	46.5	1542	**0.98**	**0.96**
all	817	808	19.1	2429	808	19.0	2413	0.99	0.99
all-solved	808	808	18.0	2273	808	17.9	2258	0.99	0.99

---

[2] The precise subset used will be made available in the companion repository of our paper.

8 Conclusion

In this paper, we defined Pandora's Multi-Variable Branching, an abstract model of the Branch-and-Bound algorithm that represents the dual gains at a particular node as a random variable with a known probability distribution. All branching candidates represent one realization of this random variable, strong branching can then be viewed as sampling this distribution to obtain a candidate with the highest possible dual gain. Equipped with this model and a family of probability distributions that model the dual gains observed on real instances, we explicitly estimate the expected reduction in the number of nodes from one more strong branching call, compare that reduction with the cost of solving the additional LPs from strong branching, and dynamically decide after each candidate evaluation whether to continue or stop strong branching. Computational experiments show that our Probabilistic Lookahead Strong Branching algorithm improves over the static lookahead rule on the MIPLIB benchmark both in time and in number of nodes, especially on hard instances, a result that could not be achieved by simply adjusting SB parameters.

The different contributions of our paper open promising venues for future research. The proposed PVB represents a variable as a single geometric mean gain, allowing us to build our lookahead strategy on the mean gains only. Future work will include designing abstract models representing unbalanced binary trees, i.e. with left and right dual gains. The proposed algorithm performs better than the default rule on average on the benchmark set but does not dominate it on all instances. Future work should investigate the criteria influencing the relative effectiveness of the two algorithms, and whether it is possible to improve over both of them based on static information available before starting to branch. The presented model is oblivious to variable histories and only fits the distribution based on the dual gains observed at the current node. An enhancement of the method could consider a Bayesian-like update of the fitted distribution, allowing the use of the dynamic lookahead earlier and with a better distribution fit. We studied distributions mixing a mass point at 0 and a continuous distribution, which captures a good fraction of the series of dual gains at any given node. This is to the best of our knowledge the first study modeling dual gains directly with a mixed distribution which appears essential from the data since many variables will yield a zero dual gain. This probabilistic model has promising applications in other parts of reliability branching, e.g., to improve the estimation of pseudocosts and to determine when pseudocosts are reliable already.

Acknowledgements. Research reported in this paper was partially supported through the Research Campus Modal funded by the German Federal Ministry of Education and Research (fund numbers 05M14ZAM, 05M20ZBM). We thank Nicolas Gast and Bruno Gaujal for discussions on online decision-making including the Pandora's box problem.

References

1. Achterberg, T.: Constraint integer programming. Ph.D. thesis, Technische Universität Berlin (2007)
2. Achterberg, T., Berthold, T.: Hybrid branching. In: van Hoeve, W.-J., Hooker, J.N. (eds.) CPAIOR 2009. LNCS, vol. 5547, pp. 309–311. Springer, Heidelberg (2009). https://doi.org/10.1007/978-3-642-01929-6_23
3. Achterberg, T., Koch, T., Martin, A.: Branching rules revisited. Oper. Res. Lett. **33**(1), 42–54 (2005)
4. Anderson, D., Le Bodic, P., Morgan, K.: Further results on an abstract model for branching and its application to mixed integer programming. Math. Program. **190**(1–2), 811–841 (2021)
5. Applegate, D., Bixby, R., Cook, W., Chvátal, V.: On the solution of traveling salesman problems (1998)
6. Bartoszyński, R., Govindarajulu, Z.: The secretary problem with interview cost. Sankhyā: Indian J. Stat. Ser. B 11–28 (1978)
7. Besancon, M., et al.: Distributions.jl: definition and modeling of probability distributions in the JuliaStats ecosystem. J. Stat. Softw. **98**(16), 1-30 (2021). https://doi.org/10.18637/jss.v098.i16
8. Bestuzheva, K., et al.: Enabling research through the SCIP optimization suite 8.0. ACM Trans. Math. Softw. **49**(2), 1–21 (2023)
9. Beyhaghi, H., Cai, L.: Pandora's problem with nonobligatory inspection: optimal structure and a PTAS
10. Bussieck, M.R., Drud, A.S., Meeraus, A.: MINLPLib'a collection of test models for mixed-integer nonlinear programming. INFORMS J. Comput. **15**(1), 114–119 (2003)
11. Conforti, M., Cornuéjols, G., Zambelli, G.: Integer Programming Models. Springer, Cham (2014)
12. Dey, S.S., Dubey, Y., Molinaro, M., Shah, P.: A theoretical and computational analysis of full strong-branching. Math. Program. 1–34 (2023)
13. Gamrath, G., et al.: The SCIP optimization suite 7.0 (2020)
14. Gamrath, G., Berthold, T., Salvagnin, D.: An exploratory computational analysis of dual degeneracy in mixed-integer programming. EURO J. Comput. Optim. **8**(3–4), 241–261 (2020)
15. Gianini, J., Samuels, S.M.: The infinite secretary problem. Ann. Probab. **4**(3), 418–432 (1976)
16. Gleixner, A., et al.: MIPLIB 2017: data-driven compilation of the 6th mixed-integer programming library. Math. Program. Comput. **13**(3), 443–490 (2021)
17. Hendel, G.: Enhancing MIP branching decisions by using the sample variance of pseudo costs. In: Michel, L. (ed.) CPAIOR 2015. LNCS, vol. 9075, pp. 199–214. Springer, Cham (2015). https://doi.org/10.1007/978-3-319-18008-3_14
18. Koch, T., Berthold, T., Pedersen, J., Vanaret, C.: Progress in mathematical programming solvers from 2001 to 2020. EURO J. Comput. Optim. **10**, 100031 (2022)
19. Le Bodic, P., Nemhauser, G.: An abstract model for branching and its application to mixed integer programming. Math. Program. **166**(1–2), 369–405 (2017)
20. Lodi, A., Tramontani, A.: Performance variability in mixed-integer programming. In: Theory Driven by Influential Applications, pp. 1–12. INFORMS (2013)
21. Turner, M., Berthold, T., Besançon, M., Koch, T.: Branching via cutting plane selection: improving hybrid branching. arXiv preprint arXiv:2306.06050 (2023)

22. Vigerske, S.: MINLPLib: a library of mixed-integer and continuous nonlinear programming instances (2018). https://www.minlplib.org. Accessed Dec 2023
23. Vigerske, S., Gleixner, A.: SCIP: global optimization of mixed-integer nonlinear programs in a branch-and-cut framework. Optim. Methods Softw. **33**(3), 563–593 (2018)
24. Weitzman, M.: Optimal Search for the Best Alternative, vol. 78. Department of Energy (1978)

Lookahead, Merge and Reduce for Compiling Relaxed Decision Diagrams for Optimization

Mohsen Nafar[(✉)] and Michael Römer

Bielefeld University, Universitätsstraße 25, 33615 Bielefeld, Germany
{mohsen.nafar,michael.roemer}@uni-bielefeld.de

Abstract. In this paper, we propose a new approach for the top-down compilation of relaxed Binary Decision Diagrams (BDDs) for Discrete Optimization: Lookahead, Merge and Reduce. The approach is inspired by the bottom-up algorithm for reducing exact BDDs in which equivalent nodes, that is, nodes with the same partial completions, are merged. In our top-down compilation approach, we apply this reduction algorithm for determining which states to be merged by constructing a lookahead layer, merging the lookahead layer nodes according to some heuristic and then deeming nodes having the same feasible completions in the lookahead BDD as approximately equivalent. Moreover, under certain structural properties we prove an upper limit on the size of the reduced layers given the size of the merged lookahead layer. In a set of preliminary computational experiments, we evaluate our approach for the 0/1 Knapsack problem, showing that the approach often achieves much stronger bounds than the traditional top-down compilation scheme.

Keywords: Decision Diagram · Discrete Optimization · Dynamic programming

1 Introduction

As recently documented in the survey [4], Decision Diagrams (DDs) are a powerful tool for discrete optimization. Given a Dynamic Programming (DP) formulation, a so-called *exact* DD can be used to compactly represent the set of feasible solutions of a discrete optimization problem. While the size of an exact DD grows exponentially in the number of decision variables, it can often be considerably compressed by applying an exact bottom-up reduction technique, which can be very useful when using the DD in a network flow reformulation [1] within a mixed-integer linear programming model. Furthermore, there are two types of approximate DDs: Relaxed DDs provide a discrete over-approximation of the solution space of a discrete optimization problem, whereas restricted DDs provide an under-approximation. Relaxed and restricted DDs can be used within a purely DD-based branch-and-bound algorithm [2] which has been successfully applied to a variety of optimization problems. In the context of a DD-based

B. Dilkina (Ed.): CPAIOR 2024, LNCS 14743, pp. 74–82, 2024.
https://doi.org/10.1007/978-3-031-60599-4_5

branch-and-bound, approximate DDs are typically compiled using the so-called top-down approach. For a given problem and a given maximum width, the quality of the bounds of the approximate DDs depends on heuristic decisions involving (i) the order in which the variables are considered, and (ii) the selection of the nodes in a DD layer to be removed (for restricted DD) or to be merged (for a relaxed DD) in case the maximum width is exceeded.

In this paper, we propose Lookahead, Merge, and Reduce, a node selection heuristic for relaxed BDDs that applies the bottom-up reduction logic for reducing exact BDDs to a partial lookahead BDD to determine approximately equivalent states. Moreover, we prove that under a certain set of conditions, the size of the reduced layer can not exceed a limit which is a function of the size of the merged lookahead layer. In a set of preliminary computational experiments with the 0/1 Knapsack problem we observed that our approach provides much stronger bounds than a standard top-down compilation procedure.

2   Decision Diagrams for Optimization

A binary decision diagram $\mathcal{D} = (\mathcal{N}, \mathcal{A})$ is a layered directed acyclic graph with node set $\mathcal{N}$ and arc set $\mathcal{A}$. The paths in $\mathcal{D}$ represent solutions to a discrete optimization problem $\mathscr{P}$ with a maximization objective function f and an n-dimensional vector of decision variables $x_1, \cdots, x_n \in \{0, 1\}$. $\mathcal{N}$ is partitioned into $n+1$ layers $\mathcal{N}_1, \ldots, \mathcal{N}_{n+1}$, where $\mathcal{N}_1 = \{\mathbf{r}\}$ and $\mathcal{N}_{n+1} = \{\mathbf{t}\}$ for a *root* $\mathbf{r}$ and a *terminal* $\mathbf{t}$. Each arc $a = (u, v)$ connects nodes of two consecutive layers and is associated with a decision $d(a)$ representing the assignment $x_u = d(a)$. This means that a path $p = (a_1, \ldots, a_n)$ starting from $\mathbf{r}$ and ending at $\mathbf{t}$ represents the solution $x(p) = (d(a_1), \ldots, d(a_n))$. We denote the set of all $\mathbf{r}$-$\mathbf{t}$ paths with $\mathcal{P}$, and we refer to the solutions to $\mathscr{P}$ represented by $\mathcal{P}$ with $\text{Sol}(\mathcal{D})$. Moreover, each arc a has length $v(a)$ and $\sum_{i=1}^{n} v(a_i)$ provides the length $v(p)$ of path p. We refer to $\mathcal{D}$ as exact if $\text{Sol}(\mathcal{D}) = \text{Sol}(\mathscr{P})$ if for each path $p \in \mathcal{P}$ we have $v(p) = f(x(p))$; then a longest path in $\mathcal{D}$ forms an optimal solution to $\mathscr{P}$.

A common approach to compile an exact DD is to provide a Dynamic Programming (DP) formulation of $\mathscr{P}$ and to compile the DD in a top-down fashion. To do so, every node u is associated with a state S_u and every arc a is associated with a state transition induced by the decision $d(a)$ associated with a. S_u is an element of the state space $\mathcal{S}$; the state $S_\mathbf{r}$ associated with the $\mathbf{r}$ is the so-called *initial state*. The state S_v of the target node v of the arc depends on the state S_u of the arc's source node as well as on d and is computed by the state-transition function $f(S_u, d)$. The objective function contribution of a decision are computed by a reward function $g(S_u, d)$. Finally, the set of out-arcs of a node u is determined by the set of feasible decisions $X(S_u)$ given state S_u. The top-down compilation then proceeds layer-by-layer until reaching layer $\mathcal{N}_n$; all arcs emanating from that layer point to the terminal node $\mathbf{t}$.

In a DD compiled in the sketched top-down fashion, any pair of nodes in a layer has different states, that is, partial paths ending in the same state point to the same node. Nonetheless, in general, layers may contain pairs of nodes which are *equivalent*, that is, nodes that have the same feasible completions or, in other words, that have the same sets of partial solutions until the terminal node. This means that the DD may contain isomorphic subgraphs which can be superimposed without affecting the solution space. An exact DD in which none of the layers contains a pair of equivalent states is called a *reduced* exact DD, and the reduction of an exact DD can be performed in linear time using a simple bottom-up algorithm [3,11]. Also see [9] for an in-depth discussion of efficient reduction algorithms for exact DDs and of a so-called relaxed reduction of a given exact DD.

Even if one can obtain a substantially smaller DD by these reduction algorithms, they require building a full exact DD which limits their usefulness to certain use cases such as embedding a DD in a network flow component of a MILP model. To deal with the exponential growth of the DD, many DD-based solution approaches such as DD-based branch-and-bound [2] rely on so-called approximate DDs that can be used to obtain upper or lower bounds for the solutions of $\mathscr{P}$. There are two types of approximate DDs: in a *restricted* DD $\mathcal{D}$, one aims at considering only promising nodes and arcs, meaning that $\mathrm{Sol}(\mathcal{D}) \subseteq \mathrm{Sol}(\mathscr{P})$, and thus, the longest path in a restricted DD provides a lower bound to $\mathscr{P}$. The second type of approximate DD, which is the one we focus on in this paper, is the *relaxed* DD, provides an upper bound: in a relaxed DD, we have $\mathrm{Sol}(\mathcal{D}) \supseteq \mathrm{Sol}(\mathscr{P})$, that is, the set of paths may contain paths associated with infeasible solutions to $\mathscr{P}$. Regarding the objective function value, every path a relaxed DD needs to satisfy $v(p) \geq f(x(p))$. In both restricted and relaxed DDs, a common approach to control the size of the DD is to impose a maximum width W for each layer in the DD. In case of a top-down compilation procedure, after having created all nodes of a given layer, one reduces its size to W by removing or merging nodes. In case of a relaxed DD, nodes are merged by redirecting the incoming arcs of the nodes to be merged to a single merged node. To ensure that no feasible completions of any of the merged nodes is lost, one requires a problem-specific merge operator $\oplus$ for the states associated with the two nodes, see [7] for a discussion of the conditions the merge operator needs to satisfy.

In any case, the strategy to select which nodes to merge is critical for the quality of the bound and thus has been subject to some amount of research. A popular and generic strategy is to sort the states according to some criterion, to keep the "best" $W - 1$ states and to merge the remaining states with the last state in the list. A classical and problem-agnostic criterion is sort in decreasing length of the partial path ending at each of the nodes, this approach will be called *sortObj* in the following. Other criteria being used for sorting the nodes take the state information associated with the nodes into account. Depending on the criterion being used, there may be nodes with the same value of the criterion; [6] propose a tie breaking strategy for that case. Instead of sorting states and creating a single large node, other authors aim at grouping nodes to merge according to some similarity measure. As an example, [8] propose to use so-called collector nodes that aim at merging states that have the same value with respect to a labeling function. A similar approach was recently used by [5] who merge nodes based on partitioning the state space for a single machine scheduling problem with release times, deadlines setup times and rejection.

3   Lookahead, Merge, and Reduce

In this section, we describe the main contribution of this paper, the LMR (i.e. Lookahead, Merge, and Reduce) approach for selecting which nodes to merge in a top-down compilation of relaxed BDDs. Later, theoretically, we derive an upper bound for one of the main parameters of LMR in a way that the relaxed BDD obtained with LMR respects a given maximum width. The key idea of LMR is to apply the concept of (completion-based) equivalence to a lookahead layer in which the lookahead layer is obtained by merging nodes based on state similarity. Specifically, in this approach, given the layer $\mathcal{N}_k$ to be relaxed, we first create a lookahead layer $\mathcal{N}_{k+1}$ whose nodes are merged according to a given node selection heuristic. Then, in analogy to the reduction of exact BDDs in which two nodes are *equivalent* if they exhibit the same set of completions in the exact BDD, we consider two nodes as *approximately equivalent* if they have the same set of completions in the partial BDD augmented by the lookahead layer. Formal definition of *approximately equivalent nodes* and accordingly *approximate equivalence class* are stated in the following.

Definition 1 (Approximately equivalent nodes). *Given an arc $a = (u, v)$, we refer to the pair $(d(a), v)$ as the decision-target (node) pair of a. Now, let $T(u)$ denote the set of the decision-target pairs of all out-arcs of u. We assume that the target nodes are part of a layer in which the nodes are relaxed/merged. Then, we refer to two nodes u and u' as approximately equivalent if $T(u) = T(u')$.*

Definition 2 (Approximate equivalence class). *A set of approximately equivalent nodes is called an approximate equivalence class.*

Figure 1 illustrates the steps to merge and bottom-up reduction procedure applied to obtain approximate equivalence classes in layer $\mathcal{N}_k$ given a layer $\mathcal{N}_{k+1}$ (on the left) and its merged version (yellow nodes on the right).

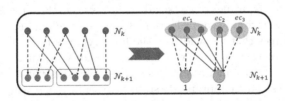

Fig. 1. Approximate equivalence classes in $\mathcal{N}_k$ according to merged $\mathcal{N}_{k+1} = \{1, 2\}$. (Color figure online)

The pseudocode of the procedure LMR is given in Algorithm 1. The parameter n_M defines the maximum number of nodes into which the lookahead layer is merged. Suppose that the width of the layer $\mathcal{N}_k$ has exceeded W, the top-down compilation is carried out to create layer $\mathcal{N}_{k+1}$, i.e. the lookahead layer, without limiting the maximum width W. Then the lookahead layer is merged into n_M

merged nodes. Then, the reduction is done to reduce/relax layer $\mathcal{N}_k$. Finally, we remove the lookahead layer from $\mathcal{D}$ and return the relaxed/reduced layer $\mathcal{N}_k$. Note that this approach can also be used to compile a restricted DD in which case we pick one of the original states in each node $u \in \mathcal{N}_k$ (i.e. select one node from every approximate equivalence class) as the representative to be kept.

Algorithm 1. LMR algorithm

1: **procedure** $LMR(DP, W, n_M)$
2:    **for** $k = 2$ TO $n + 1$ **do**
3:        **for all** $u \in \mathcal{N}_{k-1}$ **do**
4:            **for all** $d \in X(S_u)$ **do**
5:                $v = \textsc{GetOrAddNode}(\mathcal{N}_k, f(S_u, d))$
6:                $\textsc{AddArc}(u,v,d)$
7:            **end for**
8:        **end for**
9:        **if** $|\mathcal{N}_k| > W$ **then**
10:            **for all** $u \in \mathcal{N}_k$ **do**
11:                **for all** $d \in X(S_u)$ **do**
12:                    $v = \textsc{GetOrAddNode}(\mathcal{N}_{k+1}, f(S_u, d))$
13:                    $\textsc{AddArc}(u,v,d)$
14:                **end for**
15:            **end for**
16:            $\textsc{MergeLayer}(\mathcal{N}_{k+1}, n_M)$
17:            initialize empty map `ts-node`
18:            **for all** $u \in \mathcal{N}_k$ **do**
19:                **if** $T(u) \in$ `ts-node` **then**
20:                    merge u into `ts-node`$[T(u)]$
21:                **else**
22:                    `ts-node`$[T(u)] = u$
23:                **end if**
24:            **end for**
25:        **end if**
26:    **end for** **return** $\mathcal{D}$
27: **end procedure**

The choice of n_M must be done in a way that ensures that the maximum number of possible equivalence classes does not exceed W. This is because the nodes in every approximate equivalence class gets merged into exactly one node, therefore, it enforces that the width of the layer $\mathcal{N}_k$, the layer to reduced, after reduction does not exceed the given maximum width W. In the following, we theoretically prove that under a certain set of assumptions which can satisfied for a class of problems and an appropriate state representation, setting $n_M = \lfloor \frac{W+1}{2} \rfloor$ ensures that the width of the reduced layer is smaller than the given maximum width W. We then show that the 0/1 KP is an order-preserving problem.

Definition 3 (*order preserving* **problem**). *Let p_u^k be the position (index) of node u in a fixed ordering of the nodes in a layer $\mathcal{N}_k$ of a BDD of a problem $\mathcal{P}$. A problem is called an order preserving problem if there exists a total order of nodes (states) such that for any pair of nodes $u, v \in \mathcal{N}_k$ with $p_u^k < p_v^k$, the positions of their target nodes in layer $\mathcal{N}_{k+1}$ follows the same relation, i.e. $p_{u'}^{k+1} < p_{v'}^{k+1}$ where $S_{u'} = f(u, d)$ and $S_{v'} = f(v, d), \forall d \in \{0, 1\}$.*

Theorem 1. *0/1 KP problem is an order preserving problem.*

Proof. Let the nodes of a layer $\mathcal{N}_k$ in an instance of KP be ordered in an increasing order of the state values, where the state value corresponds to the weight accumulated in the knapsack so far. Let S_u be the state value of the node u and let y_k be the corresponding item of the instance which is going to be considered in the current stage of the BDD compilation, and w_{y_k} be its weight. Trivially the following holds for any pair of nodes $u, v \in \mathcal{N}_k$: $p_u^k < p_v^k \iff S_u < S_v$. Applying the transition function of the KP on nodes u and v we have that $p_{f(u,0)}^{k+1} = S_u < p_{f(v,0)}^{k+1} = S_v$ and $p_{f(u,1)}^{k+1} = S_u + w_x < p_{f(v,1)}^{k+1} = S_v + w_x$. □

Suppose the nodes in layers $\mathcal{N}_k$ and $\mathcal{N}_{k+1}$ of an order preserving problem are sorted. Assume that the layer $\mathcal{N}_{k+1}$ is merged into M merged nodes $m_1, \cdots, m_M$ such that for any pair of original nodes $u, v \in \mathcal{N}_{k+1}$ with $p_u^{k+1} < p_v^{k+1}$ and $u \in m_i, v \in m_j$ then $i \leq j$. This is the way we select nodes in the lookahead layer to merge, i.e. it is order/position preserving.

We know that the width of the layer $\mathcal{N}_k$, i.e. layer to be reduced, is at least $W + 1$ and at most $2 \cdot W$. If we can ensure that the maximum number of possible approximate equivalence classes is smaller than W, then by *pigeonhole principle* we can be sure that the width of the layer after reduction is less than the given maximum width W. Therefore, it just remains to compute the maximum possible number of approximate equivalence classes, which is the maximum number of ordered pairs of indices representing them. The following Theorem gives the maximum number of ordered pairs under certain conditions. These conditions are not random; the conditions I-III come from order preserving problems, and conditions IV-VI comes from the way we merge the lookahead layer.

Theorem 2. *Let $Q = \{q^1, \cdots, q^z\}$ be an ordered set of ordered pairs $q^l = (q_1^l, q_2^l)$, where $q_1^l, q_2^l \in \{1, \cdots, M\}, \forall l \in \{1, \cdots, z\}$, and rules I-VI hold. Then $z \leq (2 \cdot M) - 1$:*

- I. $q_1^l \leq q_2^l, \forall l \in \{1, \cdots, z\}$,
- II. $q_1^i \leq q_1^j, \forall i < j$,
- III. $q_2^i \leq q_2^j, \forall i < j$,
- IV. $q_1^1 = 1$,
- V. $q_2^z = M$,
- VI. $\forall y \in \{1, \cdots, M\}, \exists l, \textbf{ s.t. } q_1^l = y \textbf{ or } q_2^l = y \textbf{ or } q_1^l = q_2^l = y$.

Proof. Every assignment of values from $\{1, \cdots, M\}$ to coordinates of members of the set Q is equivalent to a route from an entry in the first row of the board (i.e. rule IV) in Fig. 2 and exiting from an exit of its last column (i.e. rule V). Based on the rule I we are not allowed to enter the shaded zone. Based on rules I, II, and III we can only move to "right", "down", and "cross" (i.e. right-down). Rule VI implies that in any route, each row or column must be entered at least once. The longest enter-exit route in the board corresponds to the largest number of distinct members of set Q (i.e. largest possible size of Q). The longest enter-exit route in an $M \times M$ board has a length of $(2 \cdot M) - 1$, which is entering from entry

1, 1 and moving right until reaching the entry $1, M$, and then moving down until M, M and exit from it. The length of this route is exactly $(2 \cdot M) - 1$. Thus Q cannot have more than $(2 \cdot M) - 1$ distinct members. □

Based on Theorem 1 and Theorem 2, if we apply a structured selection of nodes for merging (i.e. a structure that satisfies the rules IV, V, and VI), then if in LMR we set $n_M = \lfloor \frac{W+1}{2} \rfloor$ we can be sure that the width of no reduced layer will exceed the given maximum width W.

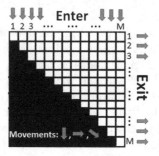

Fig. 2. Board game representation used in the proof of Theorem 2.

One of the selection to merge approaches for the 0/1 knapsack problem that follows the rules IV, V, and VI, which we used for our experiments, is called *Fixed* and is the following: let C be the capacity of the problem instance. Divide the range $[0-C]$ into n_M intervals and fix these intervals throughout the compilation. Then, merge all nodes $u \in \mathcal{N}_{k+1}$ whose state S_u belong to the same interval. Please note that there may exists other selection ways such that the resulting merging satisfies the rules IV, V, and VI. Since KP is an order preserving problem (by Theorem 1), the rules I, II, and III hold for it. Therefore, we can apply Theorem 2 and conclude that if the nodes of every lookahead layer are merged into $n_M \leq \lfloor \frac{W+1}{2} \rfloor$ merged nodes, then the maximum number of approximate equivalence classes in the reduced layer does not exceed W.

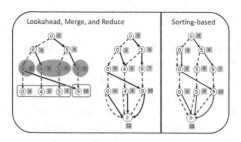

Fig. 3. Relaxed BDD compiled via LMR (left), and sortObj (right), for the example 1.

Example. Figure 3 depicts the resulting relaxed BDD compiled via LMR (left side), and sortObj (right side), where $W = 3, n_M = \lfloor \frac{W+1}{2} \rfloor = 2$), on the following 0/1 knapsack problem (KP):

$$\begin{aligned} \max \ & 4x_1 + 3x_2 + 10x_3 + 2x_4 \\ \text{s.t.} \ & 5x_1 + 4x_2 + 9x_3 + 3x_4 \le 12 \text{ and } x_i \in \{0,1\}, \forall i = 1,2,3,4 \end{aligned} \tag{1}$$

In Fig. 3, dashed and solid arcs are associated with $d = 0$ and $d = 1$. The interior labels for each node u correspond to the state values S_u; the state corresponds to the weight accumulated in the path $p(u)$ ending in u. The merge operator for merging two states S, S' in the relaxed BDD is defined as $\oplus = \min(S, S')$. The orange labels correspond to the length $v(u)$ of the partial path. Observe that in this example, the solution obtained with the LMR algorithm is optimal.

4   Computational Results

All experiments were performed on 100 instances of 0/1 KP with 200 items taken from [10]. Every approach is tested using the same variable ordering. We report the performance of two sorting-based node selection heuristics, i.e. sortObj and maxState in which nodes are sorted according to their state value, and the performances of LMR and OP (Order Preserving, without lookahead) using two different merge schemes, i.e. *Bucket* and Fixed (explained in the example). In the bucket scheme, we assume that the nodes are equally distributed into $n_M = W/2$ buckets and then every bucket is merged into a single node. The resulting gaps are shown in Fig. 4, where blue, green, black solid, black dashed, red solid, and red dashed curves correspond to maxState, sortObj, LMR-Fixed, LMR-Bucket, OP-Fixed, and OP-Bucket. In each graph, the y-axes show the gap, and the x-axes show the maximum width, BDD sizes, and run time in ms, from left to right. We set $W \in \{10, 50, 100, 200, 500\}$. A comparison between LMR-Bucket (black dashed) and OP-Bucket (red dashed) shows that LMR significantly improves upon OP. In general, the LMR algorithm consistently obtained the best gap and is much better than the classical sorting-based approaches.

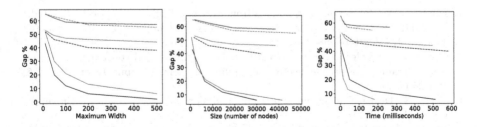

Fig. 4. Gaps from different approaches (gap vs W, gap vs size, and gap vs time)

5   Conclusion

The strength of the bounds obtained by top-down compiled approximate DDs crucially depends on the strategy to select nodes to merge or remove to respect the maximum width of each layer. In this paper, we propose a novel node selection heuristic named Lookahead, Merge, and Reduce (LMR) for compiling relaxed BDDs. LMR applies the bottom-up logic for reducing exact BDDs to a set of two consecutive layers, i.e. any layer with a width exceeding the given maximum width and its lookahead layer, to identify approximately equivalent nodes. Moreover, we prove an upper bound for the size of the reduced layer which is a function of the size of the merged lookahead layer. In a set of preliminary experiments with the 0/1 knapsack problem, we show that our approach provides much stronger bounds than a standard top-down compilation procedure.

Acknowledgements. This research was funded by the Return Programme of the Federal State of North Rhine Westphalia (NRW Rückkehrprogramm).

References

1. Behle, M.: Binary decision diagrams and integer programming (2007)
2. Bergman, D., Cire, A.A., van Hoeve, W.-J., Hooker, J.: Branch-and-bound based on decision diagrams. In: Bergman, D., Cire, A.A., van Hoeve, W.-J., Hooker, J. (eds.) Decision Diagrams for Optimization. AIFTA, pp. 95–122. Springer, Cham (2016). https://doi.org/10.1007/978-3-319-42849-9_6
3. Bryant, R.E.: Graph-based algorithms for Boolean function manipulation. IEEE Trans. Comput. **100**(8), 677–691 (1986)
4. Castro, M.P., Cire, A.A., Christopher Beck, J.: Decision diagrams for discrete optimization: a survey of recent advances. INFORMS J. Comput. **34**(4), 2271–2295 (2022)
5. de Weerdt, M., Baart, R., He, L.: Single-machine scheduling with release times, deadlines, setup times, and rejection. Eur. J. Oper. Res. **291**(2), 629–639 (2021)
6. Frohner, N., Raidl, G.R.: Towards improving merging heuristics for binary decision diagrams. In: Matsatsinis, N.F., Marinakis, Y., Pardalos, P. (eds.) LION 2019. LNCS, vol. 11968, pp. 30–45. Springer, Cham (2020). https://doi.org/10.1007/978-3-030-38629-0_3
7. Hooker, J.N.: Job sequencing bounds from decision diagrams. In: Beck, J.C. (ed.) CP 2017. LNCS, vol. 10416, pp. 565–578. Springer, Cham (2017). https://doi.org/10.1007/978-3-319-66158-2_36
8. Horn, M., Maschler, J., Raidl, G.R., Rönnberg, E.: A-based construction of decision diagrams for a prize-collecting scheduling problem. Comput. Oper. Res. **126**, 105125 (2021)
9. Perez, G.: Decision diagrams: constraints and algorithms. Ph.D. thesis, Université Côte d'Azur (2017)
10. Pisinger, D.: Where are the hard Knapsack problems? Comput. Oper. Res. **32**(9), 2271–2284 (2005)
11. Wegener, I.: Branching Programs and Binary Decision Diagrams: Theory and Applications. SIAM (2000)

LEO: Learning Efficient Orderings for Multiobjective Binary Decision Diagrams

Rahul Patel$^{(\boxtimes)}$ and Elias B. Khalil

Department of Mechanical and Industrial Engineering, University of Toronto,
Toronto, Canada
rm.patel@mail.utoronto.ca, khalil@mie.utoronto.ca

Abstract. Approaches based on Binary decision diagrams (BDDs) have recently achieved state-of-the-art results for some multiobjective integer programming problems. The variable ordering used in constructing BDDs can have a significant impact on their size and on the quality of bounds derived from relaxed or restricted BDDs for single-objective optimization problems. We first showcase a similar impact of variable ordering on the Pareto frontier (PF) enumeration time for the multiobjective knapsack problem, suggesting the need for deriving variable ordering methods that improve the scalability of the multiobjective BDD approach. To that end, we derive a novel parameter configuration space based on variable scoring functions that are linear in a small set of interpretable and easy-to-compute variable features. We show how the configuration space can be efficiently explored using black-box optimization, circumventing the curse of dimensionality (in the number of variables and objectives), and finding good orderings that reduce the PF enumeration time. However, black-box optimization approaches incur a computational overhead that outweighs the reduction in time due to good variable ordering. To alleviate this issue, we propose LEO, a supervised learning approach for finding efficient variable orderings that reduce the enumeration time. Experiments on benchmark sets from the knapsack problem with 3-7 objectives and up to 80 variables show that LEO is 3× and 1.3× faster at PF enumeration than a lexicographic ordering and algorithm configuration, respectively. Our code and instances are available at https://github.com/khalil-research/leo.

Keywords: Multiobjective optimization · Binary Decision Diagrams · Variable Ordering · Machine Learning

1 Introduction

In many real-world scenarios, one must jointly optimize over a set of conflicting objectives. Multiobjective optimization deals with the solution of such problems and has been successfully applied in novel drug design [39], space exploration

B. Dilkina (Ed.): CPAIOR 2024, LNCS 14743, pp. 83–110, 2024.
https://doi.org/10.1007/978-3-031-60599-4_6

[52,54], administrating radiotherapy [57], and supply chain network design [3], among others. In this paper, we specifically focus on multiobjective integer programming (MOIP) which have integer variables and linear constraints.

The exact solution of a multiobjective problem consists in finding the Pareto frontier (PF): the set of feasible solutions that are not dominated by any other solution, i.e., ones for which improving the value of any objective deteriorates at least one other objective. PF solutions provide the decision-maker with a set of trade-offs between the conflicting objectives. Objective-space search methods iteratively solve multiple related single-objective problems to enumerate the PF but suffer from redundant computations in which previously found solutions are encountered again or a single-objective problem turns out infeasible. On the other hand, decision-space search methods leverage branch and bound. Unlike the single-objective case where one compares a single scalar bound (e.g., in mixed-integer linear programming (MIP)), one needs to compare bound *sets* to decide if a node can be pruned; this in itself is quite challenging. Additionally, other crucial components of branch and bound such as branching variable selection and presolve are still underdeveloped, limiting the usability of this framework.

Binary decision diagrams (BDDs) have been a central tool in program verification and analysis [15,55]. More recently, however, they have been used to solve discrete optimization problems [7,8] that admit a recursive formulation akin to that of dynamic programming. BDDs leverage this structure to get an edge over MIP by efficiently encoding the feasible set into a network model which enables fast optimization. To the best of our knowledge, Bergman and Cire [6] were the first to use BDDs to solve multiobjective problems, achieving state-of-the-art results for a number of problems. The variable ordering (VO) used to construct a BDD has a significant impact on its size and consequently any optimization performed using it. However, the variable ordering question in BDD-based MOIP has not been tackled in the literature. We address this gap by designing a novel learning-based BDD variable ordering technique for faster enumeration of the PF.

We begin with the following hypothesis: VO has an impact on the PF enumeration time and an "efficient" VO can reduce it significantly. Following an empirical validation of this hypothesis, we show that such orderings can be found using black-box optimization, not directly in the (exponentially large) space of variable orderings, but rather indirectly in the space of constant-size variable scoring functions. The scoring function is a weighted linear combination of a fixed set of variable properties (or attributes), and the indirect search is in the space of possible weight combinations. However, solving the black-box optimization problem may be prohibitively time-consuming for any one instance. For this approach to be viable in practice, the time required to produce an efficient VO should be negligible relative to the actual PF enumeration time. To alleviate this issue, we train a supervised machine learning (ML) model on the orderings collected using black-box optimization. A trained model can then be used on an unseen (test) instance to predict a VO. Should such a model generalize well, it would lead to reduced PF enumeration time. We refer to our approach as LEO (Learning Efficient Orderings). Our key contributions can be summarized as follows:

1. We show that variable ordering can have a dramatic impact on solving times through a case study of the multiobjective knapsack, a canonical combinatorial problem.
2. We show how black-box optimization can be leveraged to find efficient variable orderings at scale.
3. We design a supervised learning framework for predicting variable orderings which are obtained with black-box optimization on a set of training instances. Our ML models are invariant to permutations of variables and independent of the number of variables, enabling fast training and the use of one ML model across instances of different sizes.
4. We perform an extensive set of experiments on the knapsack problem and show that LEO is 3× and 1.3× faster on average than the lexicographic ordering and the SMAC algorithm configuration tool, respectively.
5. We perform a feature importance analysis of the best class of ML models we have found, extreme gradient boosted ranking trees. The analysis reveals that: (a) a single ML model can be trained across instances with varying numbers of objectives and variables; and (b) a single knapsack-specific feature that we had not initially contemplated performs reasonably well on its own, though far worse than our ML models.

2 Preliminaries

2.1 Multiobjective Optimization

An MOIP problem takes the form $\mathcal{M} := \min_x \left\{ z(x) : x \in \mathcal{X}, \mathcal{X} \subset \mathbb{Z}_+^n \right\}$, where x is the decision vector, $\mathcal{X}$ is a polyhedral feasible set, and $z : \mathcal{X} \to \mathbb{R}^p$ a vector-valued objective function representing the p objectives. In this work, we focus on the knapsack problem with binary decision variables, hence $\mathcal{X} \subset \{0,1\}^n$. Let $[p] = \{1, ..., p\}, p \in \mathbb{Z}^+$. Consider two points $x^1, x^2 \in \mathcal{X}$ with $y^1 = z(x^1), y^2 = z(x^2)$. We say that y^1 *dominates* y^2 if $y_j^1 \leq y_j^2, \forall j \in [p]$ and $\exists \bar{j} \in [p] : y_{\bar{j}}^1 < y_{\bar{j}}^2$. Let $\mathcal{Y} = \{z(x) : x \in \mathcal{X}\}$ be the set of feasible objective values. Exact solution approaches to multiobjective optimization aim to find the set of *nondominated* objective vectors $\mathcal{Y}_N \subseteq \mathcal{Y}$, i.e., no objective vector in $\mathcal{Y}_N$ is dominated by an objective vector in $\mathcal{Y}$. This set of *nondominated* vectors is also called the Pareto frontier.

2.2 BDDs for Multiobjective Optimization

A BDD is a compact encoding of the feasible solution set of a combinatorial optimization problem that exploits a recursive formulation of the problem. Formally, a BDD is a layered acyclic graph $G = (n, \mathcal{N}, \mathcal{A}, \ell, d)$, where n is the number of variables of the multiobjective problem $\mathcal{M}$, $\mathcal{N}$ represents the set of nodes, $\mathcal{A}$ is the set of arcs, $\ell : \mathcal{N} \to [n+1]$ a mapping from a node to a layer, $d : \mathcal{A} \to \{0,1\}$ a mapping from an arc to its label.

The nodes are partitioned into $n + 1$ layers $L_1, \ldots, L_{n+1}$, where $L_l = \{u : \ell(u) = l, u \in \mathcal{N}, l \in [n + 1]\}$. The first and last layers have only one node, which is referred to as the root node $\mathbf{r}$ and terminal node $\mathbf{t}$, respectively. The width of a layer L_l and BDD G is equal to $|L_l|$ and $\max_{l \in [n+1]} |L_l|$, respectively. An arc $a := (r(a), t(a)) \in \mathcal{A}$ starts from the node $r(a) \in L_l$ and ends in node $t(a) \in L_{l+1}, l \in [n]$. It has an associated label $d(a) \in \{0, 1\}$ and a vector of values $\bar{v}(a) \in \mathbb{R}_+^p$ that represents the contribution of that arc to the p objective values.

Let $\mathcal{P}$ represent all the paths from the root to the terminal node. A path $e = (a_1, a_2, \ldots, a_n) \in \mathcal{P}$ is equal to the solution $x(e) = (d(a_1), d(a_2), \ldots, d(a_n))$, with the objective value $\bar{v}(e) = \sum_{i=1}^n \bar{v}(a_i)$. The BDD representation of $\mathcal{M}$ is exact if $\{x(e) : e \in \mathcal{P}\} = \mathcal{X}$ and valid if $\text{ND} \left(\bigcup_{e \in \mathcal{P}} \bar{v}(e) \right) = \mathcal{Y}_N$, where ND is an operator to retrieve the nondominated objective vectors. We refer the readers to Bergman et al. [5] and Bergman and Cire [6] for a detailed description of the construction of the multiobjective knapsack BDD. In what follows, we assume access to BDD construction and PF enumeration procedures and will focus our attention on the variable ordering aspect.

3   Related Work

3.1   Exact Approaches for Solving MOIP Problems

Traditional approaches to exactly solve MOIP can be divided into objective-space search and decision-space search methods. The objective-space search techniques [11,12,38,44] enumerate the PF by searching in the space of objective function values. They transform a multiobjective problem into a single-objective one either by weighted sum aggregation of objectives or transforming all but one objective into constraints. The decision-space search approaches [1,4,45,47,53] instead explore the set of feasible decisions. Both these approaches have their own set of challenges as described in the introduction. We point the reader to [27,28] for a more detailed background.

Bergman et al. [5] showcased that BDD-based approaches can leverage problem structure and can be orders of magnitude faster on certain problem classes; the use of valid network pruning operations along with effective network compilation techniques were the prime factors behind their success.

3.2   Variable Ordering for BDD Construction

Finding a VO that leads to a BDD with a minimal number of nodes is an NP-Hard problem [13,14]. This problem has received significant attention from the verification community as smaller BDDs are more efficient verifiers. Along with the size, VO also affects the objective bounds; specifically, smaller exact BDDs are able to obtain better bounds on the corresponding limited width relaxed/restricted BDDs [8].

The VO techniques for BDD construction can be broadly categorized as exact or heuristic. The exact approaches to VO [9,25,26,29], though useful for smaller cases, are not able to scale with problem size. To circumvent the issue of scalability for larger problems, heuristic VO techniques are used; the proposed methodology falls into this category. These heuristics can be general or problem-specific

[2, 8, 16, 23, 43, 48, 49] but the literature has not tackled the multiobjective optimization setting. A VO method can also be classified as either *static* or *dynamic*. Static orderings are specified in advance of BDD construction whereas dynamic orderings are derived incrementally as the BDD is constructed layer-by-layer. The latter may or may not be applicable depending on the BDD construction algorithm; see [33] for a discussion of this distinction for Graph Coloring. We focus on static orderings and discuss extensions to the dynamic setting in the Conclusion. We focus on ML-based heuristics for VO as they can be learned from a relevant dataset instead of handcrafted heuristics developed through a tedious trial-and-error process.

3.3   Machine Learning for Variable Ordering

Grumberg et al. [30] proposed a learning-based algorithm to construct smaller BDDs for model verification. In particular, they create random orders for a given instance in the training set and tag variable pairs based on their impact on the resulting BDD size. Using this data, they learn multiple pair precedence classifiers. For a new instance at test time, they query each trained pair precedence classifier to construct a precedence table. These tables are merged into one to derive the ordering. The success of this method hinges on the selective sampling of informative variable pairs and the ability to generate orders with sufficient variability in BDD size to increase the chance of observing high-quality labels. As they leverage problem-specific heuristics for selective sampling and rely on random sampling for generating labels, this method is not applicable to our setting. Specifically, we do not have a notion of how informative a given variable pair is. Additionally, random orders may not produce high variability in PF enumeration time as evidenced by our experiments.

Carbin [19] uses active learning to address the VO problem for BDDs used in program analysis with a similar goal as ours of minimizing the run time. However, certain differences make it less amenable to our setting. Specifically, the technique to generate the candidate VOs in Carbin [19] is grounded in program analysis and cannot be applied to our problem. Instead, LEO leverages bayesian optimization through SMAC in conjunction with the novel property-weight search space to generate VO candidates.

Drechsler et al. [24] use an evolutionary search approach to learn a single variable ordering heuristic for a set of instances. The learned heuristic is a sequence of BDD operations (e.g., swapping two variables in an ordering) applied to instances from circuit design and verification, where the BDD represents a boolean function that must be verified efficiently. This approach can be seen as one of algorithm configuration, which we will compare to using SMAC and ultimately outperform.

The work of Cappart et al. [18] is the first learning-based method to address VO problem for BDDs used in solving discrete optimization problems. Specifically, they learn a policy to order variables of relaxed/restricted BDDs to obtain tighter bounds for the underlying optimization problem using reinforcement learning (RL). A key component of training an RL policy is devising the reward that the agent receives after taking an action in a given state and moving to the next state. Note that we care about reducing the time to enumerate the PF, which is only available at the end of a training episode, yielding sparse rewards that inhibit the application of RL.

Karahalios and van Hoeve [33] developed an algorithm portfolio approach to select the best VO strategy from a set of alternatives when constructing relaxed decision diagrams for the (single-objective) graph coloring problem. Fundamental to a portfolio-based approach is the existence of a set of strategies, such that each one of them is better than the others at solving some problem instances. However, such an algorithm portfolio does not exist in our case and part of the challenge is to discover good ordering strategies.

There have been some recent contributions [42,56] relating to solving multiobjective problems using deep learning and graph neural networks. However, these approaches are not exact and thus beyond the scope of this paper. We discuss potential extensions of LEO to the inexact setting in the Conclusion.

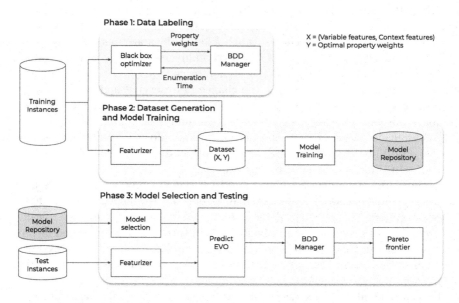

Fig. 1. Schematic of the proposed method LEO, which comprises three phases. In Phase 1, we generate property-weight labels by iteratively improving them using the interplay between "Black-box optimizer" and "BDD Manager". Phase 2 focuses on building datasets of tuples of features extracted by "Featurizer" and the best property weights obtained from Phase 1 for training learning-to-rank models. In Phase 3, "Model selection" is performed to obtain the best model and use it to predict an efficient variable ordering. Finally, we use the predicted variable ordering to construct the BDD and enumerate the Pareto frontier.

4   Methodology

We apply our technique to the multiobjective knapsack problem (MKP) which can be described as:

$$\max_{x \in \{0,1\}^n} \left\{ \left\{ \sum_{i \in [n]} a_i^p x_i \right\}_{p=1}^P : \sum_{i \in [n]} w_i x_i \leq W \right\}.$$

Here, n is the number of items/variables, w_i and a_i^p are the weight and profit corresponding to each item $i \in [n]$ and objective $p \in [P]$. Finally, the capacity of the knapsack is $W \in \mathbb{Z}_+$.

Let $\mathcal{O}$ be the set of all possible variable orderings and $\Gamma(o), o \in \mathcal{O}$ denote the PF enumeration time over a BDD constructed using ordering o. Let $o^\star \equiv \arg\min_{o \in \mathcal{O}} \Gamma(o)$ be the optimal VO. Finding $o^\star$ among all $n!$ possible permutations of n variables is intractable, so we will aim for an efficient variable ordering (EVO) o^e that is as close as possible to $o^\star$ in terms of PF enumeration time. Note that our approach is heuristic and does not come with approximation guarantees on the enumeration time of o^e relative to $o^\star$.

The proposed methodology, as depicted in Fig. 1, is divided into three phases. The objective in the first phase is to find, for each training instance, an EVO that acts as a label for supervising an ML model. In the second phase, each training instance is mapped to a set of features and an ML model is trained. Finally, we perform model selection and use the chosen model to predict EVOs, referred to as $\hat{o}^e$, that are then used to construct a BDD and compute the PF for any test instance.

4.1   Phase 1: Finding an EVO

Since finding an optimal VO that minimizes the PF enumeration time is NP-Hard, we devise a heuristic approach for finding an EVO, o^e. To find o^e for a given instance $\mathcal{I}$, we use black-box optimization as the run time $\Gamma^{\mathcal{I}}$ cannot be described analytically and optimized over. A naive approach to find o^e would be to search directly in the variable ordering space. While this might work well for tiny problems, it will not scale with the increase in the problem size as there are $n!$ possible orderings.

To alleviate this issue, we define a surrogate search space that removes the dependence on the problem size. Specifically, we introduce a score-based variable ordering where we order the variables in the decreasing order of their total score. For a given problem class, we define a set of properties $\mathcal{K}$ for its variables that capture problem structure. For example, in an MKP, the weight of an item can act as property. Table 1 lists all properties of a variable of an MKP. Let g_{ik} be the property score of variable i for some property $k \in \mathcal{K}$, $\bar{\lambda} = (\lambda_1, \cdots, \lambda_{|\mathcal{K}|})$ be the property weights in $[-1,1]^{|\mathcal{K}|}$. Then, the score of a variable i is defined as

$$s_i \equiv \sum_{k \in \mathcal{K}} \lambda_k \cdot \frac{g_{ik}}{\sum_{i \in [n]} g_{ik}}. \tag{1}$$

Table 1. Properties of a variable i of an MKP.

Property	Definition
weight	w_i
avg-value	$\sum_{p=1}^{P} a_i^p / P$
max-value	$\max\{a_i^p\}_{p=1}^{P}$
min-value	$\min\{a_i^p\}_{p=1}^{P}$
avg-value-by-weight	$\left(\sum_{p=1}^{P} a_i^p / P\right) / w_i$
max-value-by-weight	$\max\{a_i^p\}_{p=1}^{P} / w_i$
min-value-by-weight	$\min\{a_i^p\}_{p=1}^{P} / w_i$

We recover the variable ordering by sorting them in decreasing order of their score. The search for o^e is conducted in the surrogate search space $[-1, 1]^{|\mathcal{K}|}$, which only depends on $|\mathcal{K}|$ and not on the number of variables n. Note that defining the search space in this manner gives an additional layer of dynamism in the sense that two instances with the same property weights can have different variable orderings. We would like to emphasize that the idea of properties can be extended to problem classes with more than one constraint, as has been proposed in Khalil et al. [36] and Sierra-Altamiranda et al. [51] for the general integer linear programs. In fact, the current set of properties can be thought of as their special case.

With a slight abuse of notation, let $\Gamma(\bar{\lambda})$ represent the time taken to enumerate the PF using a VO obtained by property weights $\bar{\lambda}$. Given a problem instance, the black-box optimizer controls the property weights and the BDD manager computes the PF enumeration time based on the VO derived from these property weights. The black-box optimizer, in our case SMAC [41], performs Bayesian optimization to build a probabilistic model for predicting PF enumeration time over the search space. It uses this model to query a promising new incumbent VO and update the model based on its run time in an iterative manner, as depicted in Phase 1 of Fig. 1. We propose to use the variable ordering obtained from the best incumbent property weight as the label for the learning task.

4.2   Phase 2: Dataset Generation and Model Training

In this phase, we begin by generating the training dataset. We give special attention to designing our features such that the resulting models are permutation-invariant and independent of the size of the problem. Note that instead of this feature engineering approach, one could use feature learning through graph neural networks or similar deep learning techniques, see [17] for a survey. However, given that our case study is on the knapsack problem, we opt for domain-specific feature engineering that can directly exploit some problem structure and leads to somewhat interpretable ML models.

Suppose we are given J problem instances, each having n_j variables, $j \in [J]$. Let α_{ij} denote the features of variable $i \in [n_j]$ and β_j denote the instance-level context features. Using the features and the EVO computed in Phase 1, we construct a dataset $\mathcal{D} = \{(\alpha_{ij}, \beta_j, r_i(o^e{}_j) : i \in [n_j] : j \in [J]\}$. Here, o_j^e is the EVO of an instance j and $r_i : \mathbb{Z}_+^{n_j} \to \mathbb{Z}_+$ a mapping from the EVO to the rank of a variable i. For example, if $n_j = 4$ for some instance j and $o_j^e = (2, 1, 4, 3)$, then $r_1(o_j^e) = 3$, $r_2(o_j^e) = 4$, $r_3(o_j^e) = 1$, and $r_4(o_j^e) = 2$. For a complete list of variable and context features, refer to Table 5.

Learning-to-Rank. (LTR) is an ML approach specifically designed for ranking tasks, where the goal is to sort a set of items based on their relevance given a query. It is commonly used in applications such as information retrieval. We formulate the task of predicting the EVO as an LTR task and use the pointwise and pairwise ranking approaches to solve the problem.

In the **pointwise approach**, each item (variable) in the training data is treated independently, and the goal is to learn a model that directly predicts the relevance score or label for each variable. This is similar to solving a regression problem with a mean-squared error loss. Specifically, we train a model $f_\theta(\alpha_{ij}, \beta_j)$ to predict $r_i(o_j^e)$. Once the model is trained, it can be used to rank items based on their predicted scores $\hat{o}^e$. However, this approach does not explicitly consider the relationships between variables in its loss function. The **pairwise approach** aims to resolve this issue [31]. Let $\mathcal{T}_j = \{(i_1, i_2) : r_{i_1}(o_j^e) > r_{i_1}(o_j^e))\}$ be the set of all variable tuples (i_1, i_2) such that i_1 is ranked higher than i_2 for instance j. Then, the goal is to learn a model that maximizes the number of respected pairwise-ordering constraints. Specifically, we train a model $g_\phi(\cdot)$ such that the number of pairs $(i_1, i_2) \in \mathcal{T}_j$ for which $g_\phi(\alpha_{i_1 j}, \beta_j) > g_\phi(\alpha_{i_2 j}, \beta_j)$ is maximized. This approach is better equipped to solve the ranking problem as the structured loss takes into account the pairwise relationships.

4.3   Phase 3: Model Selection and Testing

We follow a two-step approach to perform model selection as the ranking task is a proxy to the downstream task of efficiently solving a multiobjective problem. Firstly, for each model class (e.g., decision trees, linear models, etc.), we select the best model based on Kendall's Tau [34], a ranking performance metric that measures the fraction of violated pairwise-ordering constraints, on instances from a validation set different from the training set. Subsequently, we pit the best models from each type against one another and select the winner based on the minimum average PF enumeration time on the validation set. Henceforth, for previously unseen instances from the test set, we will use the model selected in Phase 3 to predict the EVO and then compute the PF.

5   Computational Setup

Our code and instances are available at https://github.com/khalil-research/leo. All the experiments reported in this manuscript are conducted on a computing cluster with an Intel Xeon CPU E5-2683 CPUs. We use SMAC [41] – a black-box optimization and algorithm configuration library – for generating the labels. The ML models are built using Python 3.8, Scikit-learn [46], and XGBoost [22], and SVMRank [32]. The "BDD Manager" is based on the implementation of Bergman and Cire [6], available at https://www.andrew.cmu.edu/user/vanhoeve/mdd/.

Instance Generation: We use a dataset of randomly generated MKP instances as described in [6]. The values w_i and a_i^p are sampled randomly from a discrete uniform distribution ranging from 1 to 1000. The capacity W is set to $\lceil 0.5 \sum_{i \in I} w_i \rceil$. We generate instances with sizes

$$\mathcal{S} = \{(7, 40), (6, 40), (5, 40), (4, 50), (3, 60), (3, 70), (3, 80)\},$$

where the first and second component of the tuple specify the number of objectives and variables, respectively. For each size, we generate 1000 training instances, 100 validation instances, and 100 test instances.

5.1   Instrumenting SMAC

As a Labeling Tool: To generate EVOs for the learning-based models, we use SMAC [41]. In what follows, SmacI refers to the use of SMAC as a black-box optimizer that finds an EVO for a given training instance. Specifically, SmacI solves $\bar{\lambda}_j^e = \arg\min_{\bar{\lambda}} \Gamma_j(\lambda)$ for each instance j in the training set; we obtain o_j^e by calculating variable scores – a dot product between $\bar{\lambda}_j^e$ and the corresponding property value – and sorting variables in the decreasing order of their score.

As a Baseline: The other more standard use of SMAC, which we refer to as SmacD, is as an algorithm configuration tool. To obtain a SmacD ordering, we use SMAC to solve $\bar{\lambda}_D^e = \arg\min_{\bar{\lambda}} \mathbb{E}_{j\sim[J]} \left[\Gamma_j(\lambda) \right]$ and obtain an ordering $o_{D_j}^e$ for instance j using single property weight vector $\bar{\lambda}_D^e$. The expectation of the run time here simply represents its average over the $|J|$ training instances. Note that we get only one property weight vector for the entire dataset in the SmacD case rather than one per instance as in the SmacI case. However, we obtain an instance-specific VO when using the SmacD configuration as the underlying property values change across instances. Additional details about the SMAC setup can be found in Appendix B

5.2   Learning Models

We use linear, ridge, and lasso regression, decision trees, and gradient-boosted trees (GBT) with mean-squared error loss to build size-specific pointwise ranking models. Similarly, we train support vector machines and GBT with pairwise-ranking loss to obtain pairwise ranking models for each size. The details of the feature engineering are presented in Appendix A. We omit the context features in Table 5 when training linear size-specific models, as these features take on the same value for all variables of the same instance and thus do not contribute to the prediction of the rank of a variable. The context features are used with non-linear models such as decision trees and GBT. The model selection follows the procedure mentioned in Phase 3 of the Methodology section. The GBT-based models trained with pairwise-ranking loss perform the best across different model classes. In the experimental results that follow, the GBT-based models will be referred to as ML (trained only on variable features) and ML+C (trained on variable and context features). We also train two additional GBT models – ML+A and ML+AC – with pairwise-ranking loss, on the *union of the datasets of all sizes*. In particular, ML+A is trained with only variable features, similar to ML, whereas ML+AC also uses the instance context features like ML+C. These models achieved a Kendall's Tau ranging between 0.67 to 0.81 across all problem sizes on the validation set.

5.3 Baselines

To evaluate the performance of learning-based orderings, we compare to four baselines. **Lex** (lexicographic) uses the (arbitrary) default variable ordering in which the instance was generated. **MinWt** orders the variables in increasing order of their weight values, w_i. This is a commonly used heuristic for solving the single-objective knapsack problem. **MaxRatio** orders the variables in decreasing order of the property min-value-by-weight detailed in Table 1, which is defined as $\min\{a_i^p\}_{p=1}^P/w_i$. This rule has an intuitive interpretation: it prefers variables with larger *worst-case* (the minimum in the numerator) value-to-weight ratio. It is not surprising that this heuristic might perform well given that it is part of a $\frac{1}{2}$-approximation algorithm for the single-objective knapsack problem [21]. **SmacD**, as described in an earlier paragraph. It produces one weight setting for the property scores per dataset and can be seen as a representative for the algorithm configuration paradigm recently surveyed by Schede et al. [50].

6 Experimental Results

We examine our experimental findings through a series of questions that span the impact of VO on PF enumeration time (Q1), the performance of LEO on unseen test instances and comparison to the baselines (Q2, Q3), and a feature importance analysis of the best ML models used by LEO (Q4).

Table 2. The best and worst ratio of the average time taken by heuristic orderings to that of 5 random orderings across 250 MKP instances.

Ordering	(3, 40)	(3, 60)	(3, 80)	(5, 30)	(5, 40)	(7, 30)	(7, 40)
best	0.654	0.537	0.590	0.700	0.533	0.588	0.581
worst	1.662	2.266	1.729	1.865	2.760	2.769	1.830

Q1. Does variable ordering impact the Pareto frontier enumeration time? To test this, we compare the run time of ten heuristic orderings (based on variable properties) against the expected run time of random orderings. We estimate the expected run time of a random ordering by sampling 5 variable orderings uniformly at random from all possible $n!$ orderings and averaging their run time. Table 2 summarizes the results of this experiment. It validates two of our hypotheses: the choice of VO dramatically affects run time (as evidenced by the large gap between "best" and "worst"), and there is always some ordering that is far better than random (the "best" values are much smaller than 1). Refer to Appendix C.1 for the detailed results.

Table 3. Shifted geometric mean of the PF enumeration time on the test set containing 100 instances with a time limit of 1800 s. The number in superscript represents the count of instances that did not run successfully either due to hitting memory or time limit. The geometric mean is computed with a shift of 5.

Size	ML	ML+C	ML+A	ML+AC	SmacD	MaxRatio	MinWt	Lex
(7, 40)	50.95	[1]**47.97**	[1]53.83	[1]54.54	[1]74.24	[2]96.38	[1]69.11	[5]111.59
(6, 40)	20.90	20.59	20.06	**19.90**	25.38	26.68	25.63	49.98
(5, 40)	5.00	4.99	**4.90**	4.99	6.80	6.66	6.68	11.75
(4, 50)	**11.18**	11.34	11.50	11.39	18.12	20.33	24.66	38.75
(3, 60)	8.31	8.53	8.41	**8.24**	9.32	14.68	17.13	25.89
(3, 70)	20.46	20.80	20.79	**20.38**	22.13	39.13	50.16	77.46
(3, 80)	49.42	49.66	50.27	**48.48**	59.05	106.08	135.33	[1]202.25

Q2. How does LEO perform in comparison to baseline methods? Table 3 presents the shifted geometric mean of the PF enumeration time across different problem sizes and methods on the test set. We can observe that learning-based methods consistently outperform baselines across all problem sizes. SmacD acts as a strong baseline, being the second-best except for size (7, 40). The methods MinWt and MaxRatio have a nice connection to the single-objective knapsack problem [21], as discussed earlier; this might explain why these heuristics reduce the number of intermediate solutions being generated in the BDDs during PF enumeration (see Appendix C.3), helping the algorithm terminate quickly. They closely follow SmacD with sizes having more than 3 objectives; however, they are almost twice as worst as SmacD on instances with 3 objectives. Interestingly, when the number of objectives is larger than 5, MinWt outperforms MaxRatio (and vice versa). This underscores the relationship between the number of objectives and heuristics, i.e., one heuristic might be preferred over another depending on the structure of the problem. This also highlights why the ML method performs better than the heuristics as the feature engineering helps create an ensemble rather than using only one of them. Lastly, we observe that Lex performs worst across all sizes.

Table 4. Analyzing the impact of BDD topology on the PF enumeration time on the test set. For a given "Size" and "Method", the metric value indicates the percentage of mean performance by that "Method" compared to the Lex method. The "Nodes" ("Width") represents the percentage of the mean of the number of nodes (mean width) of the BDDs generated by "Method" to that of Lex. "Degree" denotes the percentage of the mean ratio of the in-degree of the 50 nodes with the highest in-degree in BDDs generated by "Method" to that of Lex. We focus on nodes with a high in-degree as they are key to the computational cost of enumeration. Similarly, "Checks" and "Time" indicate the percentage of the average Pareto-dominance checks and the average geometric mean of the PF enumeration time, respectively. For all metrics, values less than 100 are preferred. For example, a value of δ for "Nodes" indicates that the mean number of nodes for "Method" is $\delta/100\times$ the mean number of nodes of "Lex".

Size	(7, 40)					(3, 80)				
Method	Nodes	Width	Degree	Checks	Time	Nodes	Width	Degree	Checks	Time
ML	80.37	88.01	74.92	**38.31**	**38.98**	90.84	98.56	62.72	**19.02**	**24.09**
SmacD	70.08	79.22	149.06	64.24	59.56	92.21	99.18	79.72	25.73	28.81
MaxRatio	71.24	80.63	**66.72**	85.14	79.63	84.23	97.21	**45.26**	54.71	51.71
MinWt	**69.18**	**78.69**	151.10	59.23	55.66	**68.64**	**90.48**	114.92	70.17	65.71

We would also like to highlight that the two models trained on the union of all datasets of different sizes, ML+A and ML+AC, perform remarkably well. The main takeaway here is that our model architecture, which is independent of the number of variables in an instance, enables the training of unified models that are size-independent. Refer to Appendix C.2 for detailed statistics on PF enumeration time and performance profile (number of instances solved vs. enumeration time) for different methods.

Q3. What explains LEO's performance? Traditionally, smaller exact BDDs are sought for efficient model checking [30] or computing objective function bounds [9]. For our use case, we need to run a multicriteria shortest path algorithm on BDDs to enumerate the PF. The design of multicriteria shortest path algorithms is an active research area and several approaches with varying output-sensitive complexity bounds exist [10,20,35,40]. The complexity bound of our implementation is $O(|\mathcal{A}|P\Gamma^2)$ [20], where $|\mathcal{A}|$ is the number of arcs in the BDD, P is the number of objectives, and Γ is the maximum number of intermediate solutions generated during enumeration at a node. For a fixed P, the bound suggests that variable orderings that lead to BDDs with fewer intermediate solutions will be faster. We analyze the empirical relationship between BDD topology and PF enumeration time for various orderings in Table 4.

We can observe that MinWt leads to BDDs with the minimum Nodes and Width, whereas the BDDs generated by MaxRatio have the minimum Degree. On the other hand, ML has neither the minimum Nodes nor Degree but is the fastest in PF enumeration. This is explained by the fact that ML has the minimum number of Checks, which depends on the number of intermediate solutions generated during enumeration. ML seems to lead to a better trade-off between the number of arcs in the BDD $|\mathcal{A}|$ and the number of intermediate solutions Γ, leading to reduced PF enumeration time. Note that only optimizing the BDD topology such as the number of Nodes or the Degree may result in better-than-random but overall suboptimal performance.

Q4. How interpretable are the decisions made by LEO? LEO uses GBT as the model of choice as explained in Sect. 5.2. To obtain feature importance scores of ML models, we count the number of times a particular feature is used to split a node in the decision trees. We then normalize these scores by dividing them by the maximum score, resulting in values in $[0, 1]$ for each size.

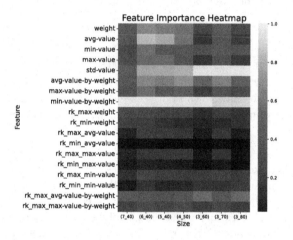

Fig. 2. Heatmap of feature importance scores normalized across different sizes. The features are described in Table 5. The property-based features assume the same names as the properties detailed in Table 1. The rank-based features of a variable have a prefix "rk_" before the heuristic ordering; refer to Table 6 for more details. Finally, the "Value SD" feature from Table 5 is named std-value in this plot.

The heatmap of feature importance scores for different sizes is provided in Fig. 2. We can note that the min-value-by-weight feature is important across all sizes, especially for cases with more than 3 objectives. In fact, the choice of VO heuristic MaxRatio was driven by feature importance scores and is a case in point of how learning-based methods can assist in designing good heuristics for optimization. Furthermore, the real-valued features are more important than the categorical rank features for problems with more than 3 objectives. For problems with 3 objectives, the std-value featureco is extremely crucial. Also, rank feature rk_max_avg-value-by-weight receives higher importance than some of the real-valued features. It is also interesting to observe that certain rank-based features are consistently ignored across all sizes, which can be used to improve the feature engineering.

In a nutshell, the heatmap of feature importance scores helps in interpreting the features that govern the decisions made by LEO, which also happens to be in alignment with a widely used heuristic to solve the single-objective knapsack problem. Refer to Appendix C.4 feature importance plots of size-independent models (ML+A and ML+AC), which showcase a similar trend.

7 Conclusion

LEO is the first machine learning framework for accelerating the BDD approach for multiobjective integer linear programming. We contribute a number of techniques and findings that may be of independent interest. For one, we have shown that variable ordering does impact the solution time of this approach. Our labeling method of using a black-box optimizer over a fixed-size parameter configuration space to discover high-quality variable orderings successfully bypasses the curse of dimensionality; the application of this approach to other similar labeling tasks may be possible, e.g., for backdoor set discovery in MIP [37]. An additional innovation is using size-independent ML models, i.e., models that do not depend on a fixed number of decision variables. This modeling choice enables the training of unified ML models, which our experiments reveal to perform very well. Through a comprehensive case study of the knapsack problem, we show that LEO can produce variable orderings that significantly reduce PF enumeration time.

There are several exciting directions for future work that can build on LEO:

- The BDD approach to multiobjective integer programming has been applied to a few problems other than knapsack [5]. Much of LEO can be directly extended to such other problems, assuming that their BDD construction is significantly influenced by the variable ordering. One barrier to such an extension is the availability of open-source BDD construction code. As the use of BDDs in optimization is a rather nascent area of research, it is not uncommon to consider a case study of a single combinatorial problem, as was done for example for Graph Coloring in [33] and Maximum Independent Set in [9].
- Our method produces a *static* variable ordering upfront of BDD construction. While this was sufficient to improve on non-ML orderings for the knapsack problem, it may be interesting to consider *dynamic* variable orderings that observe the BDD construction process layer by layer and choose the next variable accordingly, as was done in [18].
- We have opted for rather interpretable ML model classes but the exploration of more sophisticated deep learning approaches may enable closing some of the remaining gap in training/validation loss, which may improve downstream solving performance.
- Beyond the exact multiobjective setting, extending LEO to a heuristic that operates on a *restricted* BDD may provide an approximate PF much faster than full enumeration. We believe this to be an easy extension of our work.

A Feature Engineering

Table 5 lists the features for an MKP. It can be divided into two types: variable features and context features. The role of variable features is to summarize the information associated with a particular variable, whereas the context features contain the instance level information. The variable features can be further partitioned into two main categories: property-based and rank-based. The property-based features are based on properties detailed in Table 1 and rank-based features are derived by sorting variables based on property values. In summary, we have 18 variable features and 19 context features, resulting in a total of 37 features per variable. Note that our learning models are based on GBT. However, we do not use context features in all its variants. Specifically, ML and ML+A do not use them whereas ML+C and ML+AC do. Finally, we would like to highlight that similar features can be derived for the problem classes with more than one constraint as given in [36,51].

Table 5. Features for an MKP.

Type	Feature	Description	Count
Variable	Properties	The variable properties described in Table 1	7
	Rank	The ranks corresponding to the heuristic orderings described in Table 6	10
	Value SD	The standard deviation of the values	1
Context	# objectives	The number of objectives in the MKP problem	1
	# items	The number of items in the MKP problem	1
	Capacity	The capacity of the MKP	1
	Weight stats.	The mean, min., max. and std. of the MKP weights	4
	Aggregate value stats	The mean, min., max. and std. of the values of each objective	12
Total count			37

B   SMAC Setup

Initialization. For both uses of SMAC, we use the MinWt ordering as a warm-start by assigning all the property weights to zero except the weight property, which is set to -1. This reduces the need for extensive random exploration of the configuration space by providing a reasonably good ordering heuristic.

Random Seeds. As SMAC is a randomized optimization algorithm, running it with multiple random seeds increases the odds of finding a good solution. We leverage this idea for SmacI as its outputs will be used as labels for supervised learning and are thus expected to be of very high quality, i.e., we seek instance-specific parameter configurations that yield variable orderings with minimal solution times. We run SmacD with a single seed and use its average run time on the training set as a target to beat for SmacI. Since the latter optimizes at the instance level, one would hope it can do better than the distribution-level configuration of SmacD.

As such, for SmacI, we run SMAC on each instance with 5 seeds for all sizes except (5, 40) and (6, 40), for which we used one seed. We start with one seed, then average the run time of the best-performing seed per instance to that of the average enumeration time of SmacD on the training set, and relaunch SMAC with a new seed until a desired performance gap between SmacI and SmacD is achieved.

Computational Budget. In the SmacD setting, we run SMAC with a 12-hour time limit, whereas in the SmacI case the time limit is set to 20 min per instance except for sizes (3, 60), (4, 50), (5, 40), for which it is set to 5 min per instance. In both settings, SMAC runs on a 4-core machine with a 20GB memory limit. It can be observed that generating the labels can be computationally expensive. This dictates the choice of sizes in the set of instance sizes, S. Specifically, we select instance sets with an average running time of at most 100 s (not too hard) and at least 5 s (nor too easy) using the top-down compilation method described in Bergman et al. [5].

SmacI Performance Profile. Having established in Q1 that the search for high-quality variable orderings is justified, we now turn to a key component of Phase 1: the use of SMAC as a black-box optimizer that produces "label" variable orderings for subsequent ML in Phase 2. Figure 3 shows the average run time, on the training instances, of the incumbent property-weight configurations found by SMAC. These should be interpreted as standard optimization convergence curves where we seek small values (vertical axis) as quickly as possible (horizontal axis). Indeed, SMAC performs well, substantially improving over its initial warm-start solution (the MinWt ordering). The flat horizontal lines in the figure show the average run time of the single configuration found by SmacD. The latter is ultimately outperformed by the SmacI configurations on average, as desired.

C   Additional Results

C.1   Impact of Variable Ordering on PF Enumeration Time

The heuristic orderings are constructed using the variable properties. For example, the heuristic ordering called max_avg-value-by-weight sorts the variables in descending order of the ratio of a variable's average value by its weight.

Table 6 summarizes the results of this experiment. We work with 250 MKP instances of the problem sizes (3, 20), (3, 40), (3, 60), (3, 80), (5, 20), (5, 30), (5, 40), (7, 20), (7, 30), and (7, 40). The values in the table are the ratios of the average run time of a heuristic ordering to that of the 5 random orderings; values

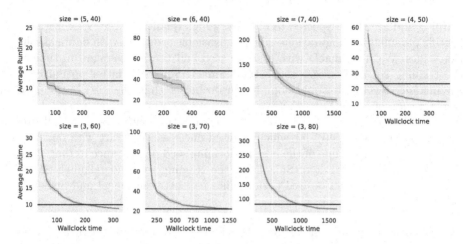

Fig. 3. SmacI performance w.r.t. wallclock time across different sizes. For a given size, we first of all find the best seed, i.e., the seed which finds an incumbent property-weight having minimum PF enumeration time, for each instance. Then for a given wallclock time, we calculate the mean and standard error of the PF enumeration time across instances for the incumbent property-weight up to that wallclock time. The mean and standard error are plotted as a blue line and blue error band around it, respectively. The horizontal black line represents the average PF enumeration time of SmacD configuration on the validation set.

smaller than one indicate that a heuristic ordering is faster than a random one, on average. The best heuristic ordering for each size is highlighted.

First, it is clear that some heuristic orderings consistently outperform the random ones across all problem sizes (min_weight, max_min-value), and by a significant margin. In contrast, some heuristic orderings are consistently worse than random. For example, the heuristic ordering max_weight, min_avg-value, and min_min-value consistently underperform when the number of variables is more than 20. Second, the choice of MinWt as a baseline was motivated by the results of this experiment as min_weight wins on most sizes as highlighted in Table 6. Altogether, this experiment validates the hypothesis that VO has an impact on the PF enumeration time and that there exists EVO that can significantly reduce this time.

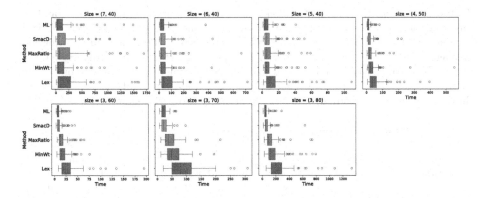

Fig. 4. Box plots of time to enumerate the PF for different sizes.

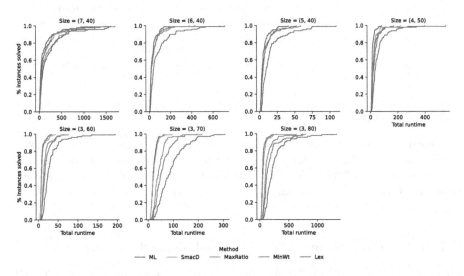

Fig. 5. Performance profile of different methods in terms of the fraction of instances solved w.r.t. time for various sizes.

C.2   Performance of LEO in Comparison to Other Baselines

The detailed version of Table 3, with additional statistics, is provided in Table 7. To complement this further, we present box plots and performance profiles in Fig. 4 and Fig. 5, respectively. Note how the PF enumeration time distribution for the ML method is much more concentrated, with a smaller median and only a few outliers compared to other methods. Note that the instances that had a timeout are omitted from this analysis. This performance improvement leads to a larger fraction of instances being solved in a smaller amount of time as highlighted in Fig. 5.

Table 6. Ratio of the average time taken by a heuristic ordering to that of 5 random orderings across 250 MKP instances across sizes specified. A ratio of less (greater) than 1 indicates that heuristic ordering performs better (worse) than random ordering, on average. The name of a heuristic ordering follows the structure ⟨sort⟩-⟨property-name⟩, where ⟨sort⟩ can either be min or max and ⟨property-name⟩ can be equal to one of the properties defined in Table 1. A value of max (min) for ⟨sort⟩ would sort the variables in the descending (ascending) order of the property ⟨property-name⟩.

Heuristic ordering ⟨sort⟩-⟨property-name⟩	Size									
	(3, 20)	(3, 40)	(3, 60)	(3, 80)	(5, 20)	(5, 30)	(5, 40)	(7, 20)	(7, 30)	(7, 40)
max_weight	**0.778**	1.147	1.378	1.317	0.815	1.865	2.760	1.165	2.769	1.830
min_weight	0.812	**0.654**	0.648	0.685	**0.769**	**0.700**	**0.533**	**0.694**	**0.588**	**0.581**
max_avg-value	0.861	0.891	0.843	0.972	0.839	0.992	0.840	0.845	1.125	0.823
min_avg-value	0.893	1.619	2.144	1.692	0.936	1.721	2.132	1.103	1.918	1.524
max_max-value	0.858	1.065	1.126	1.203	0.857	1.137	1.060	0.899	1.408	0.992
min_max-value	0.855	1.018	1.010	0.995	0.854	1.077	1.009	0.909	0.981	0.975
max_min-value	0.851	0.705	**0.537**	**0.590**	0.813	0.816	0.617	0.819	0.862	0.767
min_min-value	0.883	1.662	2.266	1.729	0.930	1.700	2.020	1.094	1.980	1.430
max_avg-value-by-weight	0.838	0.832	0.988	1.156	0.802	0.896	0.887	0.751	0.917	0.788
max_max-value-by-weight	0.837	0.837	1.037	1.192	0.804	0.871	0.824	0.732	0.790	0.735

C.3   BDD Topology Analysis

The mean cumulative number of solutions generated at a given level are given in Fig. 6 and the BDD topology analysis for all sizes is detailed in Table 8. It is evident from both of these that ML-based VO leads to fewer intermediate solutions, which results in smaller number of Checks and improved PF enumeration time.

C.4   Feature Importance Plots for Size-Independent ML Models

Figure 7 and Fig. 8 shows the feature importance plots for the ML+A and ML+AC models, respectively. The std-value and min-value-by-weight features are ranked highly by both models. The usefulness of the context features is clearly highlighted by their presence in the top 5 features of the ML+AC model.

Table 7. PF enumeration time statistics on the test set containing 100 instances with a time limit of 1800 s. Each row represents aggregate statistics for a given instance "Size" and "Method". "Count" stands for the number of instances on which the algorithm ran successfully without hitting the time or memory limits. "GMean", "Mean", "Min", "Max" and "Std" denotes the geometric mean, arithmetic mean, minimum, maximum and standard deviation of the enumeration time computed across "count" many instances. We use a shift of 5 to compute the "GMean".

Size	Method	Count	GMean	Mean	Min	Max	Std
(7, 40)	ML	100	50.95	129.26	2.26	1490.58	242.53
	ML+C	99	**47.97**	**113.28**	2.25	851.69	182.27
	ML+A	99	53.83	125.25	2.81	1130.82	209.60
	ML+AC	99	54.54	128.47	2.27	1153.07	207.21
	SmacD	99	74.24	162.30	2.95	1521.43	252.04
	MaxRatio	98	96.38	214.78	4.14	1700.44	319.32
	MinWt	99	69.11	144.82	3.16	1561.36	227.74
	Lex	95	111.59	232.30	7.66	1604.17	332.34
(6, 40)	ML	100	20.90	32.57	2.57	377.89	45.71
	ML+C	100	20.59	32.71	2.17	404.85	47.98
	ML+A	100	20.06	**31.56**	2.29	355.64	43.80
	ML+AC	100	**19.90**	32.22	2.01	427.80	49.85
	SmacD	100	25.38	40.35	2.95	434.34	55.20
	MaxRatio	100	26.68	45.75	3.05	664.88	77.28
	MinWt	100	25.63	40.67	2.91	427.06	54.98
	Lex	100	49.98	88.89	6.01	713.58	120.76
(5, 40)	ML	100	5.00	6.10	1.26	40.91	6.43
	ML+C	100	4.99	6.20	1.25	41.19	6.80
	ML+A	100	**4.90**	**6.00**	1.28	41.24	6.40
	ML+AC	100	4.99	6.34	1.06	41.29	7.29
	SmacD	100	6.80	8.62	1.38	51.22	9.34
	MaxRatio	100	6.66	8.62	1.44	58.87	10.04
	MinWt	100	6.68	8.46	1.44	52.61	9.21
	Lex	100	11.75	16.20	1.86	109.18	18.66
(4, 50)	ML	100	**11.18**	**13.53**	2.54	79.42	12.37
	ML+C	100	11.34	13.86	2.49	87.89	13.01
	ML+A	100	11.50	14.06	2.47	73.17	12.79
	ML+AC	100	11.39	13.90	2.22	77.75	12.68
	SmacD	100	18.12	24.29	4.04	218.24	30.38
	MaxRatio	100	20.33	27.03	3.24	167.95	27.71
	MinWt	100	24.66	36.13	4.67	553.15	61.12
	Lex	100	38.75	55.14	6.67	395.67	62.76
(3, 60)	ML	100	8.31	9.04	3.88	34.13	5.43
	ML+C	100	8.53	9.25	3.66	36.82	5.52
	ML+A	100	8.41	9.09	3.56	31.14	5.22
	ML+AC	100	**8.24**	**8.95**	3.44	31.99	5.36
	SmacD	100	9.32	10.34	3.53	43.58	6.96
	MaxRatio	100	14.68	16.52	5.76	64.21	11.09
	MinWt	100	17.13	19.32	5.36	75.98	12.08
	Lex	100	25.89	31.12	8.12	195.25	26.67
(3, 70)	ML	100	20.46	22.28	7.92	64.55	10.89
	ML+C	100	20.80	22.72	7.29	66.67	11.26
	ML+A	100	20.79	22.81	7.68	80.82	11.77
	ML+AC	100	**20.38**	**22.24**	7.41	65.67	10.88
	SmacD	100	22.13	24.49	7.52	96.94	13.47
	MaxRatio	100	39.13	44.43	12.01	215.63	27.95
	MinWt	100	50.16	57.56	12.17	194.07	33.10
	Lex	100	77.46	91.38	19.87	308.21	55.80
(3, 80)	ML	100	49.42	57.03	15.29	286.11	40.37
	ML+C	100	49.66	57.12	15.99	224.96	38.47
	ML+A	100	50.27	58.65	15.35	262.77	42.52
	ML+AC	100	**48.48**	**56.11**	15.43	262.57	39.60
	SmacD	100	59.05	71.67	15.66	632.92	69.80
	MaxRatio	100	106.08	129.24	29.64	741.82	110.73
	MinWt	100	135.33	168.94	30.35	794.10	141.95
	Lex	99	202.25	250.26	56.52	1332.19	204.47

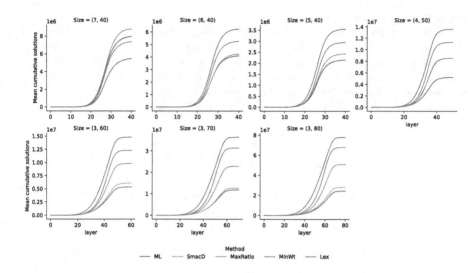

Fig. 6. Cumulative number of intermediate solutions generated at a particular layer across all sizes.

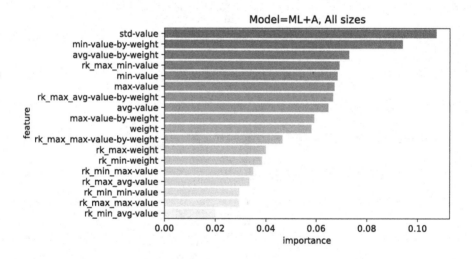

Fig. 7. Feature importance plot for ML+A model. Refer to the caption of Fig. 2 for an explanation of the feature naming conventions.

Table 8. Analyzing the impact of BDD topology on the PF enumeration time on the test set. For a given "Size" and "Method", the metric value indicates the percentage of mean performance by that "Method" compared to the Lex method. The "Nodes" ("Width") represents the percentage of the mean of the number of nodes (mean width) of the BDDs generated by "Method" to that of Lex. "Degree" denotes the percentage of the mean ratio of the in-degree of the top 50 nodes with the highest in-degree in BDDs generated by "Method" to that of Lex. Similarly, "Checks" and "Time" indicate the percentage of the average Pareto-dominance checks and the average geometric mean of the PF enumeration time, respectively. For all metrics, values less than 100 are preferred. For example, a value of δ for "Nodes" indicates that the mean number of nodes for "Method" is $\delta/100\times$ the mean number of nodes of "Lex".

Size	Method	Nodes	Width	Degree	Checks	Time
(7, 40)	ML	80.37	88.01	74.92	**38.31**	**38.98**
	SmacD	70.08	79.22	149.06	64.24	59.56
	MaxRatio	71.24	80.63	**66.72**	85.14	79.63
	MinWt	**69.18**	**78.69**	151.10	59.23	55.66
(6, 40)	ML	78.21	85.74	89.48	**39.96**	**41.82**
	SmacD	69.85	79.38	147.08	53.80	50.79
	MaxRatio	90.31	94.86	**63.16**	52.18	53.38
	MinWt	**69.85**	**79.38**	147.08	53.80	51.28
(5, 40)	ML	81.30	88.05	63.86	**35.91**	**42.51**
	SmacD	69.14	78.86	132.01	57.99	57.87
	MaxRatio	88.50	93.93	**57.97**	50.51	56.69
	MinWt	**69.14**	**78.86**	132.01	57.99	56.81
(4, 50)	ML	90.39	96.63	60.12	**23.09**	**28.84**
	SmacD	75.67	87.74	110.15	43.63	46.75
	MaxRatio	87.99	96.18	**57.78**	50.17	52.46
	MinWt	**70.19**	**84.78**	149.33	64.96	63.64
(3, 60)	ML	92.96	98.33	69.77	**20.61**	**32.12**
	SmacD	91.58	98.80	77.99	26.46	35.99
	MaxRatio	86.06	96.36	52.66	51.46	56.70
	MinWt	**68.46**	**86.59**	141.52	67.49	66.16
(3, 70)	ML	92.11	98.56	65.73	**18.26**	**26.41**
	SmacD	90.67	98.71	66.02	21.07	28.56
	MaxRatio	85.98	97.13	**49.10**	49.67	50.52
	MinWt	**70.12**	**89.70**	121.16	70.17	64.75
(3, 80)	ML	90.84	98.56	62.72	**19.02**	**24.09**
	SmacD	92.21	99.18	79.72	25.73	28.81
	MaxRatio	84.23	97.21	**45.26**	54.71	51.71
	MinWt	**68.64**	**90.48**	114.92	70.17	65.71

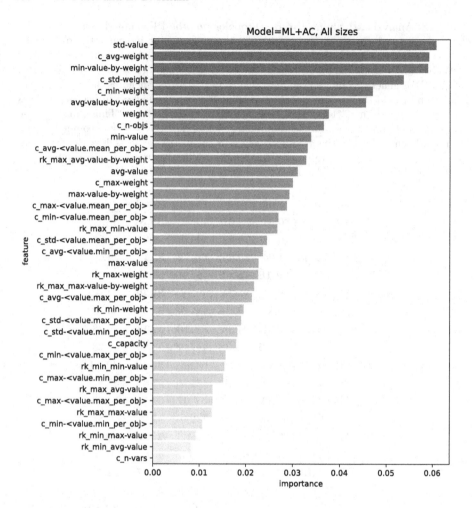

Fig. 8. Feature importance plot for ML+AC model. Features that start with "c_" are context features; see Table 5.

References

1. Adelgren, N., Gupte, A.: Branch-and-bound for biobjective mixed-integer linear programming. INFORMS J. Comput. **34**(2), 909–933 (2022)
2. Aloul, F.A., Markov, I.L., Sakallah, K.A.: Force: a fast and easy-to-implement variable-ordering heuristic. In: Proceedings of the 13th ACM Great Lakes Symposium on VLSI, pp. 116–119 (2003)

3. Altiparmak, F., Gen, M., Lin, L., Paksoy, T.: A genetic algorithm approach for multi-objective optimization of supply chain networks. Comput. Ind. Eng. **51**(1), 196–215 (2006)
4. Belotti, P., Soylu, B., Wiecek, M.M.: A branch-and-bound algorithm for biobjective mixed-integer programs. Optimization Online, pp. 1–29 (2013)
5. Bergman, D., Bodur, M., Cardonha, C., Cire, A.A.: Network models for multiobjective discrete optimization. INFORMS J. Comput. (2021)
6. Bergman, D., Cire, A.A.: Multiobjective optimization by decision diagrams. In: Rueher, M. (ed.) CP 2016. LNCS, vol. 9892, pp. 86–95. Springer, Cham (2016). https://doi.org/10.1007/978-3-319-44953-1_6
7. Bergman, D., Ciré, A.A., van Hoeve, W., Hooker, J.N.: Discrete optimization with decision diagrams. INFORMS J. Comput. **28**(1), 47–66 (2016). https://doi.org/10.1287/ijoc.2015.0648
8. Bergman, D., Cire, A.A., Van Hoeve, W.J., Hooker, J.: Decision Diagrams for Optimization, vol. 1. Springer, Cham (2016)
9. Bergman, D., Cire, A.A., van Hoeve, W.-J., Hooker, J.N.: Variable ordering for the application of BDDs to the maximum independent set problem. In: Beldiceanu, N., Jussien, N., Pinson, É. (eds.) CPAIOR 2012. LNCS, vol. 7298, pp. 34–49. Springer, Heidelberg (2012). https://doi.org/10.1007/978-3-642-29828-8_3
10. Bökler, F., Ehrgott, M., Morris, C., Mutzel, P.: Output-sensitive complexity of multiobjective combinatorial optimization. J. Multi-Criteria Decis. Anal. **24**(1–2), 25–36 (2017)
11. Boland, N., Charkhgard, H., Savelsbergh, M.: The triangle splitting method for biobjective mixed integer programming. In: Lee, J., Vygen, J. (eds.) IPCO 2014. LNCS, vol. 8494, pp. 162–173. Springer, Cham (2014). https://doi.org/10.1007/978-3-319-07557-0_14
12. Boland, N., Charkhgard, H., Savelsbergh, M.: A criterion space search algorithm for biobjective integer programming: the balanced box method. INFORMS J. Comput. **27**(4), 735–754 (2015)
13. Bollig, B., Löbbing, M., Wegener, I.: On the effect of local changes in the variable ordering of ordered decision diagrams. Inf. Process. Lett. **59**(5), 233–239 (1996)
14. Bollig, B., Wegener, I.: Improving the variable ordering of OBDDs is NP-complete. IEEE Trans. Comput. **45**(9), 993–1002 (1996)
15. Bryant, R.E.: Symbolic Boolean manipulation with ordered binary-decision diagrams. ACM Comput. Surv. (CSUR) **24**(3), 293–318 (1992)
16. Butler, K.M., Ross, D.E., Kapur, R., Mercer, M.R.: Heuristics to compute variable orderings for efficient manipulation of ordered binary decision diagrams. In: Proceedings of the 28th ACM/IEEE Design Automation Conference, pp. 417–420 (1991)
17. Cappart, Q., Chételat, D., Khalil, E.B., Lodi, A., Morris, C., Velickovic, P.: Combinatorial optimization and reasoning with graph neural networks. J. Mach. Learn. Res. **24**(130), 1–61 (2023)
18. Cappart, Q., Goutierre, E., Bergman, D., Rousseau, L.: Improving optimization bounds using machine learning: decision diagrams meet deep reinforcement learning. In: The Thirty-Third AAAI Conference on Artificial Intelligence, AAAI 2019, The Thirty-First Innovative Applications of Artificial Intelligence Conference, IAAI 2019, The Ninth AAAI Symposium on Educational Advances in Artificial Intelligence, EAAI 2019, Honolulu, Hawaii, USA, 27 January–1 February 2019, pp. 1443–1451. AAAI Press (2019). https://doi.org/10.1609/aaai.v33i01.33011443
19. Carbin, M.: Learning effective BDD variable orders for BDD-based program analysis. Technical report. Citeseer (2006)

20. de las Casas, P.M., Sedeno-Noda, A., Borndörfer, R.: An improved multiobjective shortest path algorithm. Comput. Oper. Res. **135**, 105424 (2021)

21. Chekuri, C., Fox, K.: UIUC CS 598CSC: approximation algorithms, lecture notes on Knapsack (2009). https://courses.engr.illinois.edu/cs598csc/sp2009/lectures/lecture_4.pdf. Accessed 01 July 2023

22. Chen, T., Guestrin, C.: XGBoost: a scalable tree boosting system. In: Proceedings of the 22nd ACM SIGKDD International Conference on Knowledge Discovery and Data Mining, KDD 2016, pp. 785–794. ACM, New York (2016). https://doi.org/10.1145/2939672.2939785. ISBN 978-1-4503-4232-2

23. Chung, P.Y., Hajj, I., Patel, J.H.: Efficient variable ordering heuristics for shared ROBDD. In: 1993 IEEE International Symposium on Circuits and Systems (ISCAS), pp. 1690–1693. IEEE (1993)

24. Drechsler, R., Göckel, N., Becker, B.: Learning heuristics for OBDD minimization by evolutionary algorithms. In: Voigt, H.-M., Ebeling, W., Rechenberg, I., Schwefel, H.-P. (eds.) PPSN 1996. LNCS, vol. 1141, pp. 730–739. Springer, Heidelberg (1996). https://doi.org/10.1007/3-540-61723-X_1036

25. Ebendt, R., Drechsler, R.: Exact BDD minimization for path-related objective functions. In: Reis, R., Osseiran, A., Pfleiderer, H.-J. (eds.) VLSI-SoC 2005. IIFIP, vol. 240, pp. 299–315. Springer, Boston, MA (2007). https://doi.org/10.1007/978-0-387-73661-7_19

26. Ebendt, R., Gunther, W., Drechsler, R.: Combining ordered best-first search with branch and bound for exact BDD minimization. IEEE Trans. Comput. Aided Des. Integr. Circuits Syst. **24**(10), 1515–1529 (2005)

27. Ehrgott, M.: A discussion of scalarization techniques for multiple objective integer programming. Ann. Oper. Res. **147**(1), 343–360 (2006)

28. Ehrgott, M., Gandibleux, X., Przybylski, A.: Exact methods for multi-objective combinatorial optimisation. In: Multiple Criteria Decision Analysis: State of the Art Surveys, pp. 817–850 (2016)

29. Friedman, S.J., Supowit, K.J.: Finding the optimal variable ordering for binary decision diagrams. In: Proceedings of the 24th ACM/IEEE Design Automation Conference, pp. 348–356 (1987)

30. Grumberg, O., Livne, S., Markovitch, S.: Learning to order BDD variables in verification. J. Artif. Intell. Res. **18**, 83–116 (2003)

31. Joachims, T.: Optimizing search engines using clickthrough data. In: Proceedings of the Eighth ACM SIGKDD International Conference on Knowledge Discovery and Data Mining, pp. 133–142 (2002)

32. Joachims, T.: Training linear SVMs in linear time. In: Proceedings of the 12th ACM SIGKDD International Conference on Knowledge Discovery and Data Mining, pp. 217–226 (2006)

33. Karahalios, A., van Hoeve, W.J.: Variable ordering for decision diagrams: a portfolio approach. Constraints **27**(1–2), 116–133 (2022)

34. Kendall, M.G.: A new measure of rank correlation. Biometrika **30**(1/2), 81–93 (1938)

35. Kergosien, Y., Giret, A., Neron, E., Sauvanet, G.: An efficient label-correcting algorithm for the multiobjective shortest path problem. INFORMS J. Comput. **34**(1), 76–92 (2022)

36. Khalil, E., Le Bodic, P., Song, L., Nemhauser, G., Dilkina, B.: Learning to branch in mixed integer programming. In: Proceedings of the AAAI Conference on Artificial Intelligence, vol. 30 (2016)

37. Khalil, E.B., Vaezipoor, P., Dilkina, B.: Finding backdoors to integer programs: a Monte Carlo tree search framework. In: Thirty-Sixth AAAI Conference on Artificial Intelligence, AAAI 2022, Thirty-Fourth Conference on Innovative Applications of Artificial Intelligence, IAAI 2022, The Twelveth Symposium on Educational Advances in Artificial Intelligence, EAAI 2022 Virtual Event, 22 February–1 March 2022, pp. 3786–3795. AAAI Press (2022). https://ojs.aaai.org/index.php/AAAI/article/view/20293
38. Kirlik, G., Sayın, S.: A new algorithm for generating all nondominated solutions of multiobjective discrete optimization problems. Eur. J. Oper. Res. **232**(3), 479–488 (2014)
39. Lambrinidis, G., Tsantili-Kakoulidou, A.: Multi-objective optimization methods in novel drug design. Expert Opin. Drug Discov. **16**(6), 647–658 (2021)
40. Lentz, A.: Multicriteria shortest paths and related geometric problems. Ph.D. thesis, Bordeaux (2021)
41. Lindauer, M., et al.: SMAC3: a versatile Bayesian optimization package for hyperparameter optimization (2021)
42. Liu, S., Yan, X., Jin, Y.: End-to-end pareto set prediction with graph neural networks for multi-objective facility location. In: Emmerich, M., et al. (eds.) EMO 2023. LNCS, vol. 13970, pp. 147–161. Springer, Cham (2023). https://doi.org/10.1007/978-3-031-27250-9_11
43. Lu, Y., Jain, J., Clarke, E., Fujita, M.: Efficient variable ordering using ABDD based sampling. In: Proceedings of the 37th Annual Design Automation Conference, pp. 687–692 (2000)
44. Ozlen, M., Burton, B.A., MacRae, C.A.: Multi-objective integer programming: an improved recursive algorithm. J. Optim. Theory Appl. **160**, 470–482 (2014)
45. Parragh, S.N., Tricoire, F.: Branch-and-bound for bi-objective integer programming. INFORMS J. Comput. **31**(4), 805–822 (2019)
46. Pedregosa, F., et al.: Scikit-learn: machine learning in Python. J. Mach. Learn. Res. **12**, 2825–2830 (2011)
47. Przybylski, A., Gandibleux, X.: Multi-objective branch and bound. Eur. J. Oper. Res. **260**(3), 856–872 (2017)
48. Rice, M., Kulhari, S.: A survey of static variable ordering heuristics for efficient BDD/MDD construction. University of California, Technical report, p. 130 (2008)
49. Rudell, R.: Dynamic variable ordering for ordered binary decision diagrams. In: Proceedings of 1993 International Conference on Computer Aided Design (ICCAD), pp. 42–47. IEEE (1993)
50. Schede, E., et al.: A survey of methods for automated algorithm configuration. J. Artif. Intell. Res. **75**, 425–487 (2022)
51. Sierra-Altamiranda, A., Charkhgard, H., Dayarian, I., Eshragh, A., Javadi, S.: Learning to project in multi-objective binary linear programming. arXiv preprint arXiv:1901.10868 (2019)
52. Song, Z., Chen, X., Luo, X., Wang, M., Dai, G.: Multi-objective optimization of agile satellite orbit design. Adv. Space Res. **62**(11), 3053–3064 (2018)
53. Sourd, F., Spanjaard, O.: A multiobjective branch-and-bound framework: application to the biobjective spanning tree problem. INFORMS J. Comput. **20**(3), 472–484 (2008)
54. Tangpattanakul, P., Jozefowiez, N., Lopez, P.: Multi-objective optimization for selecting and scheduling observations by agile earth observing satellites. In: Coello, C.A.C., Cutello, V., Deb, K., Forrest, S., Nicosia, G., Pavone, M. (eds.) PPSN 2012. LNCS, vol. 7492, pp. 112–121. Springer, Heidelberg (2012). https://doi.org/10.1007/978-3-642-32964-7_12

55. Wegener, I.: Branching Programs and Binary Decision Diagrams: Theory and Applications, vol. 4. SIAM (2000)
56. Wu, Y., Song, W., Cao, Z., Zhang, J., Gupta, A., Lin, M.: Graph learning assisted multi-objective integer programming. In: Advances in Neural Information Processing Systems, vol. 35, pp. 17774–17787 (2022)
57. Yu, Y., Zhang, J., Cheng, G., Schell, M., Okunieff, P.: Multi-objective optimization in radiotherapy: applications to stereotactic radiosurgery and prostate brachytherapy. Artif. Intell. Med. **19**(1), 39–51 (2000)

Minimizing the Cost of Leveraging Influencers in Social Networks: IP and CP Approaches

Felipe de C. Pereira[1]([✉]) [iD], Pedro J. de Rezende[1] [iD], and Tallys Yunes[2] [iD]

[1] Institute of Computing, University of Campinas (Unicamp), Campinas, SP, Brazil
pereira_felipe@ic.unicamp.br, pjr@unicamp.br
[2] Miami Herbert Business School, University of Miami, Coral Gables, FL, USA
tallys@miami.edu

Abstract. In this paper, we introduce and study mathematical programming formulations for the Least Cost Directed Perfect Awareness Problem (LDPAP), an NP-hard optimization problem that arises in the context of influence marketing. In the LDPAP, we seek to identify influential members of a given social network that can disseminate a piece of information and trigger its propagation throughout the network. The objective is to minimize the cost of recruiting the initial spreaders while ensuring that the information reaches everyone. This problem has been previously modeled as two different integer programming formulations that were tested on a collection of 300 small synthetic instances. In this work, we propose two new integer programming models and three constraint programming formulations for the LDPAP. We also present preprocessing techniques capable of significantly reducing the sizes of these models. To investigate and compare the efficiency and effectiveness of our approaches, we perform a series of experiments using the existing small instances and a new publicly available benchmark of 14 large instances. Our findings yield new optimal solutions to 185 small instances that were previously unsolved, tripling the total number of instances with known optima. Regarding both small and large instances, our contributions include a comprehensive analysis of the experimental results and an evaluation of the performance of each formulation in distinct scenarios, further advancing our understanding of the LDPAP toward the design of exact approaches for the problem.

Keywords: Least Cost Directed Perfect Awareness Problem ·
Influence Marketing · Social Networks · Integer Programming ·
Constraint Programming

Supported in part by grants from: *Santander Bank*, Brazil; *Brazilian National Council for Scientific and Technological Development* (CNPq), Brazil, #313329/2020-6, #314293/2023-0; *São Paulo Research Foundation* (FAPESP), Brazil, #2023/04318-7, #2023/14427-8; *Fund for Support to Teaching, Research and Outreach Activities* (FAEPEX), Brazil; *Coordination for the Improvement of Higher Education Personnel* (CAPES), Brazil – Finance Code 001.

B. Dilkina (Ed.): CPAIOR 2024, LNCS 14743, pp. 111–127, 2024.
https://doi.org/10.1007/978-3-031-60599-4_7

1   Introduction

The phenomenon of influence spreading on social networks has been modeled in combinatorial optimization since the beginning of the century [11]. One of the most studied problems within this topic is the Target Set Selection Problem (TSSP) [5], whose introduction gave rise to IP formulations [1,20,22] and a collection of variants that deal with different scenarios [2].

In the original formulation of the TSSP, the objective is to select a set of individuals of minimum size (or minimum cost) capable of starting a widespread propagation of a piece of information (news, media, opinion, etc.). Typically, it is assumed that during a propagation, inactive people only start to actively pass on the information when they receive sufficient influence from their acquaintances.

A more realistic scenario is addressed in the Perfect Awareness Problem (PAP), which was proposed in [6] and also studied in [7,13,16]. In the PAP, an additional intermediary state (being aware) is introduced. In this problem, people are considered aware of the information as soon as they are influenced for the first time. The objective is to select a minimum set of seminal spreaders so that all individuals end up in the aware state at the end of the propagation.

In [15], the Least Cost Directed Perfect Awareness Problem (LDPAP) is introduced as an extension of the PAP to consider nonreciprocal relations observed in social networks like X (formerly known as Twitter) and Instagram. The LDPAP also includes distinct degrees of influence between individuals (which neither TSSP nor PAP consider) as well as different costs for recruiting initial spreaders.

Although there are important theoretical and practical results regarding the LDPAP in the literature, which includes two integer programming (IP) formulations and a heuristic [15], there remains significant work to be done to solve the problem from an exact perspective. So far, only small instances can be solved to optimality. In this paper, we propose new formulations and exact techniques for the LDPAP that allow us to solve considerably larger instances.

Our Contributions
The main contributions of this paper for the LDPAP are:

- two novel IP formulations;
- three new constraint programming (CP) formulations;
- preprocessing techniques that reduce the size of the models;
- comparative experiments using the new and existing models on small synthetic instances and large instances built from crawled X networks.

This text is organized as follows. In Sect. 2, we formally define the LDPAP and review the literature on the problem. The new IP and CP models are presented in Sect. 3 and Sect. 4, respectively. In Sect. 5, we introduce the preprocessing techniques for these formulations. Later, in Sect. 6, we report a set of computational experiments and analyze the results with the objective of empirically evaluating the exact models. Lastly, in Sect. 7, we close the paper with concluding remarks and address future work.

2   Object of Study

Although the LDPAP has just recently been introduced into the literature, a number of theoretical and practical results are already known. Next, we formally present the LDPAP and summarize previous work.

2.1   Problem Statement

The LDPAP, as first introduced in [15], can be described as follows. Consider a portion of an online social network represented by a directed graph $D = (V, E)$, where V and E are the sets of vertices and directed edges (arcs) of D. Each vertex in V represents an individual, and each $(u, v) \in E$ indicates that u can exert influence over v. The *in-* and *out-neighborhoods* of each $v \in V$ are indicated by $N_{\text{in}}(v) = \{u \in V : (u, v) \in E\}$ and $N_{\text{out}}(v) = \{u \in V : (v, u) \in E\}$.

A set of vertices selected to first disseminate the information is called a *seed set* and it contains the *seeds*. When the seeds transmit the information to their out-neighbors, some vertices may be influenced to the point that they forward the information, triggering a propagation. In this process, each vertex assumes at least one of three possible states regarding the information being disseminated:

- *ignorant*: the vertex is not a seed and has not received the information yet;
- *aware*: the vertex is a seed or has received the information at least once;
- *spreader*: the vertex is a seed or has received the information enough times to enable it to forward the information to others.

The *cost* of selecting a vertex v as a seed is denoted by $c_v \in \mathbb{R}^+$ and the cost of a seed set $S \subseteq V$ is $c_S = \sum_{v \in S} c_v$. Moreover, the *threshold* of v, denoted by $t_v \in \mathbb{R}^+$, quantifies the need for v to be sufficiently influenced to the point where v begins to pass on the information being propagated. In other words, if v is not a seed, then v becomes a spreader only if the amount of influence received by v is at least t_v. Furthermore, the *weight* of an edge $(u, v) \in E$, denoted by $w_{u,v} \in \mathbb{R}^+$, measures the degree of influence u can exert on v.

The passage of time in a propagation is divided into rounds. In each round $\tau \geq 0$ of the propagation $\mathcal{P}_S$ started from a seed set S, the sets of vertices that are in the spreader and aware states are denoted by S_τ and A_τ, respectively. In the beginning of $\mathcal{P}_S$, $S_0 = A_0 = S$, and for each $\tau \geq 1$ we have:

$$A_\tau = A_{\tau-1} \cup \{v \in V \setminus A_{\tau-1} : |N_{\text{in}}(v) \cap S_{\tau-1}| \geq 1\};$$
$$S_\tau = S_{\tau-1} \cup \{v \in V \setminus S_{\tau-1} : \sum_{u \in N_{\text{in}}(v) \cap S_{\tau-1}} w_{u,v} \geq t_v\}.$$

Note that every vertex in the spreader state is also in the aware state, i.e., $S_\tau \subseteq A_\tau$ for any $\tau \geq 0$, but the reverse is not necessarily true. The propagation ends when $S_{\rho-1} = S_\rho$ for some $\rho \geq 1$. If all vertices are aware at the end of $\mathcal{P}_S$, i.e., $A_\rho = V$, S is called a *perfect* seed set.

Formally, in the LDPAP, we are given an instance $I = (D, c, t, w)$, where $D = (V, E)$ is a directed graph, and $c : V \to \mathbb{R}^+$, $t : V \to \mathbb{R}^+$ and $w : E \to \mathbb{R}^+$ are cost, threshold, and weight functions, respectively. The problem's objective is to find a perfect seed set of minimum cost.

2.2 Previous Work

In [15], it is shown that the LDPAP is NP-hard and cannot be approximated within a ratio of $\mathcal{O}(2^{\log^{1-\varepsilon} n})$, for any $\varepsilon > 0$, unless $\mathsf{NP} \subseteq \mathsf{DTIME}(n^{\mathrm{polylog}(n)})$. These results also hold for the PAP [6], which is a special case of the LDPAP where $(u, v) \in E$ iff $(v, u) \in E$, $w_{u,v} = 1$ for every $(u, v) \in E$, and $t_v \in \mathbb{Z}^+$, $c_v = 1$ for every $v \in V$. Moreover, the LDPAP remains NP-hard for acyclic graphs, which can be proved by a polynomial reduction from the *Hitting Set Problem* [10,15].

One may obtain a trivial lower bound for the objective function of the LDPAP by computing the sum of the costs of all vertices that are sources, if there exists any, or the least cost among all vertices, otherwise. An algorithm to obtain an alternative lower bound that is at least as good as the trivial one is proposed in [15]. The strategy is to contract the strongly connected components of the original graph to form a (probably smaller) new instance. The optimal value for the new instance provides a lower bound for the original one.

There are two IP models for the LDPAP called IP-ROUNDS and IP-ARCS, which were proposed in [15]. In that same paper, these formulations were tested with a commercial IP solver on 300 small instances containing up to 100 vertices. Within an hour of execution for each instance, the IP-ROUNDS and IP-ARCS models obtained optimal solutions for 90 and 54 instances, respectively. The authors also concluded that the IP-ARCS formulation appears to be more adequate for solving instances with sparse graphs, while the IP-ROUNDS model is a better fit for instances with graphs of intermediate and higher densities.

Due to the LDPAP's complexity, a natural way to tackle large instances is to employ heuristic algorithms. In [15], the authors introduce a heuristic for the LDPAP called RGR, which is based on the metaheuristic Reactive GRASP [19]. The RGR heuristic was tested on large instances with up to 50,000 vertices and proved to be greatly superior to a baseline greedy algorithm.

3 Integer Programming Formulations

In this section, we introduce two new IP formulations for the LDPAP, namely, IP-ARCS-POLY and IP-ORDERING. For ease of reference, we also present the two existing formulations from [15] namely, IP-ROUNDS and IP-ARCS. We remark that these models can be simplified to obtain TSSP formulations and they may be seen as adaptations from the IP models for the TSSP from [1,20,22]. Future TSSP models might lead to the design of new formulations for LDPAP.

Let $I = (D = (V, E), c, t, w)$ be an instance of the LDPAP, where $|V| = n$. Every formulation in this section contains a set $\{s_v : v \in V\}$ of binary variables such that $s_v = 1$ iff v is a seed. The models also share the same objective function (1), which minimizes the cost of the seed set.

$$\min \sum_{v \in V} c_v s_v \tag{1}$$

3.1   The Existing IP-ROUNDS Formulation

In the IP-ROUNDS model, there exists a set $\{x_{v,\tau} : v \in V, \tau \in \{0, 1, \ldots, n\}\}$ of binary variables such that $x_{v,\tau} = 1$ iff v is a spreader in round τ. Note that $n + 1$ is the maximum number of rounds it takes for a propagation to end. The rest of the formulation comprises constraints (2)–(4).

$$x_{v,0} = s_v \qquad\qquad \forall v \in V \qquad (2)$$

$$\sum_{u \in N_{\text{in}}(v)} w_{u,v} x_{u,\tau-1} + t_v x_{v,0} \geq t_v x_{v,\tau} \qquad \forall v \in V \ \forall \tau \in \{1, 2, \ldots, n\} \qquad (3)$$

$$x_{v,0} + \sum_{u \in N_{\text{in}}(v)} x_{u,n-1} \geq 1 \qquad\qquad \forall v \in V \qquad (4)$$

Constraints (2) determine that the seed set is formed by the vertices that are spreaders in round $\tau = 0$. Constraints (3) forbid each vertex v from being a spreader in round τ (i.e., make $x_{v,\tau} = 0$) if v is neither a seed ($x_{v,0} = 0$) nor receives enough influence from its in-neighbors that are spreaders in round $\tau - 1$ (i.e., $\sum_{u \in N_{\text{in}}(v)} w_{u,v} x_{u,\tau-1} < t_v$). Lastly, constraints (4) ensure that every vertex v is either a seed or has an in-neighbor that is spreader in round $\tau = n - 1$ and, consequently, is aware in round n. The IP-ROUNDS formulation has $\mathcal{O}(|V|^2)$ binary variables and $\mathcal{O}(|V|^2)$ constraints.

3.2   The Existing IP-ARCS Formulation

The IP-ARCS model has a set $\{y_{u,v} : (u, v) \in E\}$ of binary variables such that $y_{u,v} = 1$ iff u influences v during the propagation. Let $\varXi$ be the collection of all directed cycles of D. The IP-ARCS formulation includes constraints (5)–(8).

$$y_{u,v} + s_v \leq 1 \qquad\qquad \forall (u, v) \in E \qquad (5)$$

$$\sum_{i \in N_{\text{in}}(u)} w_{i,u} y_{i,u} + t_u s_u \geq t_u y_{u,v} \qquad \forall (u, v) \in E \qquad (6)$$

$$s_v + \sum_{u \in N_{\text{in}}(v)} y_{u,v} \geq 1 \qquad\qquad \forall v \in V \qquad (7)$$

$$\sum_{(u,v) \in \xi} y_{u,v} \leq |\xi| - 1 \qquad\qquad \forall \xi \in \varXi \qquad (8)$$

Constraints (5) establish that, if a vertex is a seed, it cannot be influenced by any of its in-neighbors. Constraints (6) guarantee that, for any $(u, v) \in E$, if u influences v during the propagation ($y_{u,v} = 1$), then u necessarily enters the spreader state before that happens, either by being a seed ($s_u = 1$) or by receiving an amount of influence of at least t_u. Constraints (7) enforce that each vertex is a seed or is influenced by one of its in-neighbors, and hence aware.

Lastly, (8) forbid the occurrence of a circular sequence of influences during the propagation by avoiding directed cycles in the directed subgraph of D induced by each $(u, v) \in E$ with $y_{u,v} = 1$. The IP-ARCS formulation has $\mathcal{O}(|V| + |E|)$ variables and $\mathcal{O}(|V| + |E|)$ constraints, except for (8), which can be exponential.

3.3   The New IP-ARCS-POLY Formulation

The new IP-ARCS-POLY formulation is obtained by substituting constraints (8) in IP-ARCS by an alternative set of constraints with polynomial size. The idea is similar to the subtour elimination constraints in the Miller–Tucker–Zemlin formulation for the Travelling Salesman Problem [12].

Let $\{\ell_v : v \in V\}$ be a set of integer variables with domain $\{0, 1, \ldots, n-1\}$. For each $v \in V$, ℓ_v is the length of the longest path that starts from a seed and ends at v in the directed subgraph of D induced by the edges in $\{(u, v) \in E : y_{u,v} = 1\}$. The IP-ARCS-POLY model comprises constraints (5)–(7), (9), and (10).

$$(n - 1) \cdot (1 - s_v) \geq \ell_v \qquad\qquad \forall v \in V \qquad (9)$$

$$n(y_{u,v} - 1) + 1 \leq \ell_v - \ell_u \qquad\qquad \forall(u, v) \in E \qquad (10)$$

Constraints (9) determine that if v is a seed ($s_v = 1$), then the length of the longest path that starts from a seed and ends at v is 0 (i.e., $\ell_v = 0$). On the other hand, constraints (10) enforce that if u influences v (i.e., $y_{u,v} = 1$), then the length of the longest path that starts from a seed and ends at v must be larger than the length of the longest path that starts from a seed and ends at u (i.e., $\ell_v > \ell_u$). The IP-ARCS-POLY model has $\mathcal{O}(|V| + |E|)$ binary variables, $\mathcal{O}(|V|)$ integer (non-binary) variables, and $\mathcal{O}(|V| + |E|)$ constraints.

3.4   The New IP-ORDERING Formulation

In the new IP-ORDERING formulation, there are three sets of binary variables, namely, $\{p_v : v \in V\}, \{q_v : v \in V\}, \{h_{u,v} : u, v \in V, u \neq v\}$, such that:

$p_v = 1$ iff v is not a seed, but becomes a spreader during the propagation;
$q_v = 1$ iff v is not a seed and does not become a spreader, but is in the aware state at the end of the propagation;
$h_{u,v} = 1$ iff u and v are both spreaders during the propagation, and u assumes that state before v does.

The model comprises constraints (11)–(17). The role of the h variables is to induce the order in which the vertices assume their spreader state.

$$s_v + p_v + q_v = 1 \qquad\qquad \forall v \in V \qquad (11)$$

$$\sum_{u \in N_{in}(v)} w_{u,v} h_{u,v} \geq t_v p_v \qquad\qquad \forall v \in V \qquad (12)$$

$$\sum_{u \in N_{in}(v)} (s_u + p_u) \geq q_v \qquad\qquad \forall v \in V \qquad (13)$$

$$s_u + p_u \geq h_{u,v} \qquad\qquad \forall(u, v) \in V \times V, u \neq v \qquad (14)$$

$$h_{u,v} \leq p_v \qquad\qquad \forall(u, v) \in V \times V, u \neq v \qquad (15)$$

$$h_{u,v} + h_{v,u} + q_v \leq 1 \qquad\qquad \forall(u, v) \in V \times V, u \neq v \qquad (16)$$

$$h_{i,j} + h_{j,k} \leq h_{i,k} + 1 \qquad\qquad \forall(i, j, k) \in V \times V \times V, i \neq j \neq k \qquad (17)$$

Constraints (11) guarantee that each vertex is either a seed or assumes the spreader or aware state. Constraints (12) enforce that, if v is not a seed but becomes a spreader, the amount of influence received by v from its in-neighbors that are spreaders before v must be at least t_v. Constraints (13) determine that if v is neither a seed nor becomes a spreader but is otherwise aware at the end of the propagation, then v must have at least one in-neighbor that is either a seed or becomes a spreader during the propagation.

Constraints (14) and (15) establish that, if u becomes a spreader before v does, then either u is a seed or it becomes a spreader later, and v has to become a spreader as well. Constraints (16) imply that at most one of the following can be true: u assumes the spreader state before v does; v assumes the spreader state before u does; v is neither a seed nor becomes a spreader, but becomes aware during the propagation. Lastly, (17) ensure transitivity in the ordering of the vertices that assume the spreader state during the propagation. The IP-ORDERING model has $\mathcal{O}(|V|^2)$ binary variables and $\mathcal{O}(|V|^3)$ constraints.

4   Constraint Programming Formulations

In this section, we rewrite the IP-ROUNDS, IP-ARCS-POLY, and IP-ORDERING formulations by converting some of their linear constraints into propositional logic constraints, which are more natural and adequate for a CP solver. The converted constraints represent logical implications originally formulated with big-M type coefficients. This is the case for constraints (3), (6), (9), and (10). As a result of this rewriting, we derive three CP formulations for the LDPAP, namely, CP-ROUNDS, CP-ARCS-POLY, and CP-ORDERING. Before describing them, we explain why we did not rewrite IP-ARCS.

Because of the exponentially many constraints (8) in the IP-ARCS model, one would generally use a row generation algorithm in which (8) are regarded as *lazy constraints* that get introduced only as they become violated. The CP solvers we are familiar with do not allow lazy constraints since they require all constraints to be loaded a priori, which is not doable for (8). Although it is possible to use a CP solver to repeatedly solve the problem and add violated constraints (8), we tested such an implementation and concluded it is too time consuming and not effective in practice. Some CP solvers, such as Gecode [8], include a global constraint called DAG that avoids induced directed cycles, but that constraint basically consists of using (9) and (10). Therefore, substituting DAG for (8) in IP-ARCS would simply lead to the same IP-ARCS-POLY model that we are already studying.

We now introduce the CP-ROUNDS, CP-ARCS-POLY and CP-ORDERING formulations, which share the same objective function (1).

4.1   The CP-ROUNDS Formulation

The CP-ROUNDS model is derived from IP-ROUNDS by substituting (18) for (3).

$$\text{IF } (x_{v,\tau} \text{ AND } \neg x_{v,0}) \text{ THEN } \sum_{u \in N_{\text{in}}(v)} w_{u,v} x_{u,\tau-1} \geq t_v \quad \forall v \in V \ \forall \tau \in \{1, 2, \ldots, n\} \quad (18)$$

For every vertex v, (18) say that if v is a spreader in round τ but not a seed, then the amount of influence it receives in round $\tau - 1$ is at least t_v. Like IP-ROUNDS, this model has $\mathcal{O}(|V|^2)$ binary variables and $\mathcal{O}(|V|^2)$ constraints.

4.2   The CP-ARCS-POLY Formulation

The CP-ARCS-POLY model is derived from IP-ARCS-POLY by substituting (19), (20), and (21) for (6), (9), and (10), respectively.

$$\text{IF } (y_{u,v} \text{ AND } \neg s_u) \text{ THEN } \sum_{i \in N_{\text{in}}(u)} w_{i,u} y_{i,u} \geq t_u \qquad \forall (u,v) \in E \qquad (19)$$

$$\text{IF } s_v \text{ THEN } \ell_v = 0 \qquad\qquad \forall v \in V \qquad (20)$$

$$\text{IF } y_{u,v} \text{ THEN } \ell_u < \ell_v \qquad\qquad \forall (u,v) \in E \qquad (21)$$

Constraints (19) guarantee that, for any $(u,v) \in E$, if u influences v during a propagation but u is not a seed, then u must have become a spreader by receiving enough influence. Constraints (20) establish that, if v is a seed, the longest path from a seed to v has length 0. Lastly, (21) indicate that, if u influences v, the longest path from a seed to u must be shorter than the longest path from a seed to v. The CP-ARCS-POLY model has $\mathcal{O}(|V| + |E|)$ binary variables, $\mathcal{O}(|V|)$ (non-binary) integer variables, and $\mathcal{O}(|V| + |E|)$ constraints.

4.3   The CP-ORDERING Formulation

The CP-ORDERING formulation is derived from IP-ORDERING by substituting (22) and (23) for (12) and (17), respectively.

$$\text{IF } p_v \text{ THEN } \sum_{u \in N_{\text{in}}(v)} w_{u,v} h_{u,v} \geq t_v \qquad\qquad \forall v \in V \qquad (22)$$

$$\text{IF } (h_{i,j} \text{ AND } h_{j,k}) \text{ THEN } h_{i,k} \qquad \forall (i,j,k) \in V \times V \times V, i \neq j \neq k \qquad (23)$$

Constraints (22) determine that, if a non-seed vertex v becomes a spreader, the amount of influence received by v must be at least t_v. Constraints (23) enforce transitivity on the spreading order implied by the h variables. The CP-ORDERING model has $\mathcal{O}(|V|^2)$ binary variables and $\mathcal{O}(|V|^3)$ constraints.

5   Preprocessing Methods for the LDPAP Formulations

In this section, we present useful preprocessing techniques that can be applied to the proposed models to reduce their sizes either by removing unnecessary variables or constraints, or by reducing some variable domains.

Let $I = (D = (V,E), c, t, w)$ be an instance of the LDPAP with $|V| = n$, and let $Q \subseteq V$ be the set of source vertices in D (i.e., vertices with an empty in-neighborhood). Clearly, any feasible solution for I must contain Q, otherwise the sources would not become aware during the propagation. Now, let $\mathcal{P}_Q$ be a propagation started from Q and let $\rho \geq 1$ be the last round of $\mathcal{P}_Q$. Then, $Q_\rho = Q_{\rho-1}$, where Q_i denotes the set of spreaders in the i-th round. Take $\kappa = \rho + n + 1 - |Q_{\rho-1}|$. Our first preprocessing method relies on Theorem 1.

Theorem 1. *If S is a feasible solution for I, then the propagation $\mathcal{P}_S$ started from S takes no more than κ rounds.*

Proof. Let S be a feasible solution for I so that $\mathcal{P}_S$ takes at least $\kappa + 1$ rounds. Since S is feasible, $Q \subseteq S$ and, therefore, $Q_{\rho-1} \subseteq S_{\rho-1}$. Hence, at the end of round $\rho - 1$ of $\mathcal{P}_S$, there are at least $|Q_{\rho-1}|$ spreaders and, consequently, at most $n - |Q_{\rho-1}|$ non-spreaders. Therefore, there can be at most $n - |Q_{\rho-1}| + 1$ rounds after $\rho - 1$, totaling $\rho + n + 1 - |Q_{\rho-1}| = \kappa$ rounds (because there are ρ rounds from 0 to $\rho - 1$), which contradicts the choice of S.   □

We can compute both ρ and $|Q_{\rho-1}|$ (and thus κ) by simulating $\mathcal{P}_Q$ using the `CompletePropagation` algorithm, introduced in [15], which runs in $\mathcal{O}(|V|+|E|)$ time. To do so, we keep a counter for the number of rounds and another for the number of spreaders per round.

So far in this paper, we have used $n + 1$ as an upper bound for the total number of rounds that a propagation can take. Indeed, some variables in the `IP-ROUNDS`, `IP-ARCS-POLY`, `CP-ROUNDS`, and `CP-ARCS-POLY` formulations were defined by assuming that, in the worst-case scenario, a propagation ends in round n. We now tighten the $n + 1$ upper bound to κ. In doing so, we are assuming that, in the worst case, a propagation ends in round $\kappa - 1$. Next, we describe how we adjust the models accordingly.

For the `IP-ROUNDS` and `CP-ROUNDS` models, we remove all variables $x_{v,\tau}$ with $v \in V$ and $\tau \geq \kappa$, as well as every constraint of type (3) and (18) in which they occur. Then, we substitute (24) for (4).

$$x_{v,0} + \sum_{u \in N_{in}(v)} x_{u,\kappa-2} \geq 1 \qquad\qquad \forall v \in V \qquad (24)$$

For the `IP-ARCS-POLY` and `CP-ARCS-POLY` models, we simply change the domain of the ℓ variables from $\{0, 1, \ldots, n-1\}$ to $\{0, 1, \ldots, \kappa - 2\}$.

Now, we present a preprocessing approach for the `IP-ORDERING` formulation. Recall that each binary variable $h_{u,v}$ indicates whether u assumes the spreader state before v does. Moreover, (16) and (17) guarantee that the directed graph defined by the edge set $\{(u, v) : u, v \in V, u \neq v, h_{u,v} = 1\}$ is acyclic. In other words, these constraints forbid any circular sequence of vertices with respect to the order in which they assume the spreader state.

During a propagation, however, a circular sequence of influence induced by h can only occur between vertices belonging to the same strongly connected component of D. Thus, instead of ensuring a non-circular ordering of the vertices for the entire graph, we can rewrite the constraints to focus separately on each strongly connected component of D.

We redefine the set of h variables from $\{h_{u,v} : u, v \in V, u \neq v\}$ to $\{h_{u,v} : (u, v) \in E \cup E'\}$ where E' contains edges (u, v) and (v, u) for every distinct vertices u and v that belong to the same strongly connected component of D. Next, we substitute constraints (25) to (28) for constraints (14) to (17).

$$s_u + p_u \geq h_{u,v} \qquad\qquad \forall (u,v) \in E \cup E' \quad (25)$$

$$h_{u,v} \leq p_v \qquad\qquad \forall (u,v) \in E \cup E' \quad (26)$$

$$h_{u,v} + h_{v,u} + q_v \leq 1 \qquad\qquad \forall (u,v) \in E' \quad (27)$$

$$h_{i,j} + h_{j,k} \leq h_{i,k} + 1 \quad \forall \text{ distinct } i,j,k \in V \text{ s.t. } (i,j),(j,k),(i,k) \in E' \quad (28)$$

To consider the CP-ORDERING formulation with the preprocessing approach we have described for the IP-ORDERING model, we apply the same modifications proposed above and also substitute (29) for (23).

$$\text{IF } (h_{i,j} \text{ AND } h_{j,k}) \text{ THEN } h_{i,k} \quad \forall \text{ distinct } i,j,k \in V \text{ s.t. } (i,j),(j,k),(i,k) \in E' \quad (29)$$

6 Computational Experiments

In this section, we describe the experiments we carried out with the formulations presented in Sects. 3 and 4. We used a machine equipped with an Intel® Xeon® E5-2630 v4 processor, 64 GB of RAM, and the Ubuntu 22.04.1 LTS operating system. We employed Gurobi v10.0.3 [9] as the IP solver and CP-SAT v9.7.2996 (from Google OR-Tools) [18] as the CP solver. The instances in our experiments were divided into two datasets, outlined in the next section.

We refer the reader to a publicly available repository [17] that accompanies this paper and includes the source code, problem instances, the solutions obtained, and an appendix with additional details about our experimental results.

6.1 Datasets

The first dataset, denoted by Δ_{syn}, was introduced in [15] and contains 30 synthetic instances with n vertices for each $n \in \{10, 20, \ldots, 100\}$, totaling 300 instances. The graphs in these instances were generated using a well known algorithm proposed in [3] for the creation of scale-free graphs that capture crucial characteristics of social networks. Particularly, the Δ_{syn} dataset was purposely designed to be diverse regarding the densities of the graphs, calculated as $m/(n^2 - n)$, where m is the number of edges. For each fixed n, the 30 instances in Δ_{syn} with n vertices vary from very sparse graphs to nearly complete graphs.

For each instance (D, w, c, t) in Δ_{syn}, the weight, cost, and threshold functions are given by (30), (31) and (32), which were originally presented in [15].

$$w_{u,v} = |N_{\text{out}}(u)| / \left(\sum_{w \in N_{\text{in}}(v)} |N_{\text{out}}(w)| \right) \qquad (30)$$

$$c_v = \begin{cases} 5 \cdot |N_{\text{out}}(v)|, & \text{if } |N_{\text{out}}(v)| > 0 \\ +\infty, & \text{otherwise} \end{cases} \qquad (31)$$

$$t_v = \begin{cases} \dfrac{|N_{\text{in}}(v)|}{2} \cdot \underset{u \in N_{\text{in}}(v)}{\text{median}} \{w_{u,v}\}, & \text{if } |N_{\text{in}}(v)| > 0 \\ +\infty, & \text{otherwise.} \end{cases} \qquad (32)$$

Formula (30) establishes that the amount of influence that u exerts on v is proportional to the popularity of u on the network compared to the popularity of the other in-neighbors of v. Formula (31) is based on the fact that, in 2020, influencers with up to 1 million followers on some online social networks were charging about 5 cents of a dollar per follower for a single paid post [4]. Note that when v has no influence over other users, assigning $+\infty$ as its cost prevents v from being picked as an ineffective seed.

Formula (32) determines that a vertex v begins to forward the information when it receives an amount of influence equivalent to the total influence coming from half of its in-neighbors, assuming that all of them were exerting a median level of influence on v. This is an extension of the well-studied *majority threshold function*, where, in a given unweighted and undirected graph, a non-seed vertex spreads the information when at least half of its neighbors are spreaders [5]. For more details on the generation of the Δ_{syn} dataset, we refer the reader to [15].

With the objective of evaluating the scalability of the proposed models for larger but still solvable instances, we designed a second set of instances, denoted by Δ_X, containing large graphs. The Δ_X dataset was built using a set of users of the X social network and their online interactions between August 2019 and March 2020 [21]. For each of 14 selected topics, from soccer to politics, the authors identified the users who actively posted, retweeted or quoted, and then crawled their followers that were also active users on the same topic.

These crawled relationships between "follower" and "followed" were previously used in [15] to obtain a benchmark of instances used to test a heuristic for the LDPAP (see Sect. 2.2). We first attempted to test our exact models on that benchmark, but preliminary results indicated that the formulations could not be solved within the established time limit of one hour and sometimes could not even be loaded into memory due to the very large sizes of the instances, specifically their number of edges. Thus, we were compelled to reduce the number of edges in these graphs in order to stress our models on large, yet solvable, instances.

Therefore, we designed the Δ_X dataset so that its instances contain a subset of the user relationships present in the original instances from [15]. This resulted in 14 large instances, one for each crawled topic. Next, we outline the details of the Δ_X dataset construction.

First, we grouped the users with respect to their popularity in the network, in accordance with the approach proposed in [4], where the influential users were classified according to their number of followers. To do so, we took r as the largest integer such that there exists a user with at least 2^r followers, and then partitioned the users into the following equivalence classes:

- *Group A*: at least 2^r followers;
- *Group B*: at least 2^{r-1} and fewer than 2^r followers;
- *Group C*: at least 2^{r-2} and fewer than 2^{r-1} followers;
- *Group D*: fewer than 2^{r-2} followers.

We then assumed that the information typically flows from more popular users to less popular ones or between users within the same class, except for users within the least popular Group D. Next, we constructed a graph that contains an edge (u, v), meaning that u exerts influence over v, for each pair of users u and v such that v follows u, and either u is more popular than v or they both belong to the same group A, B or C. This construction purposely forbids outgoing edges from vertices in group D, since they correspond to users with very few followers and are not significantly influential in practice. Lastly, we used (30), (31) and (32) to compute the remaining parameters required to create an instance of the LDPAP. Table 1 shows properties of the graphs that make up the instances in Δ_X, where SCCs denotes the number of strongly connected components (SCC) and LSCC is the number of vertices in the largest SCC.

Table 1. Quantifying topological characteristics of Δ_X.

Instance	Vertices	Edges	Density	SCCs	LSCC
X_{01}	6529	69385	0.00163	6344	170
X_{02}	16976	646190	0.00224	16123	719
X_{03}	29693	250162	0.00028	29515	172
X_{04}	275	1244	0.01651	250	26
X_{05}	1637	11744	0.00439	1520	99
X_{06}	6537	310694	0.00727	5725	664
X_{07}	8068	138701	0.00213	7362	572
X_{08}	7857	164710	0.00267	6994	730
X_{09}	927	5371	0.00626	884	35
X_{10}	9271	58828	0.00068	9209	50
X_{11}	3964	24296	0.00155	3903	56
X_{12}	1520	13287	0.00575	1388	104
X_{13}	993	16828	0.01708	685	173
X_{14}	1405	8181	0.00415	1370	32

6.2   Results for Instances in Δ_{syn}

In this section, we report the results for each of the 300 instances from Δ_{syn}. We configured both solvers, Gurobi and CP-SAT, to run on a single thread of execution within a time limit of 1 h for each pair of formulation and instance.

The preprocessing methods that we proposed in Sect. 5 to reduce the size of the models were employed whenever they were applicable. They proved to be much more effective than the built-in preprocessing phase inherent to the solver, so the time spent running them was fully compensated by the performance gains.

Moreover, for each instance, we provided the solvers with an initial feasible solution that corresponds to the best known perfect seed set for that instance thus far. These seed sets were obtained from experiments conducted in [15] and are publicly available in [14]. We also provided the solvers with an initial lower bound for the objective function, which was obtained by the algorithm proposed in [15] (see Sect. 2.2).

It is important to highlight that, as shown in Sect. 3, the IP-ARCS model has an exponential number of constraints (8). As a result, we followed the traditional lazy constraint strategy: whenever the IP solver found an integer solution, we performed a complete depth-first search on the directed graph induced by the integer solution and, for each cycle found, we added the corresponding violated constraint to the formulation.

We remark that, in [15], the IP-ROUNDS and IP-ARCS models were tested with the Δ_{syn} dataset and found optimal solutions for 90 and 54 instances, respectively, totalling 93 instances solved.

Regarding the experiments conducted in the present paper, Table 2 shows the number of provably optimal solutions obtained by each model, where the instances are grouped by their number of vertices (recall that there are 30 instances for each value of $|V|$).

Table 2. Number of solved instances from Δ_{syn}.

Formulation	$|V|$										Total
	10	20	30	40	50	60	70	80	90	100	
IP-ROUNDS	30	28	9	4	3	3	3	3	4	4	91
IP-ARCS	30	9	4	3	2	2	1	1	1	1	54
IP-ARCS-POLY	30	14	5	3	2	2	1	1	1	1	60
IP-ORDERING	30	26	6	3	2	2	1	1	1	1	73
CP-ROUNDS	30	30	30	30	30	30	30	29	22	17	278
CP-ARCS-POLY	30	22	9	5	3	3	1	2	1	1	77
CP-ORDERING	30	30	30	28	12	6	6	2	3	2	149

We first observe that the results for the IP-ROUNDS and IP-ARCS models were similar to the ones obtained in [15], since they solved almost the exact same instances they did before. The IP-ARCS-POLY model had slightly superior performance than its exponential-size version, IP-ARCS, by solving 6 more instances with 20 and 30 vertices. The IP-ORDERING formulation outperformed both IP-ARCS and IP-ARCS-POLY formulations by solving a total of 73 instances. Among the IP models, IP-ROUNDS remained the best formulation for the Δ_{syn} dataset, solving a total of 91 instances.

When it comes to the CP formulations, we observe that each model outperformed its own IP version. More specifically, the CP-ORDERING and CP-ROUNDS models solved substantially more instances than their IP counterparts for each group of instances (with at least 20 vertices). Indeed, the CP-ORDERING and CP-ROUNDS models were the ones that solved the largest numbers of instances, 149 and 278, respectively, surpassing all the IP models by a large margin. Next, we analyze the results obtained by these two formulations in more detail.

Because the CP-ROUNDS model was able to solve all 149 instances solved by the CP-ORDERING model, we compare the performance of these formulations by the time it took them to find these optimal solutions. Table 3 reports statistics on the running times for those 149 instances.

From the median values in Table 3, we see that both models solved at least half of the 149 instances in less than 2 s. However, the average times indicate that the CP-ROUNDS formulation was about 38 times faster than the CP-ORDERING model. The intervals between minimum and maximum running times, together with the standard deviations, suggest that the CP-ROUNDS model had more stable running times that grew more gradually.

Table 3. Running times (in seconds) for the 149 instances solved by both CP-ROUNDS and CP-ORDERING formulations.

Formulation	Min	Max	Median	Average	Std Dev
CP-ROUNDS	0.006	56.572	0.617	3.622	8.453
CP-ORDERING	0.004	3362.744	1.450	138.469	458.432

Regarding the 129 instances solved by the CP-ROUNDS model but not by CP-ORDERING (because it exceed the 1-hour time limit), CP-ORDERING obtained an average optimality gap of 78.31%, with a standard deviation of 24.26. On the other hand, the CP-ROUNDS formulation proved optimality within 469 s on average, with a standard deviation of 746.

With respect to the 22 instances for which all formulations exceeded the time limit, the IP-ROUNDS and CP-ROUNDS models obtained the smallest optimality gaps, on average, namely 92.60% and 92.83%, respectively.

6.3   Results for Instances in Δ_X

In this section, we report the results of the experiments conduced with the Δ_X dataset. We maintained the same experimental settings used for the Δ_{syn} set (see Sect. 6.2), with the exceptions noted below.

We configured the solvers to run on at most 10 threads of execution for each pair of formulation and instance. We also provided the solvers with initial feasible solutions that were obtained by running the only existing heuristic for the LDPAP [15] for 10 min per instance. Due to the large sizes of the instances from Δ_X, the combinatorial lower bound proposed in [15] could not be computed. Instead, we provided the solvers with trivial lower bounds (see Sect. 2.2).

Table 4 reports the optimality gaps obtained for each formulation within the 1-hour time limit. Entries containing the '−' symbol indicate the run could not be completed due to lack of memory.

Table 4. Optimality gaps (%) obtained for the instances in Δ_X.

Formulation	X_{01}	X_{02}	X_{03}	X_{04}	X_{05}	X_{06}	X_{07}	X_{08}	X_{09}	X_{10}	X_{11}	X_{12}	X_{13}	X_{14}
IP-ROUNDS	88.87	–	–	0.00	0.00	–	–	–	0.00	–	20.22	31.32	**85.28**	0.00
IP-ARCS	88.87	**93.63**	0.00	0.00	0.00	96.06	76.36	52.46	0.00	64.31	0.00	31.32	98.73	0.00
IP-ARCS-POLY	88.87	98.65	0.00	0.00	0.00	96.06	76.36	52.46	0.00	64.31	0.00	31.32	98.73	0.00
IP-ORDERING	88.80	–	0.00	0.00	0.00	–	–	–	0.00	0.00	0.00	31.32	98.73	0.00
CP-ROUNDS	–	–	–	0.00	0.00	–	–	–	0.00	–	0.00	0.00	98.73	0.00
CP-ARCS-POLY	**88.46**	98.65	0.00	0.00	0.00	96.06	0.00	0.00	0.00	64.31	0.00	31.32	98.73	0.00
CP-ORDERING	–	–	–	0.00	0.00	–	–	–	0.00	0.00	0.00	0.00	–	0.00

We first analyze the results for the smallest instances from Δ_X with respect to the number of vertices in their largest SCCs (see Table 1). Instances X_{04}, X_{05}, X_{09}, and X_{14} were solved by each of the formulations in at most 200 s per execution, although most of these executions took only a few seconds.

Instance X_{10} was solved only by the IP-ORDERING and CP-ORDERING models in 532 and 17 s, respectively. Instance X_{11} was solved by all models, except for IP-ROUNDS, and optimal solutions were obtained in less than 2 min, except for CP-ROUNDS, which took 18 min. Instance X_{12} was solved by the CP-ROUNDS and CP-ORDERING models in 278 and 73 s, respectively. The CP-ORDERING model was the only one capable of solving all the instances mentioned so far, always needing less than 2 min to prove optimality.

The remaining instances are larger than the previous ones with respect to their largest SCCs. In this case, the executions with CP-ORDERING were halted due to lack of memory, most likely due to the large number of constraints (29).

Instance X_{03} was solved by the CP-ARCS-POLY, IP-ORDERING, IP-ARCS and IP-ARCS-POLY models, taking less than 3 min for the first three. Instances X_{07} and X_{08} were solved only by the CP-ARCS-POLY model in less than 3 min. The remaining instances, namely X_{01}, X_{02}, X_{06}, and X_{13}, were not solved by any model. For these instances, the smallest optimality gaps were achieved, most of the time, by either the IP-ARCS or the CP-ARCS-POLY formulation.

In conclusion, for the Δ_X dataset, the CP-ORDERING formulation had the best overall performance on instances with relatively small strongly connected components. For instances whose LSCCs had more than 104 vertices, CP-ORDERING ran into memory limitations, and for these cases the IP-ARCS and CP-ARCS-POLY formulations were the most effective ones.

7 Concluding Remarks and Future Work

In this paper, we study different formulations and preprocessing approaches for the LDPAP. We propose two new IP models and three new CP formulations. We also design and perform experiments to test the existing and new models on a set of small synthetic instances and a new collection of large instances obtained from the X social network.

Based on our results, the new CP-ROUNDS formulation is more suitable, both in terms of efficiency and effectiveness, to solve LDPAP instances with up to 100 vertices. With this model, we managed to increase the number of known solved instances from the Δ_{syn} dataset from 93 to 278 (out of 300). For this group of instances, every CP model outperformed its IP counterpart.

Regarding the 14 large instances from the Δ_X dataset, we learned that the CP-ROUNDS model does not scale well with respect to the number of vertices in the graph. Our results suggest that CP-ORDERING is the most efficient and effective formulation for large instances with fairly small SCCs. For instances containing larger SCCs, the IP-ARCS and CP-ARCS-POLY formulations appear to be better suited.

As for future research directions, we believe that both the CP-ROUNDS and CP-ORDERING formulations can be employed in the development of heuristics for the LDPAP based on large neighborhood search or even on techniques grounded in the decomposition of large instances into smaller ones. For the second case, the smaller instances can, in turn, be quickly solved with these models depending on their topological characteristics. Also, it is our view that the possibility of applying Benders decomposition to the IP-ARCS model is worth investigating.

Disclosure of Interests. The authors have no competing interests to declare that are relevant to the content of this article.

References

1. Ackerman, E., Ben-Zwi, O., Wolfovitz, G.: Combinatorial model and bounds for target set selection. Theoret. Comput. Sci. **411**(44), 4017–4022 (2010). https://doi.org/10.1016/j.tcs.2010.08.021
2. Banerjee, S., Jenamani, M., Pratihar, D.K.: A survey on influence maximization in a social network. Knowl. Inf. Syst. **62**(9), 3417–3455 (2020). https://doi.org/10.1007/s10115-020-01461-4
3. Bollobás, B., Borgs, C., Chayes, J., Riordan, O.: Directed scale-free graphs. In: Proceedings of the Fourteenth Annual ACM-SIAM Symposium on Discrete Algorithms, pp. 132-139. SODA '03, Society for Industrial and Applied Mathematics, USA (2003)
4. Campbell, C., Farrell, J.R.: More than meets the eye: the functional components underlying influencer marketing. Bus. Horiz. **63**(4), 469–479 (2020). https://doi.org/10.1016/j.bushor.2020.03.003
5. Chen, N.: On the approximability of influence in social networks. SIAM J. Discret. Math. **23**(3), 1400–1415 (2009). https://doi.org/10.1137/08073617X
6. Cordasco, G., Gargano, L., Rescigno, A.A.: Active influence spreading in social networks. Theoret. Comput. Sci. **764**, 15–29 (2019). https://doi.org/10.1016/j.tcs.2018.02.024
7. Gautam, R.K., Kare, A.S., Bhavani, S.D.: Centrality measures based heuristics for perfect awareness problem in social networks. In: Morusupalli, R., Dandibhotla, T.S., Atluri, V.V., Windridge, D., Lingras, P., Komati, V.R. (eds.) MIWAI 2023. LNCS, vol. 14078, pp. 91–100. Springer, Cham (2023). https://doi.org/10.1007/978-3-031-36402-0_8
8. Gecode Team: Gecode — generic constraint development environment (2023). http://www.gecode.org

9. Gurobi Optimization, LLC: Gurobi Optimizer Reference Manual (2023). https://www.gurobi.com

10. Karp, R.M.: Reducibility Among Combinatorial Problems, pp. 85–103. Springer US, Boston, MA (1972). https://doi.org/10.1007/978-1-4684-2001-2_9

11. Kempe, D., Kleinberg, J., Tardos, E.: Maximizing the spread of influence through a social network. In: Proceedings of the Ninth ACM SIGKDD International Conference on Knowledge Discovery and Data Mining, pp. 137-146. KDD 2003, ACM, New York, NY, USA (2003). https://doi.org/10.1145/956750.956769

12. Miller, C., Tucker, A., Zemlin, R.: Integer programming formulation of traveling salesman problems. J. ACM **7**(4), 326–329 (1960). https://doi.org/10.1145/321043.321046

13. Pereira, F.C.: A computational study of the Perfect Awareness Problem. Master's thesis, University of Campinas, Brazil (2021). http://hdl.handle.net/20.500.12733/1641217

14. Pereira, F.C., de Rezende, P.J.: The Least Cost Directed Perfect Awareness Problem – Benchmark Instances and Solutions. Mendeley Data, V2 (2023). https://doi.org/10.17632/xgtjgzf28r

15. Pereira, F.C., de Rezende, P.J.: The least cost directed perfect awareness problem: complexity, algorithms and computations. Online Soc. Netw. Media **37–38** (2023). https://doi.org/10.1016/j.osnem.2023.100255

16. Pereira, F.C., de Rezende, P.J., de Souza, C.C.: Effective heuristics for the perfect awareness problem. Procedia Comput. Sci. **195**, 489–498 (2021). https://doi.org/10.1016/j.procs.2021.11.059, proceedings of the XI Latin and American Algorithms, Graphs and Optimization Symposium

17. Pereira, F.C., de Rezende, P.J., Yunes, T.: Minimizing the cost of leveraging influencers in social networks: IP and CP approaches - complementary data. Mendeley Data, V2 (2023). https://doi.org/10.17632/tkk5pdswty

18. Perron, L., Didier, F.: CP-SAT (Google OR-Tools) (2023). https://developers.google.com/optimization/cp/cp_solver/

19. Prais, M., Ribeiro, C.C.: Reactive grasp: an application to a matrix decomposition problem in TDMA traffic assignment. Informs J. Comput. **12**(3), 164–176 (2000). https://doi.org/10.1287/ijoc.12.3.164.12639

20. Raghavan, S., Zhang, R.: A branch-and-cut approach for the weighted target set selection problem on social networks. Informs J. Optimiz. **1**(4), 304–322 (2019). https://doi.org/10.1287/ijoo.2019.0012

21. Schweimer, C., et al.: Generating simple directed social network graphs for information spreading. In: Proceedings of the ACM Web Conference 2022, pp. 1475–1485. WWW 2022, ACM, New York, USA (2022). https://doi.org/10.1145/3485447.3512194

22. Shakarian, P., Eyre, S., Paulo, D.: A scalable heuristic for viral marketing under the tipping model. Soc. Netw. Anal. Min. **3**(4), 1225–1248 (2013). https://doi.org/10.1007/s13278-013-0135-7

Learning Deterministic Surrogates
for Robust Convex QCQPs

Egon Peršak$^{(\boxtimes)}$ and Miguel F. Anjos

University of Edinburgh, Edinburgh, UK
E.Persak@sms.ed.ac.uk

Abstract. Decision-focused learning is a promising development for contextual optimisation. It enables us to train prediction models that reflect the conditional sensitivity and uncertainty structure of the problem. However, there have been limited attempts to extend this paradigm to robust optimisation. We propose a double implicit layer model for training prediction models with respect to robust decision loss in uncertain convex quadratically constrained quadratic programs (QCQP). The first layer solves a deterministic version of the problem, the second layer evaluates the worst case realisation for an uncertainty set centred on the observation given the decisions obtained from the first layer. This enables us to learn model parameterisations that lead to robust decisions while only solving a simpler deterministic problem at test time. Additionally, instead of having to solve a robust counterpart we solve two smaller and potentially easier problems in training. The second layer (worst case problem) can be seen as a regularisation approach for predict-and-optimise by fitting to a neighbourhood of problems instead of just a point observation. We motivate a reformulation of the worst-case problem for a case of uncertainty sets that would otherwise lead to trust region problems. Both layers are typically strictly convex in this problem setting and thus have meaningful gradients almost everywhere. We demonstrate an application of this model on simulated experiments. The method is an effective regularisation tool for decision-focused learning for uncertain convex QCQPs.

Keywords: Differentiable Optimisation · Quadratic Programming · Robust Optimisation · Robust Predict-and-Optimise

1 Introduction

Decision problems are seldom static. Foresighted planning requires us to assert some judgement or prediction of the future to ground the hypothesis space of our decisions. Fortunately, it is common for processes relevant to decision making to have a reasonable degree of predictability either in the form of trends or seasonality. Unfortunately, the dynamics of these processes tend to be stochastic, are often non-linear, potentially non-stationary, and have been observed insufficiently for robust inference. This is a problem as decision quality can be sensitive

© The Author(s), under exclusive license to Springer Nature Switzerland AG 2024
B. Dilkina (Ed.): CPAIOR 2024, LNCS 14743, pp. 128–140, 2024.
https://doi.org/10.1007/978-3-031-60599-4_8

to differences between predicted problem parameters and true problem parameters. Even a small prediction error may render a solution highly suboptimal or even infeasible.

The classical solution to this from the mathematical optimisation community has been robust optimisation. That is to find a solution that has the best worst case realisation across a neighbourhood of problems defined by an uncertainty set. Determining an appropriate contextual uncertainty set is at least as difficult as the prediction problem and is beyond the scope of this work, we assume we have a method which is sufficient for our coverage or regularisation needs. A more recent development that aims to tackle the sensitivity of prediction defined optimisation is decision-focused learning. Instead of attempting to minimise some measure between the predicted and realised problem parameters it aims to minimise the resulting decision loss. The decision loss trained models reflect the underlying conditional sensitivity and uncertainty structure of the problem. The predictions effectively trick the optimisation model into making more robust solutions in an environment with sparsely observed contextual realisations.

There are two key shortcomings with the decision-focused learning approach. It is computationally demanding as computing the gradients for decision losses requires differentiating through an optimisation process which typically requires at least one computation of an optimal solution. In this work we focus on quadratic convex optimisation problems for which we can differentiate the optimality conditions for gradient computation. Note that a wealth of methods has been developed for differentiable optimisation [16]; other approaches for differentiating across $arg\min$ may be more appropriate, but this is beyond the scope of this work. Decision-focused learning therefore scales poorly with increasing levels of the computational difficulty of optimisation problems.

The prediction problem for optimisation is typically very high dimensional, as we need to predict the vectors and matrices that define an optimisation problem. As observations are limited, this means we are necessarily dealing with a lot of empirical uncertainty on top of aleatoric uncertainty and are faced with a few-shot learning problem. While standard decision-focused learning ameliorates the sensitivity issue, it does not resolve it without having many observations in similar contexts. Even in data-rich cases the underlying aleatoric uncertainty means that predictions based on minimising empirical regret will overfit to observed realisations for a local context. A potential solution is taking the decision losses from the robust problem centred on the observation. In its classical form this exacerbates the computational issue as the robust counterpart of an optimisation problem is at best (with elegantly designed uncertainty sets) in the same class of computational difficulty, but larger. In general, uncertainty sets have to be carefully designed to preserve tractability and their robust counterparts usually belong to more computationally challenging problem families [5]. The nature of the decision problem may mean that we have insufficient time to obtain a robust solution at test time.

To address these shortcomings we propose a model which learns parameterisations for surrogate deterministic programs whose solutions approximate robust solutions. This means we have to separate the deterministic version of the program from the worst case problem. In the worst case problem we are trying to find a combination of the worst case outcome and constraint violation for a given decision. The worst case problem does not have to be solved at test time as it is effectively a regularisation tool which penalises decisions based on their local worst-case outcome. The regularisation explicitly accounts for the local sensitivity structure for each observation. In effect, we are trying to solve the inverse problem of what parameters for a deterministic problem would achieve robust solutions for a given uncertainty set.

We narrow our attention to uncertain convex QCQP. This research focus is motivated by two properties. The first is that for strictly convex QCQPs the optimisation output in terms of parameters is continuous and thus subdifferentiable everywhere and differentiable almost everywhere. The second is that QCQPs may have optimal solutions in the interior of the feasible set and therefore the surrogate models can either learn in the direction of the constraints or objective. In contrast, linear programs have outputs which are piecewise constant requiring some form of regularisation in the objective function for decision-focused learning, for example a 2-norm on the decision vector [21] to compute meaningful gradients. This turns the problem into a second order cone program (SOCP), which is the complexity for robust linear programming with ellipsoidal uncertainty. Furthermore, the solution to a robust linear program may not be at a vertex meaning a linear surrogate will potentially have trouble approximating it, or in the case of fixed constraints will not be able to approximate it at all. We are confident our approach can be extended to more general convex optimisation frameworks, but it is beyond the scope of this work. Since general (convex and non-convex) QCQPs tend to be tackled with relaxations that reformulate them into SDP extending robust decision losses to uncertain SDPs is a promising future direction. MAXSAT, a common problem in logical reasoning, has a convex relaxation in the form of an SDP and previous research has utilised this fact to develop decision-focused learning for satisfiability solvers [20]. Robust decision losses for SDPs would therefore enable effective regularisation for logical reasoning problems.

The remainder of this work is structured as follows. Section 2 is a non-exhaustive overview of the prior work in decision-focused learning, robust convex optimisation, and online methods for QCQPs. Section 3 contains the contribution of this work by presenting the two implicit layer approach for deterministic surrogates and why it works. Section 4 presents synthetic experiments demonstrating the usefulness of robust decision losses. The work is concluded with a discussion of potential applications, and future directions research in Sect. 5.

2   Background

The field of contextual stochastic optimisation concerns itself with decision making under uncertainty in a data driven way. A paradigm which has seen much excitement in this community is integrating the prediction and optimisation step, which has been variously termed as predict-and-optimise, decision-focused learning, and end-to-end learning. From a differentiable programming standpoint it sees optimisation problems as just another differentiable component in a mapping of data context to decisions. The field of deep learning sees it as an example of an implicit layer.

Three main directions have been developed to extract gradients from optimisation layers: using the implicit function theorem to differentiate the optimality conditions [2,3,9,21], using a mathematically sensible surrogate loss function [10,19], or using a surrogate differentiable optimiser [6,12]. A host of techniques have been developed and they differ in what types of uncertain optimisation problems they are applicable for; for an overview we recommend section 5 of [16]. Notably few deal with constraint uncertainty. Differentiable optimisation has been leveraged to learn linear objective surrogate models for non-linear objective combinatorial optimisation models [11], which are otherwise largely intractable. The non-linear objective effectively acts as a loss function for the decisions produced by the surrogate model. Robust decision losses were first proposed by [17]. Their analysis shows that for decision-focused learning applied to uncertain optimisation problems empirical regret does not always equal expected regret and is therefore an ineffective loss surrogate. Though the analysis is done for linear programs with uncertain objectives, the same argument can be extended for general uncertain convex optimisation problems. They propose three losses that are better approximates of expected regret. Using a surrogate loss (SPO+ [10], Perturbed Fenchel-Young Loss [6]) function approach they modify the target decision given the observation $x^*(c)$ to the optimal decision to the robust problem $x^*_{RO}(c)$. This requires a pass through the robust problem in training, but they choose an uncertainty set for which the robust counterpart is at the same level of complexity. Our approach extends the concept of robust decision losses to uncertain convex QCQPs using an implicit differentiation approach.

QCQP are optimisation problems where the objective and all constraint functions ≤ 0 are quadratics of the form $f(\mathbf{x}) = \mathbf{x}^T A\mathbf{x} + \mathbf{b}^T\mathbf{x} + c$. If A is positive semidefinite (PSD) the problem is convex and efficiently solvable using the interior point method for SOCP [15]. In the special case of $A = 0$ this reduces to a linear programming problem. QCQPs have applications in fields such as machine learning, finance, robotic control, and energy systems.

There are well known results for uncertain QCQP using robust optimisation. A robust QCQP with ellipsoidal uncertainty has an exact SDP equivalent [5]. There are convex uncertainty sets for which the robust counterpart is known to be NP-hard, such as an intersection of ellipsoids [5]. For large robust QCQP the resulting SDP formulation may be too large for existing solvers to handle so a host of methods have been developed that iteratively solve the deterministic problem to approximately solve the robust one [4,13]. Our approach is of similar iterative nature and in the case of only having one context it can be seen as an approximate solution method for robust QCQPs.

3   Learning Deterministic Surrogates

3.1   The Optimisation Problem

We propose a double implicit layer strategy for learning deterministic surrogates for robust QCQPs. These problems take the form of:

$$
\begin{aligned}
\mathbf{x}^* \in \arg\min_{\mathbf{x}} \; & \mathbf{x}^T Q \mathbf{x} + \mathbf{c}^T \mathbf{x} + q \\
\text{s.t. } & \mathbf{x}^T A_i \mathbf{x} + \mathbf{b}_i^T \mathbf{x} + \gamma_i \leq 0 \quad \forall i \in 1, ..., n \\
& (A_i, \mathbf{b}_i, \gamma_i) \in \mathcal{U}_i \quad\quad\quad\; \forall i \in 1, ..., n
\end{aligned}
\tag{1}
$$

Unlike most predict-and-optimise work where the uncertainty is only in the objective we assume the uncertainty is only in the constraints and that it is constraint-wise. This is more general as a problem with an uncertain objective can trivially be reformulated as an equivalent problem with a certain objective and uncertain constraints. The constraints in this problem can be equivalently written as:

$$
\max_{(A_i, \mathbf{b}_i, \gamma_i) \in \mathcal{U}_i} \{ \mathbf{x}^T A_i \mathbf{x} + \mathbf{b}_i^T \mathbf{x} + \gamma_i \} \leq 0 \quad \forall i \in 1, ..., n
\tag{2}
$$

We assume Q, A_i are positive semi-definite, though as we later point out this assumption can be relaxed for A_i in our method for certain types of uncertainty sets. We want to determine a set of problem parameters

$$
\hat{\mathcal{P}} = \{ \hat{Q}, \hat{\mathbf{c}}, \hat{q}, (\hat{A}_i, \hat{\mathbf{b}}_i, \hat{\gamma}_i) | \forall i \in 1, ..., n \}
$$

such that fixing the parameters in problem (1) yields an optimal solution $\mathbf{x}^*_{det}$ which satisfies all worst case constraints (2) and approximately minimises the objective. The purpose of this is two-fold: for a single problem it represents a solution method for robust QCQPs with convex uncertainty sets, and more importantly, when learned with context $\mathbf{z}_j$ it enables decision-focused learning to predict problem parameterisations robustly.

Given some context (covariates) $\mathbf{z}_j$ we want to predict a set of parameters $\hat{\mathcal{P}}_j$ that yields a decision $\mathbf{x}^*_{det,j}$ which is feasible and minimises a version of problem (1). Note that we do not predict $\hat{Q}$ or $\hat{A}_i$ directly, instead we predict some matrix of the same size $\hat{M}_i$ and set the problem values as $\hat{M}_i^T \hat{M}_i$ to ensure they are PSD. The version of the problem is based on observed realisations of parameters $\mathcal{P}_j$ and some fixed or conditional uncertainty set $\mathcal{U}_j(\mathcal{P}_j)$ centred on the realisation. How the uncertainty set is determined is beyond the scope of this work, but we assume that a robust method is available. The shape and scale of the uncertainty set can be understood as determining the degree of regularisation of the uncertain optimisation problem.

3.2   Double Implicit Layer Model

Predict-and-Optimise. For this prediction task we propose a two block neural network (NN) composed of a prediction block (PR) parameterised by parameters

θ_P and a robust optimisation block (ROB). The prediction block can consist of any standard NN components, so long as it is capable of multioutput prediction (since we have to predict vectors and matrices in this case).

$$\mathrm{PR}(\mathbf{z}_j; \theta_P) = \hat{\mathcal{P}}_j \qquad (3)$$

ROB consists of two implicit layers. The first layer ROB_1 solves a deterministic version of the problem where the inputs are the fixed/predicted parameters from the prediction block and outputs the corresponding optimal decisions.

$$\mathrm{ROB}_1(\hat{\mathcal{P}}_j): \quad \mathbf{x}^*_{det,j} \in \arg\min_{\mathbf{x}} f(x, \hat{\mathcal{P}}_j)$$
$$\text{s.t. } g_i(x, \hat{\mathcal{P}}_j) \leq 0 \quad \forall i \in 1, ..., n \qquad (4)$$

where f, g_i are the corresponding quadratic objective and constraint functions written concisely for convenience. We use the belongs to notation since the optimal solution may not be unique, but in practice the layer will output a unique optimal solution. In the standard predict-and-optimise set-up this output would be used to compute a variant of decision loss, typically some regret based metric like empirical regret: $f(\mathrm{ROB}_1(\hat{\mathcal{P}}_j), \mathcal{P}_j) - f(\mathrm{ROB}_1(\mathcal{P}_j), \mathcal{P}_j)$. Note that empirical regret as stated here does not include a consideration of uncertain constraints. We desire feasibility across all cases, but once it is achieved the only value of importance is the objective. We deal with constraint uncertainty by constructing loss functions that provide gradients toward feasibility.

Worst Case Layer: In ROB we augment this by adding a second implicit layer ROB_2 which evaluates the worst case outcome given an uncertainty set. Since the uncertainty is in the constraints we want to penalise constraint violation in the objective function. Designing constraint penalties depends on the form of the uncertainty set. The worst case problem for robust QCQPs is known to be tractable for simple ellipsoid uncertainty sets. We look at two forms of defining ellipsoid uncertainty sets for the PSD quadratic term coefficient $A_i = P_i^T P_i$. Note that PSD matrices can always be expressed in this form using Cholesky factorisation. The uncertainty sets can either be defined in terms of A_i or P_i.

The quadratic functions that describe the constraints of the problem are of the form $\mathbf{x}^T A_i \mathbf{x} + \mathbf{v}^T \mathbf{x} + s$. The first term can be restated in a lifted form $A_i \bullet \mathbf{X}$ where $\mathbf{X} = \mathbf{x}\mathbf{x}^T$, and $\bullet$ represents the Frobenius inner product $\mathrm{tr}(A_i^T \mathbf{X})$, the sum of the element-wise product. This means that if the uncertainty set is convex and defined in terms of A_i, for fixed $\mathbf{x}$ the objective function is linear. A problem arises in the case of defining ellipsoidal uncertainty sets in terms of $P_i \in \{P_i | P_i = P_{i,0} + \sum_{i=k}^n u_k P_{i,k}, \|\mathbf{u}\|_2 \leq 1\}$. The worst case constraint (2) becomes a maximisation of a convex function problem, also known as a trust region problem, which is non-convex and therefore generally intractable. However, this problem has some hidden convexity and can be restated as an equivalent SDP using the S-lemma [5,7]. The worst case constraint in (2) is equivalent to the existence of a scalar l such that a specific matrix S_i composed of affine transformations of the elements defining the uncertainty set, l, and the decision vector is positive semi-definite.

Case P_i: We propose the following problem to determine the degree of worst case constraint violation in case P_i:

$$\text{ROB}_2(\mathbf{x}^*_{det,j}, \mathcal{P}_j; \mathcal{U}_j) : \mathbf{S}^*_j \in \arg\max_{\mathbf{l},\mathbf{S}} \sum_{i=1}^n \lambda_{min}(S_i)$$

$$\text{s.t. } S_i = S(\mathcal{U}_{j,i}(\mathcal{P}_j), \mathbf{x}^*_{det,j}, l_i) \tag{5}$$

$$l_i \in \mathbb{R}$$

$$\forall i \in 1, ..., n$$

where $S(\cdot)$ is the linear matrix function that composes the parameters from the uncertainty set $(\mathcal{P}_j, \mathcal{U}_j)$, the fixed decisions $\mathbf{x}^*_{det,j}$, and the l variable into the feasibility condition matrix. Denote the collection of S_i matrices for each of the constraints as $\mathbf{S}$ and the corresponding vector of l_i as $\mathbf{l}$ where a pair (S_i, l_i) reflects the feasibility condition components for each constraint i. We fix the decision vector as the output of ROB_1 and assume the uncertainty set components are fixed, so only the decision variable l_i determines the matrix S_i. Our formulation of ROB_2 for the case of P_i converts the feasibility condition of the existence of an l_i such that $S_i \succeq 0$ into a concave penalty. For the decision to be feasible we need to determine if an l_i exists such that S_i is PSD. This is equivalent to showing that there exists a $S(l_i)$ such that the smallest eigenvector λ_{min} of S_i is non-negative. We test for the existence of such an l_i by maximising the smallest eigenvalue. The function λ_{min} is concave and tractable in a maximisation problem. Note that the optimisation problem (5) is separable constraint-wise.

If the problem is feasible for $\mathbf{x}^*_{det,j}$ then the smallest eigenvalue of each of the solutions $\mathbf{S}^*_{j,i}$ in (5) is greater or equal 0. If the fixed decision is infeasible the negative eigenvalues of $\mathbf{S}^*_{j,i}$ reflect a form of distance to feasibility for the fixed decision from ROB_1. We can therefore use it as a penalty for uncertain constraint violation as it gives us a direction toward feasibility. The fixed decision is also feasible if (5) is unbounded and we manage this exception by setting all penalties to 0. The loss function in this case is:

$$\mathcal{L}(\mathbf{x}^*_{det,j}, \mathbf{S}^*_j) = f(\mathbf{x}^*_{det,j}, \mathcal{P}_j) + \sum_{i=1}^n \left[\tau_i \left(\sum \text{ReLU}(-\lambda(\mathbf{S}^*_{j,i})) \right) \right] \tag{6}$$

where $\lambda(\cdot)$ is a differentiable function $\lambda : \mathbb{R}^{m \times m} \to \mathbb{R}^m$ that returns the eigenvalues of a matrix and $\tau_i \geq 0$ is a penalty coefficient. Note that all penalty coefficients in this work are set manually, further work is needed to determine what good values for them are heuristically and how to sensibly set them systematically. The right term penalises any negative eigenvalues in $\mathbf{S}^*_j$ which reflects constraint violation. If there is no constraint violation gradients will only be taken with respect to the deterministic problem. ROB_2 can be understood as a test of feasibility and since it is an optimisation problem we can use differentiable optimisation to extract gradients toward feasibility. The peculiar formulation owes to the fact that the initial equivalent condition to robust constraint feasibility is an existence statement.

Case A_i: Our formulation of ROB_2 for the case of an A_i ellipsoid is simpler. For a fixed $\mathbf{x}^*_{det,j}$ the worst case problem has a linear objective function with a SOCP constraint. The robust counterpart for such a problem is also easier (SOCP) than in case P_i (SDP). We reflect the magnitude of constraint violation and provide a direction to feasibility by applying Lagrangian relaxation to the uncertain constraints. We assign a penalty coefficient of $\beta_i > 0$ to the value of each constraint as we want to maximise the violation. The robustness penalty problem for this case is:

$$
\text{ROB}_2(\mathbf{x}_{det,j}^*, \mathcal{P}_j; \mathcal{U}_j) : \mathcal{P}_{wc}^* \in \arg\max_{\mathcal{P}_{wc}} \sum_{i=1}^n \beta_i g_i(\mathbf{x}_{det,j}^*, \mathcal{P}_{wc,i})
$$
$$
\text{s.t. } \mathcal{P}_{wc,i} \in \mathcal{U}_{j,i}(\mathcal{P}_j) \quad \forall i \in 1, ..., n
\tag{7}
$$

The decision variables in this case are the uncertain matrix, vector, and scalar parameters that define the QCQP. The optimal solution to this problem is the worst case realisation in terms of constraint violation. The corresponding loss function is:

$$
\mathcal{L}(\mathbf{x}_{det,j}^*, \mathcal{P}_{wc,j}^*) = f(\mathbf{x}_{det,j}^*, \mathcal{P}_j) + \sum_{i=1}^n \text{ReLU}(\beta_i g_i(\mathbf{x}_{det,j}^*, \mathcal{P}_{wc,j}^*))
\tag{8}
$$

The penalty for constraint violation in the objective has one obvious flaw, it awards being more in the interior of the feasible set. To correct for that we apply a ReLU activation function to the penalty terms so that the constraint violation penalty is only applied if the current solution is infeasible for that constraint. In both cases at test time we simply deactivate ROB_2 to get the deterministic surrogate solution.

The simplicity of case A_i warrants a rethinking of how uncertainty sets $\mathcal{U}$ should be defined in this method for larger instances. We can define the uncertainty set as any convex set that is compatible with the subfamilies of conic programming. In this setting, given that we have a linear objective we do not necessarily have to restrict ourselves to PSD quadratic coefficient terms. This means that we can learn convex QCQP surrogates for uncertain problems that have potentially non-convex quadratic realisations. Moreover, defining uncertainty sets within this framework does not require specialised mathematical understanding of how to tractably reformulate a problem.

3.3   The Learning Problem

The backwards pass relies on finding the gradient of the loss function with respect to θ_P. Using the chain rule we can decompose gradient calculation as:

$$
\frac{\partial \mathcal{L}}{\partial \theta_P} = \left[\frac{\partial \mathcal{L}}{\partial \mathbf{x}_{det,j}^*} + \frac{\partial \mathcal{L}}{\partial \text{ROB}_2^*} \frac{\partial \text{ROB}_2(\mathbf{x}_{det,j}^*, \mathcal{P}_j; \mathcal{U}_{j,i})}{\partial \mathbf{x}_{det,j}^*} \right] \frac{\partial \text{ROB}_1(\hat{\mathcal{P}}_j)}{\partial \hat{\mathcal{P}}_j} \frac{\partial \text{PR}(\mathbf{z}_j; \theta_P)}{\partial \theta_P}
\tag{9}
$$

There are two terms in this expression that are not easy to evaluate: the partial derivatives for the ROB layers. These can be obtained using an appropriate method for differentiable optimisation. In our case, since we are dealing with conic representable problems with uncertain constraints the method of differentiating through a conic program [2] is appropriate. The method obtains the derivative by implicitly differentiating the conic programme's homogeneous self-dual embedding and is conveniently implemented in the package cvxpylayers [1]. Note that cvxpylayers has been observed to struggle with large problems as differentiating the KKT conditions has worst-case cubic complexity [18] in the number of variables and constraints. Our experiments were conducted on small instances and the model would scale poorly to large problems. One potential way of scaling our method is by leveraging ADMM, which has been adapted to efficiently deal with large differentiable quadratic programs [8].

While a differentiation method for conic representable optimisation problems exists, it does not mean that we have meaningful gradients everywhere. In the case of linear programming the optimal value as a function of problem parameters is a piece-wise constant discontinuous function. This means that we have a gradient of 0 at any point that is differentiable. We know that for strictly convex QP the output in terms of parameters is continuous and theorem 1 from [3] shows that the implicit function is differentiable almost everywhere and sub-differentiable everywhere. Quadratic programs can be equivalently reformulated as linear objective QCQPs so the same theorem applies. In our setup ROB_1 is a QCQP, assuming that all coefficient matrices are PD making the problem strictly convex theorem 1 from [3] applies to ROB_1.

In ROB_2 the nature of the implicit function depends on the definition of the uncertainty set. In the case A_i all constraint functions of the problem are strictly convex with a linear objective. This means that a change in the direction or the orientation of the linear objective will result in a continuous change of the optimal solution. In case P_i the optimisation problem is to maximise a smallest eigenvalue function and a set of linear matrix equality constraints describing the matrix. This is equivalent to minimising the lowest eigenvalue of the negative of the matrix, which is known to be a convex problem that can be cast as an SDP [14]. The optimal output of SDPs is typically continuous in terms of problem parameters and should have meaningful gradients almost everywhere.

4   Experiments

We conducted two small synthetic experiments to demonstrate a proof of concept[1]. The first experiment shows that the algorithm approximates the robust counterpart solution on a single, no-context instance problem with a case P_i uncertainty set. The second is an application to a simulated contextual optimisation problem with a case A_i uncertainty set where we compare the results with a non-regularised decision-focused learning method.

---

[1] Preliminary code available at: https://github.com/EgoPer/Deterministic-Surrogates-for-Uncertain-Convex-QCQP.

Experiment 1. We replicate problem (1) with an ellipsoid uncertainty set for $P_i, \mathbf{b}_i, \gamma_i$ where $P_i^T P_i = A_i$. The quadratic coefficient in the objective Q and scalar q were set to zero. This recreates the problem studied in [5] for which there is a widely known tractable convex robust counterpart (RC). We randomly generate the problem coefficients at five problem sizes with five constraints, each with an uncertainty set defined by four points. It is first solved using the robust counterpart. We randomly initialise the deterministic surrogate problem (SUR) and apply gradient descent in the form of Adam for two hundred steps. We select the feasible solution with the lowest objective value as the output of this process. During optimisation the solution $\mathbf{x}^*_{det,j}$ oscillates between being feasible (no constraint violation penalty gradients) and infeasible. Figure 1 shows the narrow proportional gap between the known RC optimal solution and the approximation using robust surrogates. So far we have evidenced that our algorithm approximately converges to an exact algorithm two orders of magnitude slower. What is promising is that ROB_1 for this problem is resolved two orders of magnitude faster than the robust counterpart, which reflects the difference in computational difficulty between QCQP and SDP. Note that our method is much slower than just solving the RC and that it is not designed to be used on problems without context. This is just an empirical demonstration that our method approximates the robust solution.

Problem size	RC opt value	SUR opt value	proportion gap after 200
10	-0.1949	-0.1816	0.0683
20	-0.5704	-0.5402	0.0529
30	-0.5077	-0.4809	0.0528
40	-0.5866	-0.5479	0.0660
50	-0.6000	-0.5608	0.0653

Fig. 1. Comparing Deterministic Surrogates with Robust Counterparts

Experiment 2. The real benefit of this method is in contextual settings where one may be inclined to apply decision-focused learning. We design a problem with uncertain constraints to demonstrate the shortcomings of vanilla decision losses. In experiment 2 we generate a contextual QCQP with an uncertain quadratic constraint. We generate a fixed correlation matrix C based on randomly generated positive eigenvalues. We generate context $z_{j,i} \sim \text{Unif}(0.1, 1)$ where j indexes over the hundred samples we generate and i over the number of contextual covariates (in our case four). We then generate conditional variances $\sigma^2_{j,i}$ as the inverse of convex combinations of the context $z_{j,i}$, the weights for which are also randomly generated. Let D_j be a matrix for which the diagonal are the generated conditional variances. We define conditional covariance matrices as $\Sigma_j = D_j^{\frac{1}{2}} C D_j^{\frac{1}{2}}$. These conditional covariance matrices are used to parameterise a conditional Wishart distribution with a degree of freedom of fifty from which we sample the uncertain PSD matrices for our problem and then divide them by the degree of freedom. The obtained values should reflect the distribution of empirical covariance matrix estimates we would observe for an underlying process generated by $\mathcal{N}(0, \Sigma_j)$. Our optimisation problem is:

$$\mathbf{x}^*_{det,j} \in \arg\min_{\mathbf{x}} -\mathbf{c}^T\mathbf{x}$$
$$\text{s.t. } \mathbf{x}^T \Sigma_j \mathbf{x} \leq r \tag{10}$$
$$\mathbf{x}^T \mathbf{1} \leq 1, \quad \mathbf{x} \geq 0$$

where $\mathbf{c} \geq 0$ is a fixed cost vector and r is some defined level of acceptable variance-denominated risk. A possible interpretation of this problem is an investment manager deciding what bonds to hold to maturity to maximise yield while making sure risk levels do not force liquidation. Since Σ_j is uncertain a deterministic solution to this problem could be infeasible.

We train a simple two layer neural network to predict $\hat{c}_j, \hat{\Sigma}_j$ based on input z_j using vanilla decision losses and robust decision losses obtained with the method of deterministic surrogates. We define our regularising uncertainty set as:

$$\mathcal{U}_j(\Sigma_j) = \{\Sigma_j + A | A = A^T, ||A||_F \leq 1\}$$

Note that this set was designed somewhat arbitrarily and research on what good uncertainty set designs for decision-focused learning regularisation are is needed. The two prediction models (normal and robust decision losses) are trained on seventy samples and tested on thirty. Note that we use loss function (8) in both the vanilla and robust decision loss case. In the vanilla decision loss case the worst case outcome variable is set as the observed covariance. We ran a simple experiment which was randomly initialised. For the given initialisation the robust decision loss produces predictions which cause feasible decisions in all thirty test cases, whereas vanilla decision loss produces predictions which cause decisions which are infeasible in thirteen out of the thirty test cases. The average optimal value without the robust decision loss regularisation was 37.3% lower, but infeasible for nearly half of the cases. These values are obviously very sensitive to initialisation, but in general we observed that robust decision losses yield more feasible test cases. This demonstrates the generalisation risk of only training with respect to point observations of the problem instead of considering a neighbourhood. The code provided can easily be adapted for more experimentation, we just demonstrate a proof of concept.

5   Conclusion

We establish a method for obtaining robust decision losses for uncertain convex QCQP. This is achieved by casting the problem as determining a deterministic surrogate that contextually approximates the robust solution to an uncertain problem. The second optimisation layer enables us to obtain gradients with respect to the worst case in a neighbourhood of problem realisations. Robust decision losses can be understood as a regularisation method as they enable us to obtain relevant gradients even when observed problem realisations are biased to locally more sensitive decisions. A key component in the success of modern deep learning has been the propagation of regularisation technologies such as dropout, skip connections, and layer normalisation. Decision-focused learning will require the development and validation of its own portfolio of regularisation (generalisation-improving) technologies before it becomes an established approach that is used in practice. This work is an attempt to contribute to that portfolio for a broad class of convex optimisation problems.

From the perspective of contextual robust optimisation the principal advantage of this method is the speedup at test time achieved by separating the robust problem into two optimisation problems. At test time it means we can approximate robust solutions in the time it takes to optimise deterministic problems. At train time instead of solving one larger robust counterpart, two smaller ones are solved. ROB_1 is a QCQP, whereas ROB_2 is in the same family as the robust counterpart, but with fewer decision variables which can be faster to solve on average. This approach promises approximately robust solutions at deterministic solving times which may prove valuable in low latency decision making.

We see promising research directions in exploring how different choices of uncertainty sets for robust decision losses affect decision-focused learning for QCQP, applying the method to real-world problems, and extending the methodology to uncertain SDP.

Disclosure of Interests. The authors have no competing interests to declare that are relevant to the content of this article.

References

1. Agrawal, A., Amos, B., Barratt, S., Boyd, S., Diamond, S., Kolter, Z.: Differentiable convex optimization layers. Adv. Neural Inf. Process. Syst. (2019)
2. Agrawal, A., Barratt, S., Boyd, S., Busseti, E., Moursi, W.M.: Differentiating through a cone program. arXiv preprint arXiv:1904.09043 (2019)
3. Amos, B., Kolter, J.Z.: Optnet: differentiable optimization as a layer in neural networks. In: International Conference on Machine Learning, pp. 136–145. PMLR (2017)
4. Ben-Tal, A., Hazan, E., Koren, T., Mannor, S.: Oracle-based robust optimization via online learning. Oper. Res. **63**(3), 628–638 (2015)
5. Ben-Tal, A., Nemirovski, A.: Robust convex optimization. Math. Oper. Res. **23**(4), 769–805 (1998)
6. Berthet, Q., Blondel, M., Teboul, O., Cuturi, M., Vert, J.P., Bach, F.: Learning with differentiable pertubed optimizers. Adv. Neural. Inf. Process. Syst. **33**, 9508–9519 (2020)
7. Bertsimas, D., Brown, D.B., Caramanis, C.: Theory and applications of robust optimization. SIAM Rev. **53**(3), 464–501 (2011)
8. Butler, A., Kwon, R.H.: Efficient differentiable quadratic programming layers: an ADMM approach. Comput. Optim. Appl. **84**(2), 449–476 (2023)
9. Donti, P., Amos, B., Kolter, J.Z.: Task-based end-to-end model learning in stochastic optimization. Adv. Neural Inf. Process. Syst. **30** (2017)

10. Elmachtoub, A.N., Grigas, P.: Smart predict, then optimize. Manag. Sci. **68**(1), 9–26 (2022)
11. Ferber, A.M., et al.: Surco: learning linear surrogates for combinatorial nonlinear optimization problems. In: International Conference on Machine Learning, pp. 10034–10052. PMLR (2023)
12. Kong, L., Cui, J., Zhuang, Y., Feng, R., Prakash, B.A., Zhang, C.: End-to-end stochastic optimization with energy-based model. Adv. Neural. Inf. Process. Syst. **35**, 11341–11354 (2022)
13. Kroer, C., Ho-Nguyen, N., Lu, G., Kılınç-Karzan, F.: Performance evaluation of iterative methods for solving robust convex quadratic problems. In: 10th NIPS Workshop on Optimization for Machine Learning, Dec. vol. 8 (2017)
14. Mengi, E., Yildirim, E.A., Kilic, M.: Numerical optimization of eigenvalues of hermitian matrix functions. SIAM J. Matrix Anal. Appl. **35**(2), 699–724 (2014)
15. Nesterov, Y., Nemirovskii, A.: Interior-point polynomial algorithms in convex programming. SIAM (1994)
16. Sadana, U., Chenreddy, A., Delage, E., Forel, A., Frejinger, E., Vidal, T.: A survey of contextual optimization methods for decision making under uncertainty. arXiv preprint arXiv:2306.10374 (2023)
17. Schutte, N., Postek, K., Yorke-Smith, N.: Robust losses for decision-focused learning. arXiv preprint arXiv:2310.04328 (2023)
18. Sun, H., Shi, Y., Wang, J., Tuan, H.D., Poor, H.V., Tao, D.: Alternating differentiation for optimization layers. arXiv preprint arXiv:2210.01802 (2022)
19. Vlastelica, M., Paulus, A., Musil, V., Martius, G., Rolínek, M.: Differentiation of blackbox combinatorial solvers. In: International Conference on Learning Representations. ICLR'20, May 2020. https://openreview.net/forum?id=BkevoJSYPB
20. Wang, P.W., Donti, P., Wilder, B., Kolter, Z.: Satnet: bridging deep learning and logical reasoning using a differentiable satisfiability solver. In: International Conference on Machine Learning, pp. 6545–6554. PMLR (2019)
21. Wilder, B., Dilkina, B., Tambe, M.: Melding the data-decisions pipeline: decision-focused learning for combinatorial optimization. In: Proceedings of the AAAI Conference on Artificial Intelligence, vol. 33, pp. 1658–1665 (2019)

Strategies for Compressing the Pareto Frontier: Application to Strategic Planning of Hydropower in the Amazon Basin

Zhongdi Qu[1]($\boxtimes$), Marc Grimson[1], Yue Mao[1], Sebastian Heilpern[2],
Imanol Miqueleiz[2], Felipe Pacheco[2], Alexander Flecker[2], and Carla P. Gomes[1]

[1] Department of Computer Science, Cornell University, Ithaca, USA
{zq84,mg2425,ym277}@cornell.edu, gomes@cs.cornell.edu
[2] Department of Ecology and Evolutionary Biology, Cornell University, Ithaca, USA
{s.heilpern,im298,felipe.pacheco,asf3}@cornell.edu

Abstract. The development of ethical AI decision-making systems requires considering multiple criteria, often resulting in a large spectrum of partially ordered solutions. At the core of this challenge lies the Pareto frontier, the set of all feasible solutions where no solution is dominated by another. In previous work, we developed both exact and approximate algorithms for generating the Pareto frontier for tree-structured networks. However, as the number of criteria grows, the Pareto frontier increases exponentially, posing a significant challenge for decision-makers. To address this challenge, we propose various strategies to efficiently compress the Pareto frontier, including an approximation method with optimality and polynomial runtime guarantees. We provide detailed empirical results on the strategies' effectiveness in the context of strategic planning of the hydropower expansion in the Amazon basin. Our strategies offer a more manageable approach for navigating Pareto frontiers.

Keywords: Multi-objective optimization · Approximation algorithms · Hierarchical clustering

1 Introduction

In recent years, there has been a growing interest in developing AI decision-support systems that can evaluate trade-offs based on multiple criteria, moving away from the conventional single-objective systems. This shift is particularly important when considering more ethical AI decision-making systems that align

This project is partially supported by the Eric and Wendy Schmidt AI in Science Postdoctoral Fellowship, a Schmidt Futures program; the National Science Foundation (NSF); the National Institute of Food and Agriculture (USDA/NIFA); the Air Force Office of Scientific Research (AFOSR); and the Cornell Atkinson Center for a Sustainable Future (ACSF).

B. Dilkina (Ed.): CPAIOR 2024, LNCS 14743, pp. 141–157, 2024.
https://doi.org/10.1007/978-3-031-60599-4_9

Fig. 1. (a) Existing (red) and proposed (yellow) hydropower dams in the Amazon basin and (b) Rio Santiago, a free-flowing river in the Andean Amazon with large hydropower dams in planning stages (Alvaro del Campo/The Field Museum). (Color figure online)

with multiple human values [24]. It is especially crucial to consider multiple criteria in computational sustainability [13], where balancing economic, environmental, and societal objectives is essential for achieving the Sustainable Development Goals (SDGs) [23].

Multi-objective optimization is computationally challenging. At the core of this challenge lies the Pareto frontier: the set of solutions in a multi-dimensional space representing the trade-offs among different potentially conflicting objectives. In other words, when optimizing for multiple objectives, the result is often a large spectrum of partially ordered solutions. The Pareto frontier is therefore the set of solutions that are not dominated by any other solution. Our previous work focused on developing both exact and approximate algorithms to compute the Pareto frontier in tree-structured networks [1,3,12].

Our research has been motivated by the need for strategic planning of hydropower expansion in the Amazon basin. Hydropower plays a critical role in current and future renewable energy strategies globally. The variation in project sizes and the diverse characteristics of river systems highlight the need for a deeper understanding of the trade-offs between hydropower capacity and ecosystem services. This understanding is key in evaluating dam portfolios across the Amazon river network, where hydropower projects have been proposed at over 350 locations (Fig. 1). Multicriteria optimization is crucial in identifying dam portfolios that balance social-environmental costs with energy production. However, our multiobjective optimization approaches often yield solutions consisting of millions of portfolios. As the number of criteria increases, the Pareto frontier grows exponentially, presenting a substantial challenge for decision-makers. The disparity between our computational approaches producing a vast number of Pareto-optimal solutions and the practical needs of decision-making in dam expansion is a significant hurdle for policymakers striving to construct dams with minimal environmental impact while achieving energy goals. Therefore, innovative approaches that effectively compress the number of optimal Pareto portfolios are critical to finding practical and realistic solutions.

Our Contributions: To facilitate navigating the Pareto frontier, herein we propose different approaches: **(1) A representation of the Pareto frontier,** which consists of a subset of solutions from the Pareto frontier with coverage guarantees and can be generated in time polynomial in the size of the frontier; **(2) An approximation of the Pareto frontier,** based on a dynamic-programming-based strategy, with optimality guarantees and polynomial runtime guarantees; and **(3) An estimation of the Pareto frontier,** based on a dynamic-programming-based strategy, with optimality guarantees, without polynomial runtime guarantees, but with good empirical performance. **(4)** We provide detailed **empirical results** analyzing the trade-offs of our different strategies against various baselines, **in the context of strategic planning of the hydropower expansion in the Amazon basin.** Our strategies offer more manageable ways for navigating Pareto frontiers.

2   Related Works

To solve unstructured multi-objective optimization problems, genetic algorithms have been widely used, including the family of Non-dominated Sorting Genetic Algorithms (NSGA [22], NSGA-II [8], and NSGA-III [7]) and Multi-objective Evolutionary Algorithm Based on Decomposition (MOEA/D) [28]. However, when it comes to problems with an underlying structure, like the tree-structured river network for the planning of hydropower dams, these algorithms usually are not competitive with algorithms that take advantage of that structure [26]. Moreover, genetic algorithms rarely provide theoretical guarantees on optimality or runtime: so far the theoretical analysis of these algorithms has been restricted to relatively simplistic and few objectives [9–11,29].

Our work fits into a series of research that exploits the underlying tree-structured river network to approximate the Pareto frontier for the planning of hydropower in the Amazon basin. [26] first proposed the dynamic-programming-based algorithm to find the exact Pareto frontier and a fully polynomial time approximation scheme (FPTAS) to approximate the frontier. Following works [3,14,15] further improved the methods through techniques including divide-and-conquer, expansion, compression, and affine transformation. The methods we propose here can be fully incorporated into the developed approaches.

Our methods employ hierarchical clustering techniques. The idea of leveraging clustering in multi-objective optimization to improve algorithm performance or to help interpret the Pareto frontier has been explored before, but mostly in the context of genetic algorithms. In [6,19,27,30] clustering algorithms, including hierarchical clustering, have been used to discover the population structure and aid in parent selection and offspring retention. Clustering helps with discovering solutions that are distributed more widely and uniformly.

Our work relates to Binary Decision Diagrams (BDDs) [4]. To solve a multi-objective discrete optimization problem, the BDD method uses decision diagrams to represent exactly the feasible set of the problem and then uses a multicriteria shortest path algorithm for finding the set of non-dominated solutions [4]. However, the size of the diagram could grow exponentially. Approximate decision diagrams that have a polynomial limit on the size have been developed [2]. A crucial difference between this method and ours is that BDD assumes linear separability in the objective functions, whereas, for our problem domain, the objective functions are non-separable and are dependent upon all decisions.

3 Preliminaries

3.1 Multi-objective Optimization

A multi-objective optimization problem consists of optimizing several often conflicting objectives simultaneously. Therefore, typically there does not exist a single solution that optimizes all the objectives at the same time. Accordingly, we give the definitions of optimality in the multi-objective scenario.

Pareto Dominance. Without loss of generality, assume we are maximizing d objectives at the same time. For a solution π, $z(\pi) = (z^1(\pi), \ldots, z^d(\pi))$ is the values of the d objectives. A solution π dominates another solution π', written as $z(\pi) \succ z(\pi')$, if and only if for all $1 \leq i \leq d, z^i(\pi) \geq z^i(\pi')$, and there exists $1 \leq j \leq d$ such that $z^j(\pi) > z^j(\pi')$.

Pareto Frontier. Our goal in the multi-objective optimization problem is to find the set of non-dominated solutions, which we define to be the Pareto frontier: let S be the set of all feasible solutions, the Pareto frontier is $\{\pi \in S | z(\pi) \nsucc z(\pi'), \forall \pi' \in S\}$.

In practice, the size of the Pareto frontier may be exponential even for a fixed number of objectives. As a result, finding or interpreting the entire frontier may be computationally expensive. Therefore, a more realistic goal is to find a good approximation or representation of the Pareto frontier.

ϵ-approximation. Given a Pareto frontier P, a set of solutions S ϵ-approximates P if and only if for every $\pi \in P$, there exists a solution $\pi' \in S$ such that $z^i(\pi') \geq (1 - \epsilon)z^i(\pi)$ for all $1 \leq i \leq d$, and S is found in polynomial time.

Note that an ϵ-approximation is found in polynomial time. When a set with such optimality guarantee is found in superpolynomial time, we call it an ϵ-*estimation*.

γ-representation. Given a Pareto frontier P, a subset of the frontier $P' \subseteq P$ γ-represents P if and only if for every $\pi \in P$, there exists a solution $\pi' \in P'$ such that $z^i(\pi') \geq (1 - \gamma)z^i(\pi)$ for all $1 \leq i \leq d$.

Note that a crucial difference between ϵ-approximation/estimation and γ-representation is that, while a solution in the ϵ-approximation/estimation set is not necessarily Pareto-optimal, a solution in the γ-representation is always Pareto-optimal since the γ-representation set is a subset of the Pareto frontier.

3.2 Strategic Planning of Hydropower in the Amazon Basin

The Problem. Construction of hydropower dams provides electricity but can cause significant adverse environmental impacts including disruption of fish

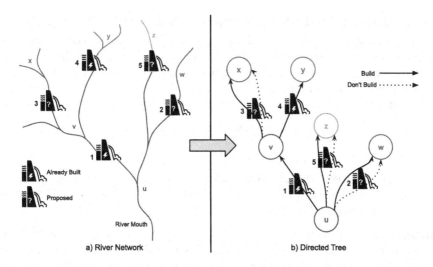

Fig. 2. Converting (a) a river network to (b) a directed multi-edged tree. Contiguous sections of the river uninterrupted by dam sites become nodes in the tree. Dam sites, both proposed and already built, are edges that connect upstream and downstream portions of the river, with a unique edge per decision.

migration routes and greenhouse gas emissions. Planning of the hydropower dam placements requires the balancing of energy production and ecosystem impacts. Accordingly, given a set of proposed dam sites, a solution is a subset of the dams to be built and our goal is to find a set of solutions that approximates or represents the Pareto frontier with respect to the following 6 objectives: **(1)** hydropower generation, **(2)** connectivity (the total length of the un-obstructed stream segments that a fish can travel starting from the river mouth without passing any dam site), **(3)** sediment (the amount of sediment and nutrients transported to the river mouth every year considering the fact that each dam traps a certain percentage of total sediment from upstream), **(4)** biodiversity (the overall impact on the fish population caused by dam construction), **(5)** degree of regulation (the total degree of flow regime alteration caused by dam construction), and **(6)** greenhouse gas emissions (the total greenhouse gas emissions caused by dam construction).

The Algorithm. Previous works [14,15,26] model the river network as a multi-edged directed tree structure (see Fig. 2). In the multi-edged directed tree representation, each edge represents a possible decision at a potential dam site, and its two vertices are respectively the river regions directly upstream and downstream of the site. Thus, each pair of parent/child nodes may have one or more edges depending on the number of decisions relevant to a given dam location. Every node v in the tree is associated with a non-negative node reward r_v^i for each objective i. Each edge is represented by (u, v, j) with parent and child nodes u and v and index j to distinguish the edge from sibling edges. Additionally, each edge is associated with a non-negative edge reward s_{uvj}^i and a non-negative transfer coefficient p_{uvj}^i, for each objective i. A solution (or partial solution) is defined as a spanning tree of the multi-edged tree (or partial spanning tree of a sub-tree). The i-th objective value of a partial solution at a leaf node v, π_v is its corresponding reward, i.e., $z^i(\pi_v) = r_v^i$. The i-th objective value of a partial solution at a non-leaf node u, π_u is defined recursively:

$$z^i(\pi_u) = r_u^i + \sum_{(u,v,j)\in\pi} s_{uvj}^i + p_{uvj}^i z^i(\pi_v) \tag{1}$$

Based on the tree-structure formulation, [26] proposed a dynamic programming algorithm that can find the exact Pareto frontier, based on the crucial observation, proven in [26], that

Theorem 1. *Let u be a node in the tree and $u_1, \ldots, u_k$ be its children. Any Pareto-optimal partial solution at u can be constructed by combining one Pareto-optimal partial solution from child u_i for each $i \in [1, \ldots, k]$ and the choice of edges connecting u and u_i.*

As a result, the algorithm recursively computes the Pareto-optimal partial solutions from leaf nodes to the root. At each node u, the algorithm comes up with the candidate solutions by combining the Pareto-optimal partial solutions at u's children. Then, the algorithm discards any dominated solutions to obtain the Pareto-optimal partial solutions at u.

Given that the size of the frontier could be exponential, they also proposed a fully polynomial-time approximation scheme (FPTAS) that approximates the Pareto frontier within an arbitrarily small ϵ and runs in time polynomial in the size of the instance and $1/\epsilon$. The FPTAS introduces a hyperparameter $K_u^i = \epsilon r_u^i$ for each node u and each objective i, and defines the rounded objective value $\hat{z}^i(\pi_u)$ recursively as

$$\hat{z}^i(\pi_u) = r_u^i + \left\lfloor \frac{\sum_{(u,v,j)\in\pi} s_{ujv}^i + p_{uvj}^i \hat{z}^i(\pi_v)}{K_u^i} \right\rfloor K_u^i. \tag{2}$$

In [26], it was proven that the Pareto frontier on tree-structured networks can be ϵ-approximated, namely:

Theorem 2. *Let P_s be the set of (partial) Pareto-optimal solutions for a node s and $\hat{P}_s$ be the set of (partial) Pareto-optimal solutions computed via the dynamic programming algorithm using the rounded objective function 2. We must have $\hat{P}_s$ is an ϵ-approximation of P_s.*

4   A Representation of the Pareto Frontier

Given a Pareto frontier P, the problem of finding the γ-representation of P is to find a subset P' of P such that for every solution $\pi \in P$, there exists a solution $\pi' \in P'$ such that $z^i(\pi') \geq (1 - \epsilon)z^i(\pi)$ for all $1 \leq i \leq d$. To this end, we have designed an algorithm based on hierarchical clustering [16,20]. Hierarchical clustering takes a set of data points and seeks to build a hierarchy of clusters of the data points. It has been widely applied to fields including taxonomy [21], bioinformatics [17,25], and social network analysis [18]. The agglomerative version of the algorithm starts with each data point as a separate cluster, and pairs of clusters are greedily merged as one moves up in the hierarchy. To decide which clusters should be combined, a measure of distance between sets of data points is required. Typically, this measure includes a distance metric between single points of the data set and a linkage method that specifies the distance of two sets as a function of the pairwise distances between the data points across the two sets. What distance metric to use depends on the underlying application, and some examples include the Euclidean distance and the Hamming distance. The linkage method, on the other hand, influences the shape of the clusters. For example, complete linkage, i.e., the distance of two sets is the maximum distance between any two data points across the sets, tends to produce more spherical clusters than single linkage, where the minimum distance is used. For our case, Euclidean distance between the objective values normalized to $[0, 1]$, and average linkage, i.e., the distance of two sets is the average distance between the pairs of data points across the sets, are used.

The algorithm to find γ-representation

- Input: A Pareto frontier $P = \{\pi_1, \ldots, \pi_n\}$, and a parameter γ.
- Output: A subset of the Pareto frontier $P' \subseteq P$ such that $\forall \pi \in P$ there is a $\pi' \in P'$ such that $z^i(\pi') \geq (1 - \gamma)z^i(\pi)$ for all $1 \leq i \leq d$.

1. Perform hierarchical clustering on P:
   (a) For each objective, normalize the objective values to $[0, 1]$.
   (b) Initialize $C_1 = \{\pi_1\}, C_2 = \{\pi_2\}, \ldots, C_n = \{\pi_n\}$, and $\mathcal{C} = \{C_1, \ldots, C_n\}$.
   (c) Find the two clusters in $\mathcal{C}$, C_i and C_j, with the smallest distance, as defined by Euclidean distance and average linkage, among all pairs of clusters.
   (d) $C_{|\mathcal{C}|+1} = C_i \cup C_j, \mathcal{C} = \mathcal{C} \cup \{C_{|\mathcal{C}|+1}\} - \{C_i\} - \{C_j\}$.
   (e) Repeat steps 1(c) to 1(d) until $\mathcal{C} = \{P\}$.
2. Run Algorithm 1 on the final cluster $\{P\}$.

Theorem 3. *The runtime of the algorithm to find γ-representation on a Pareto frontier P with $|P| = n$ is $O(n^3)$ and the algorithm returns a set $P' \subseteq P$ such that for every solution $\pi \in P$, there exists a solution $\pi' \in P'$ such that $z^i(\pi') \geq (1 - \gamma)z^i(\pi)$ for all $1 \leq i \leq d$.*

Proof. The for loop from line 3 to line 11 of Algorithm 1 makes sure that Coverage($\{P\}$) is indeed a γ-representation of P. The time complexity for doing hierarchical clustering on P is $O(n^3)$. The hierarchy of clusters can be represented as a binary tree where a node u having children l and r means clusters l and r are merged to form cluster u. The leaves of this binary tree are the individual solutions in P. Therefore, the size of the tree is $2n - 1$. In the worst case, Coverage($\{P\}$) will traverse every cluster in the tree and look at every solution in the clusters. Thus, Coverage($\{P\}$) runs in $O(n^2)$.

Algorithm 1: Coverage

Data: A cluster of solutions $C = C_i \cup C_j$ where C_i and C_j have been merged in the hierarchical clustering process to form C.

Result: A subset C' of C such that for all $\pi \in C$ there is a $\pi' \in C'$ such that $z^i(\pi') \geq (1 - \gamma)z^i(\pi)$ for all $1 \leq i \leq d$

1   $\pi' \leftarrow rand(C)$;   /* $rand(C)$ returns a random sample from C. */
2   $failed \leftarrow$ **False**;
3   **foreach** $\pi \in C$ **do**
4   **for** $1 \leq i \leq d$ **do**
5   **if** $z^i(\pi') < (1 - \gamma)z^i(\pi)$ **then**
6            $failed \leftarrow$ **True**;
7   **break**;
8   **end**
9   **end**
10   **if** $failed$ **then break** ;
11   **end**
12   **if** $failed$ **then return** Coverage(C_i) $\cup$ Coverage(C_j) ;
13   **else return** $\{\pi'\}$;

---

5   An Estimation of the Pareto Frontier

The method described in Sect. 4 works as a post-processing step after the Pareto frontier has been discovered. Alternatively, we consider incorporating the representation method into the dynamic programming algorithm proposed in [26] to estimate the Pareto frontier. The algorithm models the river network as a tree structure and recursively computes the Pareto-optimal partial solutions from the leaf nodes to the root of the tree. At every node, candidate solutions are formed by combining the Pareto-optimal partial solutions at the node's children. Dominated partial solutions are then discarded. We argue that if we apply the γ-representation algorithm after the pruning of the dominated solutions at some levels of nodes in the tree, then we have an estimation of the Pareto frontier, with optimality guarantees, but not necessarily polynomial runtime guarantees.

Note that for a tree T_u rooted at node u, we call the level of node u level 1, the level of u's children level 2, etc. We apply the γ-representation algorithm to L levels of the nodes in T_u, which means that to all the nodes from level L to level 1, after the dominated partial solutions have been discarded, we run the algorithm to find the γ-representation of the Pareto-optimal partial solutions and use the representation set, instead of all the Pareto-optimal partial solutions, to assemble the solutions at the node's parent (Fig. 3b).

Theorem 4. *Consider a node u in a run of the dynamic programming algorithm proposed in [26], and the subtree rooted at u, T_u. Suppose to L levels of the nodes in T_u we apply the γ-representation algorithm, then at node u, we obtain a set of solutions S_u such that for every Pareto-optimal partial solution at u, π_u, there is a solution $\bar{\pi}_u \in S_u$ such that $z^i(\bar{\pi}_u) \geq (1 - \gamma)^L z^i(\pi_u)$ for all $1 \leq i \leq d$.*

(a) tree structure   (b) 2 levels of the nodes

Fig. 3. Strategy for applying the γ-representation on the tree. (a) shows the underlying tree structure, with (b) showing the representation applied to 2 levels of the nodes. Nodes in red are nodes that have the γ-representation applied. (Color figure online)

Proof. We prove the theorem by induction. For $L = 0$ and $L = 1$, the statements are direct results of Theorems 1 and 3 respectively. Suppose the statement is true for $L' = L - 1 \geq 1$, and consider a Pareto-optimal partial solution π_u at u. By Theorem 1 and Eq. 1, we have that for all $1 \leq i \leq d$,

$$z^i(\pi_u) = r^i_u + \sum_{(u,v,j) \in \pi_u} s^i_{uvj} + p^i_{uvj} z^i(\pi_v)$$

where π_v is a Pareto-optimal partial solution at node v. By the induction hypothesis, the algorithm has found at v a set of partial solutions S_v that includes a solution π'_v such that $z^i(\pi'_v) \geq (1-\gamma)^{(L-1)} z^i(\pi_v)$ for all $1 \leq i \leq d$. By substituting π_v with π'_v, and combining all the π'_v with the same edges as in π_u, we obtain a partial solution π'_u such that for all $1 \leq i \leq d$

$$z^i(\pi'_u) = r^i_u + \sum_{(u,v,j) \in \pi_u} s^i_{uvj} + p^i_{uvj} z^i(\pi'_v) \geq (1-\gamma)^{(L-1)} z^i(\pi_u)$$

When the dynamic programming algorithm prunes the solutions at node u, either π'_u is kept as a Pareto-optimal partial solution, or π'_u is discarded because the algorithm has found a π''_u such that for all $1 \leq i \leq d$, $z^i(\pi''_u) \geq z^i(\pi'_u) \geq (1-\gamma)^{(L-1)} z^i(\pi_u)$. Either way, after pruning at node u, we must have kept a partial solution $\tilde{\pi}_u$ such that $z^i(\tilde{\pi}_u) \geq (1-\gamma)^{(L-1)} z^i(\pi_u)$. Then when we apply the algorithm to find the γ-representation at node u, by Theorem 3, we will be guaranteed to have in the representation set S_u a partial solution $\bar{\pi}_u$ such that for all $1 \leq i \leq d$, $z^i(\bar{\pi}_u) \geq (1-\gamma) z^i(\tilde{\pi}_u) \geq (1-\gamma)^L z^i(\pi_u)$.

Applying Theorem 4 to the root node, we obtain the following Lemma.

Lemma 1. *Suppose during a run of the dynamic programming algorithm proposed in [26], the algorithm to find the γ-representation is applied to L levels of the nodes, then at the root node we obtain a set S such that for every Pareto-optimal solution $\pi \in P$, there exists a solution $\pi' \in S$ such that for all $1 \leq i \leq d$, $z^i(\pi') \geq (1-\gamma)^L z^i(\pi)$.*

6   An Approximation of the Pareto Frontier

The algorithm described in Sect. 5 finds an estimation of the Pareto frontier with optimality guarantee, but the algorithm could run in exponential time since at each node there are a potentially exponential number of partial solutions to consider. To mitigate that problem, [26] has applied a rounding technique to the exact dynamic programming algorithm and the result is an FPTAS that ϵ-approximates the Pareto frontier P. We argue that if we apply the γ-representation algorithm to some levels of the nodes in the FPTAS, we obtain a further compressed approximation of the Pareto Frontier, with optimality and polynomial runtime guarantees.

Theorem 5. *Consider a node u in a run of the FPTAS proposed by [26] with parameters ϵ and $K_v^i = \epsilon r_v$. If for L levels of the subtree T_u rooted at u, we apply to the nodes the γ-representation algorithm in the rounded objectives, and the parameters L, γ, and ϵ satisfy that if $L > 1$, then $(1 - \gamma)^{L-1} + \epsilon \leq 1$, then at u we obtain a set S_u such that for every Pareto-optimal partial solution at u, π_u, there is a solution $\tilde{\pi}_u \in S_u$ such that for all $1 \leq i \leq d$, we have $z^i(\tilde{\pi}_u) \geq (1 - \gamma)^L(1 - \epsilon)z^i(\pi_u)$.*

Proof. We first prove again by induction that Theorem 4 still holds for the rounded objectives. The base cases for $L = 0$ and $L = 1$ are direct consequences of Theorem 2 and Theorem 3. We prove the induction step where $L > 1$. Consider the root node u and its children again. Write $\sum_{(u,v,j)\in\pi_u} s_{uvj}^i + p_{uvj}^i \hat{z}^i(\pi_v)$ as $NK_u^i + R$ for some non-negative integer N and some non-negative real number $R < K_u^i$. Then by Eq. 2 we have

$$\hat{z}^i(\pi_u) = r_u^i + \left\lfloor \frac{NK_u^i + R}{K_u^i} \right\rfloor K_u^i = r_u^i + NK_u^i$$

Similarly as in the proof for Theorem 4, by substituting all the π_v's with their γ-representations and choosing the same edges between u and v, we obtain a partial solution at u, π_u' such that

$$\hat{z}^i(\pi_u') \geq r_u^i + \left\lfloor \frac{(1 - \gamma)^{L-1}(NK_u^i + R)}{K_u^i} \right\rfloor K_u^i > r_u^i + ((1 - \gamma)^{L-1}N - 1)K_u^i$$

Then substituting $K_u^i = \epsilon r_u^i$, we get

$$\hat{z}^i(\pi_u') - (1 - \gamma)^{L-1}\hat{z}^i(\pi_u)$$
$$> r_u^i + (1 - \gamma)^{L-1}N\epsilon r_u^i - \epsilon r_u^i - (1 - \gamma)^{L-1}r_u^i - (1 - \gamma)^{L-1}N\epsilon r_u^i$$
$$= r_u^i(1 - \epsilon - (1 - \gamma)^{L-1}) \geq 0$$

i.e., $\hat{z}^i(\pi_u') \geq (1-\gamma)^{L-1}\hat{z}^i(\pi_u)$. Similarly to the proof of Theorem 4, after discarding the dominated solutions at u, we are guaranteed to be left with a $\tilde{\pi}_u$ such that $\hat{z}^i(\tilde{\pi}_u) \geq (1-\gamma)^{L-1}\hat{z}^i(\pi_u)$. Then after running the γ-representation algorithm at node u, we have in the representation set a $\bar{\pi}_u$ such that $\hat{z}^i(\bar{\pi}_u) \geq (1-\gamma)^L \hat{z}^i(\pi_u)$. Given that $\hat{z}^i(\pi_u) \geq (1-\epsilon)z^i(\pi_u)$, proved in [26], and $z^i(\bar{\pi}_u) \geq \hat{z}^i(\bar{\pi}_u)$, as a result of taking the floor operation, we have that

$$z^i(\bar{\pi}_u) \geq \hat{z}^i(\bar{\pi}_u) \geq (1-\gamma)^L \hat{z}^i(\pi_u) \geq (1-\gamma)^L(1-\epsilon)z^i(\pi_u)$$

Lemma 2. *Suppose during a run of the FPTAS proposed in [26] with parameters ϵ and $K_v^i = \epsilon r_v$, for L levels of the tree, we apply to the nodes the γ-representation algorithm in the rounded objectives, and the parameters L, γ, and ϵ satisfy that if $L > 1$, then $(1-\gamma)^{L-1} + \epsilon \leq 1$, then the algorithm in time $O((\frac{n}{\epsilon})^{3d})$ returns a set S such that for every solution $\pi \in P$, there is a solution $\pi' \in S$ such that for all $1 \leq i \leq d$, we have $z^i(\pi') \geq (1-\gamma)^L(1-\epsilon)z^i(\pi)$.*

Proof. Applying Theorem 5 to the root node, we obtain the optimality guarantee. The FPTAS with the the γ-representation algorithm incorporated still runs in polynomial time: [26] has shown that at each node u, there are $O((\frac{n_u}{\epsilon})^d)$ partial solutions to consider, where n_u is the number of nodes in T_u, so by Theorem 3, running the γ-representation algorithm on u takes $O((\frac{n_u}{\epsilon})^{3d})$. [26] has further shown that the runtime to compute all the solutions at u is $O((\frac{n_u}{\epsilon})^{2d})$. If the γ-representation algorithm is run at u, then the total runtime at u becomes $O((\frac{n_u}{\epsilon})^{3d})$. At the root node, the total runtime is $O((\frac{n}{\epsilon})^{3d})$, where n is the number of nodes in the tree.

7   Experiments

We report experimental results on using the γ-representation algorithm to find representations, estimations and approximations of Pareto frontiers for hydropower planning in the Amazon River. To accelerate the experiments, we apply the γ-representation algorithm to independent chunks of solutions in parallel. The parallelized algorithm preserves the theoretical guarantees but might return representation sets bigger than using the non-parallelized version, since clustering and choice of representative points are local to each chunk. The bigger the chunks, the slower the algorithm runs but the less the impact on the size of the representation set. For all our experiments, we used chunks of size 50000 and distribute them across 12 threads.

Representation - Table 1 shows the results of running the γ-representation algorithm as described in Sect. 4 to find representation sets of exact Pareto frontiers for different values of γ and different criteria on the full Amazon. The number of solutions decreases substantially as γ increases. To evaluate the quality of the representation sets, we calculate their hypervolumes using the framework introduced in [5] and compare them with the baseline where $\gamma = 0$, i.e., the entire exact Pareto frontier. Specifically, we normalize each objective value $z_i(\pi)$ to $[0, 1]$ by scaling it to $z_i(\pi)^* = \frac{|z_i(\pi) - z_i(\pi)_{\text{worst}}|}{|z_i(\pi)_{\text{best}} - z_i(\pi)_{\text{worst}}|}$, where $z_i(\pi)_{\text{best}}$ and $z_i(\pi)_{\text{worst}}$ are the maximum and minimum (or minimum and maximum, if the criterion is to be minimized) that can be achieved across the whole feasible solution space, i.e. building all dams or building no dam in our case. Then, we compute for each representation set the hypervolume of the objective space dominated by the solutions in the set, with zero vector as the reference point. In general, a greater hypervolume indicates a better quality. As γ increases, the hypervolume decreases, but not significantly. For example, the hypervolume decrease for the biggest γ, i.e., $\gamma = 0.1$, ranges from 1.3% to 9.7% for the different criteria, while the reduction in the number of solutions ranges from 946 to 1620 folds.

We also report the runtime for each experiment. We see that running the γ-representation algorithm requires extra processing time after the Pareto frontier has been found. The increase in runtime is polynomial, as proved in Sect. 4.

Estimation - Tables 2, and 3 contain the results of running the γ-representation algorithm at different levels of the tree during the dynamic programming

Table 1. Representing the two-criteria Pareto frontier for the full Amazon river for energy (E), connectivity (C), and greenhouse gas emission (G). The Pareto frontier is found by the exact dynamic programming algorithm, so, the reported solutions, including the representations, are guaranteed to be exactly Pareto-optimal.

Criteria	γ	Number of Solutions	Hypervolume	Runtime (s)
EC	0	33127	0.833	7.2098
	0.001	4594	0.833	49.4176
	0.01	311	0.832	27.5256
	0.1	35	0.822	25.1264
EG	0	58762	0.807	305.5793
	0.001	11090	0.803	427.3345
	0.01	1007	0.802	358.49
	0.1	60	0.792	349.2219

Table 2. γ-representation at different levels when optimizing for energy and connectivity for the full Amazon.

γ	Level	Optimality Guarantee	Number of Solutions	Hypervolume	Runtime (s)
0	N/A	1	33127	0.833	7.2098
0.001	1	1	4594	0.833	49.4176
	2	0.998	1996	0.832	24.4195
	3	0.997	1795	0.832	19.5634
0.01	1	1	311	0.832	27.5256
	2	0.980	108	0.815	1.8864
	3	0.970	102	0.815	1.3826
0.1	1	1	35	0.822	25.1264
	2	0.810	11	0.702	0.8607
	3	0.729	10	0.677	0.5394

algorithm as described in Sect. 5. Running the γ-representation algorithm for more levels of the tree shrinks the size of the solution set drastically. On the other hand, the qualities of the estimations as measured by hypervolume when compared with the exact Pareto frontier are 8.3% to 18.7% worse for the most aggressive setting ($\gamma = 0.1$ and $L = 3$), where decreases of more than 2000 folds in the sizes of the solution sets are observed.

Moreover, when the γ-representation algorithm is applied to level 2 and 3 nodes, the number of partial solutions at those nodes decreases too. The decreases at the intermediate nodes help with reducing the runtime, since on the smaller levels, fewer combinations of partial solutions need to be considered. Overall, even though the estimation algorithm is not guaranteed to run in polynomial time, empirically we observe that it has good runtime performance.

Note that when $L = 1$, the process is equivalent to finding the exact Pareto frontier and then applying the γ-representation algorithm on the exact full frontier. Therefore, the resulting solutions are still all Pareto-optimal.

Approximation - Table 4 displays the results of the γ-representation algorithm at different levels of the tree during the FPTAS as described in Sect. 6 for the

Table 3. γ-representation at different levels when optimizing for energy and greenhouse gas emission for the full Amazon.

γ	Level	Optimality Guarantee	Number of Solutions	Hypervolume	Runtime (s)
0	N/A	1	58762	0.807	305.5793
0.001	1	1	11090	0.803	427.3345
	2	0.998	8114	0.807	493.4981
	3	0.997	8285	0.807	518.3726
0.01	1	1	1007	0.802	358.49
	2	0.980	725	0.798	33.6507
	3	0.970	798	0.799	33.5599
0.1	1	1	60	0.792	349.2219
	2	0.810	32	0.74	6.9024
	3	0.729	25	0.74	5.9851

Table 4. γ-representation when optimizing three criteria (energy, connectivity, and sediment) for the full Amazon. The Pareto frontiers are approximated by running the FPTAS with $\epsilon = 0.005$.

γ	Level	Optimality Guarantee	Number of Solutions	Hypervolume	Runtime
0	N/A	0.995	4279265	0.535	25652
0.001	1	0.994	295516	0.535	31287
0.01	1	0.985	6202	0.506	39833
0.1	1	0.896	98	0.480	34146

Table 5. γ-representation at different levels of the tree when optimizing six criteria (energy, connectivity, sediment, degree of regulation, biodiversity, and greenhouse gases) for the Marañón, a sub-basin of the Amazon. The Pareto frontiers are approximated by running the FPTAS with $\epsilon = 0.2$.

γ	Level	Optimality Guarantee	Number of Solutions	Hypervolume	Runtime (s)
0	N/A	0.8	700791	0.320	1383
0.005	1	0.796	315808	0.320	4358
	2	0.792	24554	0.313	2920
	3	0.788	22772	0.311	2002
0.1	1	0.72	15803	0.316	3264
	2	0.648	4345	0.308	1747
	3	0.583	3362	0.307	730
0.2	1	0.64	2690	0.311	2898
	2	0.512	325	0.299	1891
	3	0.410	205	0.298	640

full Amazon. We see that the effect of γ-representation is preserved when run on top of the approximation using rounded objectives. Notably for optimizing three criteria on the full Amazon, with $\gamma = 0.1$ we can decrease the number of solutions from over 4 million to 98, while the hypervolume only decreases by 10.3%.

Figure 4 plots baselines (exact or approximated Pareto frontier) and their representations, estimations, or approximations from applying the γ-representation algorithm at different levels of the tree for 2 and 3 criteria for the full Amazon. We see that the representation set is well-distributed across the Pareto frontier. They are sparser on the ends of the frontier because there we have small values for at least one of the objectives, making the $(1 - \gamma)$ bound easier to achieve.

To analyze the γ-representation algorithm for a larger number of objectives, we have also experimented with optimizing six objectives for the Marañón, a sub-basin of the Amazon. The choice of the smaller basin allows us to run more objectives, in a reasonable amount of time. The results are reported in Table 5. For a large number of objectives, applying the γ-representation algorithm also results in a significant decrease in the number of solutions. Similarly, the decreased number of solutions at the intermediate nodes has improved the runtime.

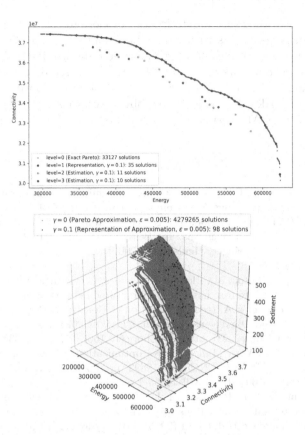

Fig. 4. Top panel: Exact Pareto frontier (# solutions: 33127) and its representation (# solutions: 35) and estimations (# solutions for level = 2: 11, # solutions for level = 3: 10) from applying the γ-representation algorithm at different levels of the tree for energy and connectivity for the full Amazon with $\epsilon = 0$ and $\gamma = 0.1$. Bottom panel: Approximated Pareto frontiers for energy, connectivity, and sediment for the full Amazon with $\epsilon = 0.005, \gamma = 0$ (# solutions: 4279265) and $\epsilon = 0.005, \gamma = 0.1$ (# solutions: 98)

8   Conclusion

We propose a clustering-based algorithm to find a **representation** set from the Pareto frontier with a coverage guarantee, which runs in time polynomial in the size of the frontier. We also consider two different strategies for incorporating the representation algorithm into a dynamic-programming-based approach: an **approximation** strategy, with polynomial runtime and optimality guarantee, and an **estimation** strategy with optimality guarantee and good empirical runtime performance, but without polynomial runtime guarantee. The three methods provide different ways to compress the Pareto frontier, resulting in solution sets significantly smaller than the full Pareto frontier, which are

γ-representations of the exact Pareto frontier or close to the frontier. Our main goal is to equip policymakers with streamlined approaches for effectively navigating Pareto frontiers, thus facilitating a more efficient decision-making process. Moreover, we hope our work will catalyze further research on the computation and visualization of Pareto frontiers. Multi-objective Pareto optimization is key to understanding trade-offs among various objectives, thus playing a pivotal role in the development of AI decision-support systems for informed decision-making.

References

1. Almeida, R.M., et al.: Reducing greenhouse gas emissions of amazon hydropower with strategic dam planning. Nat. Commun. **10**(1), 1–9 (2019)
2. Andersen, H.R., Hadzic, T., Hooker, J.N., Tiedemann, P.: A constraint store based on multivalued decision diagrams. In: Bessière, C. (ed.) CP 2007. LNCS, vol. 4741, pp. 118–132. Springer, Heidelberg (2007). https://doi.org/10.1007/978-3-540-74970-7_11
3. Bai, Y., Shi, Q., Grimson, M., Flecker, A., Gomes, C.P.: Efficiently approximating high-dimensional pareto frontiers for tree-structured networks using expansion and compression. In: Cire, A.A. (eds.) Integration of Constraint Programming, Artificial Intelligence, and Operations Research. CPAIOR 2023. LNCS, vol. 13884, pp. 1–17. Springer, Cham (2023). https://doi.org/10.1007/978-3-031-33271-5_1
4. Bergman, D., Cire, A.A.: Multiobjective optimization by decision diagrams. In: Rueher, M. (ed.) CP 2016. LNCS, vol. 9892, pp. 86–95. Springer, Cham (2016). https://doi.org/10.1007/978-3-319-44953-1_6
5. Cao, Y., Smucker, B.J., Robinson, T.J.: On using the hypervolume indicator to compare pareto fronts: applications to multi-criteria optimal experimental design. J. Stat. Plan. Inference **160**, 60–74 (2015). https://doi.org/10.1016/j.jspi.2014.12.004, https://www.sciencedirect.com/science/article/pii/S0378375814002006
6. Chen, W., Ishibuchi, H., Shang, K.: Clustering-based subset selection in evolutionary multiobjective optimization. In: 2021 IEEE International Conference on Systems, Man, and Cybernetics (SMC), pp. 468–475. IEEE (2021)
7. Deb, K., Jain, H.: An evolutionary many-objective optimization algorithm using reference-point-based nondominated sorting approach, part i: solving problems with box constraints. IEEE Trans. Evol. Comput. **18**(4), 577–601 (2013)
8. Deb, K., Pratap, A., Agarwal, S., Meyarivan, T.: A fast and elitist multiobjective genetic algorithm: nsga-ii. IEEE Trans. Evol. Comput. **6**(2), 182–197 (2002)
9. Doerr, B., Qu, Z.: A first runtime analysis of the nsga-ii on a multimodal problem. IEEE Transactions on Evolutionary Computation (2023)
10. Doerr, B., Qu, Z.: From understanding the population dynamics of the nsga-ii to the first proven lower bounds. In: Proceedings of the AAAI Conference on Artificial Intelligence, vol. 37, pp. 12408–12416 (2023)
11. Doerr, B., Qu, Z.: Runtime analysis for the nsga-ii: Provable speed-ups from crossover. In: Proceedings of the AAAI Conference on Artificial Intelligence. vol. 37, pp. 12399–12407 (2023)
12. Flecker, A.S., et al.: Reducing adverse impacts of amazon hydropower expansion. Science **375**(6582), 753–760 (2022)
13. Gomes, C., et al.: Computational sustainability: computing for a better world and a sustainable future. Commun. ACM **62**(9), 56–65 (2019)

14. Gomes-Selman, J.M., Shi, Q., Xue, Y., García-Villacorta, R., Flecker, A.S., Gomes, C.P.: Boosting efficiency for computing the pareto frontier on tree structured networks. In: van Hoeve, W.-J. (ed.) CPAIOR 2018. LNCS, vol. 10848, pp. 263–279. Springer, Cham (2018). https://doi.org/10.1007/978-3-319-93031-2_19
15. Grimson, M., et al.: Scaling up pareto optimization for tree structures with affine transformations: Evaluating hybrid floating solar-hydropower systems in the amazon. In: Proceedings of the AAAI Conference on Artificial Intelligence (submitted)
16. Johnson, S.C.: Hierarchical clustering schemes. Psychometrika **32**(3), 241–254 (1967). https://doi.org/10.1007/bf02289588, http://dx.doi.org/10.1007/BF02289588
17. Murtagh, F., Legendre, P.: Ward's hierarchical agglomerative clustering method: which algorithms implement ward's criterion? J. Classif. **31**(3), 274–295 (2014). https://doi.org/10.1007/s00357-014-9161-z, http://dx.doi.org/10.1007/s00357-014-9161-z
18. Newman, M.E.J.: Finding community structure in networks using the eigenvectors of matrices. Phys. Rev. E **74**(3) (2006). https://doi.org/10.1103/physreve.74.036104, http://dx.doi.org/10.1103/PhysRevE.74.036104
19. Sahraei, S., Asadzadeh, M.: Cluster-based multi-objective optimization for identifying diverse design options: application to water resources problems. Environ. Model. Softw. **135**, 104902 (2021)
20. Sibson, R.: Slink: an optimally efficient algorithm for the single-link cluster method. Comput. J. **16**(1), 30–34 (1973). https://doi.org/10.1093/comjnl/16.1.30, http://dx.doi.org/10.1093/comjnl/16.1.30
21. Sokal, R.R.: Numerical taxonomy. Sci. Am. **215**(6), 106–117 (1966). http://www.jstor.org/stable/24931358
22. Srinivas, N., Deb, K.: Muiltiobjective optimization using nondominated sorting in genetic algorithms. Evol. Comput. **2**(3), 221–248 (1994)
23. United Nations General Assembly: Transforming our world: the 2030 agenda for sustainable development (2015). https://sdgs.un.org/2030agenda
24. Vamplew, P., Dazeley, R., Foale, C., Firmin, S., Mummery, J.: Human-aligned artificial intelligence is a multiobjective problem. Ethics Inf. Technol. **20**, 27–40 (2018)
25. Wei, D., Jiang, Q., Wei, Y., Wang, S.: A novel hierarchical clustering algorithm for gene sequences. BMC Bioinform. **13**(1) (2012). https://doi.org/10.1186/1471-2105-13-174, http://dx.doi.org/10.1186/1471-2105-13-174
26. Wu, X., et al.: Efficiently approximating the pareto frontier: hydropower dam placement in the amazon basin. In: Proceedings of the AAAI Conference on Artificial Intelligence, vol. 32 (2018)
27. Zhang, H., Song, S., Zhou, A., Gao, X.Z.: A clustering based multiobjective evolutionary algorithm. In: 2014 IEEE Congress on Evolutionary Computation (CEC), pp. 723–730. IEEE (2014)
28. Zhang, Q., Li, H.: MOEA/D: a multiobjective evolutionary algorithm based on decomposition. IEEE Trans. Evol. Comput. **11**(6), 712–731 (2007)
29. Zheng, W., Liu, Y., Doerr, B.: A first mathematical runtime analysis of the Non-Dominated Sorting Genetic Algorithm II (NSGA-II). In: Conference on Artificial Intelligence, AAAI 2022. AAAI Press (2022). preprint at https://arxiv.org/abs/2112.08581
30. Zhou, S., et al.: A multi-objective evolutionary algorithm with hierarchical clustering-based selection. IEEE Access **11**, 2557–2569 (2023)

Improving Metaheuristic Efficiency for Stochastic Optimization by Sequential Predictive Sampling

Noah Schutte[1]([✉]) [ID], Krzysztof Postek[2] [ID], and Neil Yorke-Smith[1] [ID]

[1] Delft University of Technology, Delft, The Netherlands
{n.j.schutte,n.yorke-smith}@tudelft.nl
[2] Delft, The Netherlands

Abstract. Metaheuristics are known to be effective in finding good solutions in combinatorial optimization, but solving stochastic problems is costly due to the need for evaluation of multiple scenarios. We propose a general method to reduce the number of scenario evaluations per solution and thus improve metaheuristic efficiency. We use a sequential sampling procedure exploiting estimates of the solutions' expected objective values. These values are obtained with a predictive model, which is founded on an estimated discrete probability distribution linearly related to all solutions' objective distributions; the probability distribution is continuously refined based on incoming solution evaluation. The proposed method is tested using simulated annealing, but in general applicable to single solution metaheuristics. The method's performance is compared to descriptive sampling and an adaptation of a sequential sampling method assuming noisy evaluations. Experimental results on three problems indicate the proposed method is robust overall, and performs better on average than the baselines on two of the problems.

Keywords: single solution metaheuristics · stochastic optimization · sequential sampling · prediction-based search

1 Introduction

Stochastic optimization problems gain interest because most real-world problems involve uncertainty. Some practical examples of stochastic combinatorial problems are operating room scheduling [38], renewable energy applications [37] and transportation [27]. We define this class of problems as follows.

Definition 1. *Given a finite set of solutions X, random variables ω and a real-valued objective function f, we define a **Stochastic Combinatorial Optimization Problem (SCOP):** $\min_{x \in X} \mathbb{E}_{\omega}[f(x, \omega)]$.*

Noting that the expectation could be replaced by another risk functional like an α-quantile; we will limit ourselves to the expectation for clarity.

K. Postek—Independent Researcher.

B. Dilkina (Ed.): CPAIOR 2024, LNCS 14743, pp. 158–175, 2024.
https://doi.org/10.1007/978-3-031-60599-4_10

SCOPs are significantly more complex than deterministic combinatorial optimization problems. While both aim at finding an optimal solution, in SCOPs solutions need to be evaluated against uncertainty. Uncertainty does not impact the feasibility of a SCOP's solution, as the set of feasible solutions X we consider is independent of the uncertain parameter ω. However, uncertainty affects the solution's objective value, as it is obtained by evaluating each uncertain parameter realization. Hence SCOPs have additional complexity from solution evaluation.

Several modelling frameworks exist to tackle SCOPs. Exact modelling frameworks, like mathematical or dynamic programming [6,36], are designed to guarantee finding an optimal solution. However, due to the complexity of general SCOPs, these frameworks are often limited to small problems or unrealistic modelling assumptions. Non-exact methods, on the other hand, are particularly successful in finding good solutions in reasonable time. Specifically, *metaheuristics* – general non-exact frameworks that utilize heuristics – have shown to be effective for both deterministic and stochastic problems as they efficiently search the solution space [7,16]. However, metaheuristics for SCOPs are still limited in performance due to facing the challenge of evaluating found solutions against uncertainty [7]. This is in practice often done using simulation, which can be expensive and therefore slows down the search [15].

There is significant research on metaheuristics for specific SCOPs [7,16]. For example, Schutte et al. [31] improve the efficiency of adaptive large neighbourhood search for a stochastic surgery scheduling problem. However, there is little work on methods that improve metaheuristic efficiency for SCOPs in general. This paper advances this research area with three specific contributions:

- A generic method that improves the efficiency of metaheuristics when solving SCOPs. The method utilizes prior information in a memory-efficient way and uses a predictive model to estimate the solutions' expected objectives and to estimate the variance of this estimate.
- The notion of a general output scenario distribution. This provides a means to compare solution quality. We provide two approaches to estimate this distribution.
- Application of the method in combination with simulated annealing to demonstrate its effectiveness on three different problems.

2   Background

2.1   Stochastic Optimization

We focus on the discrete stochastic optimization problems, as in Definition 1. Within this framework it is assumed that the distribution of the uncertain parameter ω is known. Given a tractable mathematical formulation of the SCOP, the model can be solved efficiently with a guaranteed optimal solution. However, for a wide range of problems these exact methods become computationally too expensive [7,12].

One reason for this is that uncertainty makes evaluating both objectives and constraints complicated as it potentially requires integration over continuous random variable ω. To circumvent this, discretization is often used to make SCOPs tractable [8]. A common approach for this is *sample average approximation* [18] in which random samples are being drawn from the continuous probability distribution. These scenarios are assumed as being equally likely and the SCOP then has a deterministic equivalent formulation as an approximation, where each component of each uncertain variable is replaced by all potential scenarios. This still poses a challenge because to compute the average objective value, the objective function $f(x, \omega)$ in Definition 1 still needs to be evaluated for each scenario ω.

This evaluation is challenging because the objective function f need not even be a closed-form function. For example, it can be the optimal value of another optimization problem as in the case of a common class of SCOPs – two-stage problems. There, the goal is to find an optimal *here-and-now* decision before knowing the uncertain parameters' realizations, while the *wait-and-see* decisions are optimized after the uncertain parameters are realized. In other situations f cannot even be mathematically modelled. This is because processes and systems are often modelled by simulation and therefore, in practical SCOPs, evaluating f is often done by a simulation. *Simulation–optimization* thus studies optimization methods under the assumption that $f(x, \omega)$ is a black box. The main challenge there is to limit the number of simulations, as they can be expensive [1]. Because of the above two examples, we assume the following:

Assumption 1. *Objective function f in Definition 1 does not have a closed form, implying that the expected objective $\mathbb{E}_\omega[f(x, \omega)]$ also does not have a closed form.*

Assumption 1 implies that when ω has a continuous distribution, we need to approximate the expected objective using a discrete distribution. In the sequel, Ω is denoted as the set that includes all potential realizations of ω.

2.2   Metaheuristics for Stochastic Optimization

Metaheuristics are designed to find good solutions fast and, within a limited time, they often outperform exact methods. Simulated annealing, tabu search and genetic algorithms are effective metaheuristics trying to find better solutions by modifying earlier-found solutions. The modified solutions are the potential *moves* to make and they are evaluated before further moves are made. This procedure repeats itself until a final solution is returned. Initially designed for solving deterministic problems, metaheuristics have also shown to be effective for SCOPs [7,16]. Stochastics makes the objective evaluation computationally more expensive for a solution x as, based on Assumption 1 and Definition 1, we need to evaluate $f(x, \omega)$ for every scenario $\omega \in \Omega$ to find the exact expectation. The more expensive the evaluation of f and the higher the computational costs of a move (creating a new solution), the slower the metaheuristic search becomes. We show in Example 1 that even for problems with a simple f the required additional evaluations can increase the computational costs significantly. We first properly introduce simulated annealing.

Algorithm 1. Simulated Annealing

Input: budget b_0, initial temperature t_0, initial solution x
Output: final solution x

1: Let $t \leftarrow t_0$, $b \leftarrow b_0$
2: **while** $b > 0$ **do**
3:     Reduce budget b
4:     Update temperature t
5:     Get move x' and evaluate x'
6:     **if** $\mathbb{P}_{\text{accept}}(x, x', t)$ high enough **then**
7:         $x \leftarrow x'$
8: **return** solution x

Simulated annealing (SA) is an archetypal metaheuristic, outlined high-level in Algorithm 1. SA is a *single solution* metaheuristic keeping a single *current* solution in memory at all times, as opposed to *population-based* metaheuristics like genetic algorithms. Depending on the quality of a move and the value of a parameter t called the temperature, the new *candidate* solution is accepted or rejected. After this a new move is made in the following iteration. A higher temperature encourages exploration, i.e., the algorithm is more likely to accept solutions, even if they are worse. A lower temperature encourages exploitation, as it only accepts better solutions. Over time, the temperature is decreased within the SA procedure. Given this definition of SA, we now present Example 1.

Example 1. Consider a project scheduling problem in which the goal is to minimize the total makespan of a project. The project consists of set A with n activities that need to be scheduled. The scheduling of activities is constrained by some resources that the activities require. This is known as the Resource Constrained Project Scheduling Problem (RCPSP). A deterministic solution to this problem is most commonly defined as a list of activity start times $(s_{a_1}, \ldots, s_{a_n}) : a_i \in A, s_{a_i} \in \mathbb{R}_+$. If the duration of activities is uncertain however, the actual start times are unknown upfront. Because of this a solution becomes a policy. We define a solution as a *priority list* of activities $(a_1, \ldots, a_n) : a_i \in A$. This list is the order in which activities will start, such that an activity can never start before another activity that is higher on the priority list [11]. Now we use SA to find a good solution. As a move, we swap the place of two activities in the priority list. The evaluation consists of building a realized schedule $f(x, \omega)$ for each scenario $\omega \in \Omega$ and taking the average of the obtained makespans. Swapping two activities is a single operation. Building a realized schedule is n operations, as it consists of scheduling n activities. This is done $|\Omega|$ times. Hence the evaluation is about $|\Omega| * n$ times as expensive.

2.3 Sampling

Recently there has been a resurgence of interest in sampling due to its relevance in sequential decision-making problems in general and reinforcement learning (RL) in particular [9]. The *multi-armed bandit problem* is a sub-problem of RL in which the decision maker has to decide between n actions, each with an unknown but fixed probability distribution of rewards [21]. This problem specifically deals with the trade-off of exploration (new action) and exploitation (best-rewarded action). In our problem setting, the metaheuristic is tasked with solving the sequential decision problem of evaluating at each iteration a new scenario for a current solution, or moving to a new solution: each action maps to a solution and the obtained reward is the evaluated solution objective for some scenario. This is a similar problem setting to RL, but not exactly the same: in our setting the fixed probability distribution of rewards is still unknown, but the fixed probability distribution of uncertain variables that determine the rewards is known. We can exploit this by deciding which samples to draw, i.e., which scenarios to evaluate. In the next section we present two baselines that use this principle. Note that RL in general can be used as a metaheuristic alternative or as a hybrid [32], but typically requires a lot of data and is therefore expected to be inferior in a setting with (relatively) expensive solution evaluations.

3 Baseline Sampling Methods

3.1 Descriptive Sampling

Our central question is how to obtain high-quality solutions with the meta-heuristic evaluating fewer scenarios per solution. In this section we provide two baselines.

The first idea to decrease the number of evaluations is to create a smaller set of scenarios Ω when using sample average approximation. The drawback is that the set might not be representative anymore for the underlying distribution or data set. A method that creates a more representative set of scenarios compared to random sampling is *descriptive sampling* [29]. When a set of m scenarios is constructed, descriptive sampling creates m equally spaced (by probability density) realizations for each random variable. The scenarios are then constructed by selecting a random permutation for all possible realizations for each random variable, where each realization corresponds to a scenario.

One of the drawbacks of an approach like descriptive sampling is that it creates a representative set based on the distribution of the random variable ω itself. We call this the *input distribution*, which is different from the *output distribution* defined as the distribution of $f(x, \omega)$ given a solution $x \in X$. This distinction is important as we are optimizing over this objective, and we need the approximated output distribution to be representative of the actual output distribution if we want to make good optimization decisions. We show with Example 2 that a representative set for the input distribution is not necessarily a representative set for the output distribution.

Example 2. Consider a RCPSP as described in Example 1, with 3 activities $A = \{a, b, c\}$. Due to limited resources at most 2 activities can be scheduled at the same time. The duration of activity a is fixed at $d_a = 2$, and the duration of b and c are independent and uniformly distributed $d_b, d_c \sim U(0, 4)$. Our goal is to find the priority list for which the expected makespan is minimized. Consider descriptive sampling with two scenarios for realizations of (d_b, d_c). Out of 4 possible descriptive samples (consisting of durations in $\{1, 3\}$), we assume we get $\omega = (1, 1)$ and $\nu = (3, 3)$. Since d_b and d_c have the same distribution and the project will start with two activities at the same time, effectively there are two different solutions possible: (a, b, c) and (b, c, a). We have the following relationships between the expected makespans of the solutions:

$$\mathbb{E}_{\{d_b, d_c \sim U(0,4)\}}[f((b, c, a), (d_b, d_c))] = 3\tfrac{1}{3} \quad < \quad \mathbb{E}_{\{d_b, d_c \sim U(0,4)\}}[f((a, b, c), (d_b, d_c))] = 3\tfrac{1}{2},$$

$$\mathbb{E}_{\{(d_b, d_c) \in \{\omega, \nu\}\}}[f((b, c, a), (d_b, d_c))] = 4 \quad > \quad \mathbb{E}_{\{(d_b, d_c) \in \{\omega, \nu\}\}}[f((a, b, c), (d_b, d_c))] = 3\tfrac{1}{2}.$$

This shows that while the true optimal solution is (b, c, a), descriptive sampling would return (a, b, c) as optimal.

3.2 Sequential Sampling

A second idea aimed at efficient evaluation is to use *sequential sampling* within the metaheuristic. In sequential sampling there is not a given set or number of scenarios sampled, but after every sample a decision is made to continue sampling or not. In the case of *population-based* metaheuristics, multiple new solutions are encountered at the same time. This means that the sequential sampling strategy becomes more elaborate, as there is not just one solution for which the sample decision needs to be made. For example, Bartz-Beielstein et al. [5] apply sequential sampling to particle swarm optimization and Groves and Branke [14] apply sequential sampling to an evolutionary algorithm.

Similarly, sequential sampling has been applied with SA under noise, where it is assumed that an objective evaluation has added Gaussian noise [2]. In this case, without a predetermined number of evaluations per solution, a solution will only be accepted or rejected when enough evaluations have been done to make this decision. Therefore, at any point there will be three potential decisions: accept the solution, reject the solution or evaluate on more scenarios. Bulgak and Sanders [10] used this principle to accept/reject when the solution difference between candidate and current solution is significant. This is inefficient however, as it does not adhere to the fundamental *detailed balance equation* that deterministic SA is based upon. This balance equation ensures that SA reaches equilibrium at each temperature level if given sufficient time, such that it converges in probability to the optimal solution [17]. Ball et al. [2] impose the detailed balance equation conditions at this decision level, under which they maximize the acceptance probability per sample, which is a measure for the efficiency of the algorithm. The authors propose the following acceptance probability:

$$\mathbb{P}_{\text{accept}}(c_n, t) = \min(1, e^{-2(c_n + \sigma^2/(2t))(c_{n-1} + \sigma^2/(2t))/\sigma^2}), \tag{1}$$

where c_n is the cumulative performance difference after n samples:

$$c_n = \sum_{\omega \in \Omega_n} f(x, \omega) - f(x', \omega), \tag{2}$$

Algorithm 2. Sequential Difference Sampling

Input: current solution x, candidate solution x'
Output: new current solution x

1: Let $n \leftarrow 0$, $c_n \leftarrow 0$, $\Omega_x \leftarrow \emptyset$
2: **while** *true* **do**
3:     $n \leftarrow n + 1$
4:     Sample $\omega \in \Omega \backslash \Omega_x$
5:     $\Omega_x \leftarrow \Omega_x \cup \omega$
6:     Evaluate $f(x, \omega), f(x', \omega)$
7:     $c_n \leftarrow c_{n-1} + f(x, \omega) - f(x', \omega)$
8:     Determine $\mathbb{P}_{\text{accept}}(c_n, t)$
9:     **if** $\mathbb{P}_{\text{accept}}(c_n, t) > u, u \sim U(0, 1)$ **then**
10:         **return** solution x'
11:     **else if** $c_n < 0$ **then**
12:         **return** solution x

---

and Ω_n is the set of n samples. Similarly they derive an optimal rejection rule to apply if the solution is not accepted. However, since this rule is dependent on an unknown prior distribution of futures moves, a more simple but effective rejection rule is proposed: Reject if $c_n < 0$, i.e.,

$$\mathbb{P}_{\text{reject}}(c_n, t) = \mathbb{1}_{\{c_n < 0\}}. \tag{3}$$

If the move is neither accepted nor rejected, another sample is taken, and this process is repeated as can be seen in Algorithm 2. We fit this procedure to our methodology as described in Sect. 5.

4   Sequential Predictive Sampling

Our goal is to increase efficiency by reducing the number of scenarios evaluated. We aim to support this goal by utilizing prior information: function evaluations of the earlier-found solutions. In most metaheuristics, solution information is only kept if their corresponding solutions are still considered promising. By contrast, Prudius and Andradóttir [25] introduce an averaging framework in which solutions are evaluated partially during the search: an estimate of the expected objective per solution is stored as the average of all obtained objectives, which makes it possible to improve this estimate by more solution evaluations. Our method, however, does not keep all solution evaluations in memory, as it only deems the more recent solutions evaluations interesting.

The proposed methodology is summarized as follows. Given a new solution, two random scenarios from Ω are evaluated (two is the minimum number that gives us relative information). Based on this evaluation, a predictive model is set up and returns an estimate of the expected objective as well as the variance of this estimate. Based on this, an accept, reject or sample decision can be made in the overarching metaheuristic. The predictive model takes a *general scenario output distribution* as input. This distribution is estimated based on evaluations done on previously found solutions.

Commonly, a solution's expected objective would be determined by evaluating over all scenarios in some set of scenarios Ω. Although we want expected objectives to be representative over all uncertainty (i.e., the whole Ω), evaluating every solution on every scenario is not necessary. As we argue below, this is because not all evaluations are independent.

A fundamental assumption behind any search algorithm is that some solutions are better than others:

Assumption 2. *Some solutions are better than others:* $\exists x, y \in X$ *such that*

$$\mathbb{E}_\omega[f(x, \omega)] < \mathbb{E}_\omega[f(y, \omega)].$$

Similarly, we assume that some scenarios are uniformly better than others, recognizing that uncertainty can have a positive or a negative impact:

Assumption 3. *Some scenarios are better than others: Considering x as a discrete uniformly distributed random variable with as support solution space X, we assume that for some $\omega, \nu \in \Omega$ it holds that*

$$\mathbb{E}_x[f(x, \omega)] < \mathbb{E}_x[f(x, \nu)].$$

Example 3. Due to the structure of a scheduling problem, a realization with long activity durations across all activities is worse than one with short activity durations. Take for examples the two scenarios obtained in Example 2, defined as the two durations of uncertain activities b and c: $\omega = (1, 1)$ and $\nu = (3, 3)$. Given solution (b, c, a), we get as makespans: $f((b, c, a), \omega) = 3 < f((b, c, a), \nu) = 5$. More generally for this problem, if one scenario ν *dominates* another scenario ω, the makespan can never be better, i.e.,

$$d_i^\nu \geq d_i^\omega \quad \forall i \implies f(x, \nu) \geq f(x, \omega) \quad \forall x \in X.$$

This holds for all problems where the impact of the uncertain parameter is similarly related to the objective function (linearly for some solutions).

Based on Assumption 2 and 3 we go one step further. The fact that both solutions and scenarios can be better or worse, makes it realistic to assume some dependency between evaluations that are based on these solutions and scenarios. When an evaluation $f(x, \omega)$ is perceived as 'good', it is likely that x and ω are both relatively good as well.

4.1   General Scenario Output Distribution Estimation

Given our reasoning on dependence of evaluations, we introduce a linear dependence assumption that we consider to hold with a margin of error that is small enough to help us compare evaluations. Even though the assumption will most often not hold exactly, it will help us steer the solution evaluations.

We first define two uniform discrete distributions that we will use to relate evaluations: *solution output distribution* E and *general scenario output distribution* D with as sets of atoms:

$$E_X := \{e_x : x \in X, e_x := \mathbb{E}_\omega[f(x,\omega)]\},$$
$$D_\Omega := \{d_\omega : \omega \in \Omega, d_\omega := \mathbb{E}_x[f(x,\omega)]\}.$$

Assumption 4. *There exist linear relationships between solutions given a scenario, i.e., for any $x, y \in X$:*

$$f(x,\omega)/f(y,\omega) = e_x/e_y \quad \forall \omega \in \Omega,$$

where the linear relation is decomposed as a fraction e_x/e_y. Further, there exist linear relationships between scenarios given a solution, i.e., for any $\omega, \nu \in \Omega$:

$$f(x,\omega)/f(x,\nu) = d_\omega/d_\nu \quad \forall x \in X,$$

where the linear relation is decomposed as a fraction d_ω/d_ν.

Lemma 1. *Given Assumption 4, we have $\forall x, y \in X$, $\forall \omega, \nu \in \Omega$:*

$$f(x,\omega)/f(y,\nu) = e_x d_\omega/(e_y d_\nu) \Rightarrow \forall c \in \mathbb{R} : f(x,\omega) = c\,e_x d_\omega \wedge f(y,\nu) = c\,e_y d_\nu.$$

We note that constant c in Lemma 1 is irrelevant in comparing evaluations and we therefore omit it from here on. Given Assumption 4, Lemma 1 ensures we can have a rank-1 decomposition of matrix F, $F = ed^T$, where e and d denote the vectorized estimates of sets E_X and D_Ω. In practice, Lemma 1 only holds when the objective function is linear in the uncertain parameters, which is unlikely when f is considered to not have a closed-form (Assumption 1). When Assumption 4 holds, the stochastic problem effectively becomes a deterministic problem where the solution quality can be compared exactly when all solutions are evaluated on the same, single scenario $\omega \in \Omega$. Therefore, Assumption 4 serves as a strong baseline. It ensures that when uncertainty is (close to) linear in the evaluation function, the method performs similar to assuming the problem is deterministic. Now that we have defined general scenario output distribution D, we present a method to estimate it. We will use linearity throughout such that the estimator is perfect if the rank-1 decomposition $F = ed^T$ exists.

Bayesian inference is popular in estimating distributions [13]. However, the general scenario output distribution can have any kind of shape and therefore we refrain from assuming a prior distribution. A prior distribution can be especially troublesome since observing only a few evaluations per solution will update the posterior only slightly for each new solution. Furthermore, solution evaluations are not random draws from the same distribution. Because of this we use discrete linear updates. Given an output scenario distribution D and Assumption 4, we can update this distribution assuming the evaluations from the last obtained solution are most representative of the general scenario output distribution. Denoting x as the last evaluated solution, we use:

$$d_\omega \leftarrow rf(x,\omega) \frac{\sum_{\omega \in \Omega_x} d_\omega}{\sum_{\omega \in \Omega_x} f(x,\omega)} + (1-r)d_\omega, \quad \forall \omega \in \Omega_x, \tag{4}$$

where Ω_x is the set of scenarios on which x is evaluated and r is the rate by which the current values are replaced, which makes the updates smoother. When using SA, this update can be done after each iteration. Note that if the rank-1 decomposition is equal to the actual evaluation matrix, i.e., $\boldsymbol{F} = \boldsymbol{ed}^T$, these updates ensure the general scenario output distribution has a linear relationship with all solution specific scenario output distributions. This leads to a perfect predictor given a linear predictive model, hence we will use this model class.

4.2   Predictive Model

Now we have defined an approach to obtain a general scenario output distribution D with Eq. 4, we use this distribution to compare obtained solutions during the metaheuristic search. Since we are developing a sequential sampling method, this comparison will include specifying a accept, reject or sample decision as introduced in Sect. 3.2. When evaluating a solution x, we set up the following predictive model:

$$f(x,\omega) = \beta_x d_\omega + \varepsilon_{x,\omega}, \tag{5}$$

where $\varepsilon_{x,\omega}$ is an error term and we estimate β_x by the ordinary least squares estimator that minimizes squared errors of the predictive model:

$$\hat{\beta}_x := \operatorname*{argmin}_{\beta_x} \sum_{\omega \in \Omega_x} (f(x,\omega) - \beta_x d_\omega)^2 = \frac{\sum_{\omega \in \Omega_x} d_\omega f(x,\omega)}{\sum_{\omega \in \Omega_x} d_\omega^2}.$$

Given the predictive model from Eq. 5 with $\beta_x = \hat{\beta}_x$, we get an estimate of the expected objective of solution x:

$$\hat{\mu}_x := \hat{\mathbb{E}}_\omega[f(x,\omega)] = \frac{1}{|\Omega|} \Big(\sum_{\omega \in \Omega_x^c} \hat{\beta}_x d_\omega + \sum_{\omega \in \Omega_x} f(x,\omega) \Big),$$

where $\Omega_x^c = \Omega \setminus \Omega_x$. An estimate of the objective value by itself is useful, but additionally we also get a measure of the uncertainty of the prediction by estimating the variances as:

$$\widehat{\operatorname{Var}}(\mu_x) = \frac{1}{|\Omega|^2} \Big(\big(\sum_{\omega \in \Omega_x^c} d_\omega \big)^2 \operatorname{Var}(\hat{\beta}_x) + |\Omega_x^c| \operatorname{Var}(\varepsilon_x) \Big).$$

$$\widehat{\operatorname{Var}}(\varepsilon_x) = \frac{1}{|\Omega_x| - 1} \sum_{\omega \in \Omega_x} (f(x,\omega) - \hat{\beta}_x d_\omega)^2.$$

$$\operatorname{Var}(\hat{\beta}_x) = \widehat{\operatorname{Var}}(\varepsilon_x) / \sum_{\omega \in \Omega_x} d_\omega^2.$$

Given this setup, we know the expected objective of a solution x has a Student's t-distribution with $|\Omega_x| - 1$ degrees of freedom, shifted by mean $\hat{\mu}_x$ and scaled by

---

Algorithm 3. Sequential Predictive Sampling

Input: current and candidate solution x, x', estimation parameters $\hat{\mu}_x, \hat{\sigma}_x^2$, scenario set Ω, evaluated scenario set Ω_x, general scenario output distribution D, temperature t

Output: new current solution x

1: Update D ▷ using Equation 4
2: **while** *true* **do**
3:     Select $y \in \{x, x'\}$ ▷ with $\mathbb{P}(y = x) = |\Omega_x|/(|\Omega_x| + |\Omega_{x'}|)$
4:     Sample $\omega \in \Omega \backslash \Omega_y$ ▷ at random
5:     $\Omega_y \leftarrow \Omega_y \cup \omega$
6:     Evaluate $f(y, \omega)$
7:     Update $\hat{\mu}_y$ and $\hat{\sigma}_y^2$ ▷ using predictive model in Equation 5
8:     Determine $\mathbb{P}_{\text{accept}}(x, x', t)$, $\mathbb{P}_{\text{reject}}(x, x', t)$ ▷ using Equation 1, 3
9:     **if** $\mathbb{P}_{\text{accept}}(x, x', t) > u, u \sim U(0, 1)$ **then**
10:         **return** solution x'
11:     **else if** $\mathbb{P}_{\text{reject}}(x, x', t) > u, u \sim U(0, 1)$ **then**
12:         **return** solution x

---

estimated variance $\widehat{\text{Var}}(\mu_x)$, assuming error term $\varepsilon_{x,\omega}$ is normally distributed. Hence, we effectively have an estimated variance of: $\hat{\sigma}_x^2 := \widehat{\text{Var}}(\mu_x)\frac{|\Omega_x|-1}{|\Omega_x|-3}$.

The use of a linear model ensures that if the evaluations' relationships are linear, the predictive model can predict perfectly: $\hat{\sigma}_x^2 = 0$.

4.3 Full *SeqPre* Procedure

We summarize the full procedure in Algorithm 3. The comments therein show the exact implementation of the procedure for the experiments. We note that in the case of SA, the procedure replaces algorithmic steps 5–7 of Algorithm 1. All the input variables are updated and kept in memory during the search. Note that in Step 3 we always select x' if it does not have any evaluations yet. Note also that the while loop (steps 2–12) is guaranteed to terminate since scenario set Ω is finite and therefore when all scenarios are evaluated $\hat{\sigma}_y^2 = 0$ and $\mathbb{P}_{\text{accept}}(x, x', t), \mathbb{P}_{\text{reject}}(x, x', t) \in \{0, 1\}$.

5 Experiments

We examine the performance of our proposed method on three problems. Experiments were implemented in Python and performed on a single 2.0 GHz CPU running Ubuntu 20.4 with 32 GB RAM. Source code is available on GitHub [30].

5.1 General Experimental Configuration

Each method uses the same underlying SA procedure as outlined in Algorithm 1. To compare solution objectives we give each search algorithm a fixed budget

b_0 of evaluations. This is common practice for search algorithms as it shows efficiency independent of the underlying machine specifications [20]. We also denote runtime in the results, however we want to emphasize that the experimental problems have relatively cheap evaluations (non-extensive simulations) which makes the overhead greater than it would be for more practical problems. As annealing schedule, geometric annealing is used with a fixed final temperature for fair comparison between methods:

$$t \leftarrow t_0(t_f/t_0)^{(1-b/b_0)}.$$

The final temperature is set at a small value $t_f = 0.01$. The initial temperature is:

$$t_0 = -h \lceil \min_{x \in X} f(x, \mathbb{E}_{\omega \in \Omega}[\omega]) \rceil / \ln(p),$$

where $\lceil \cdot \rceil$ rounds up to the closest power of 10 and $h = 0.1$, $p = 0.5$ are such that initially solutions that are 10% worse are accepted with a probability of 0.5.

5.2   Methods Compared

We use three benchmarks that evaluate a fixed number of scenarios at each iteration. One of them evaluates the single scenario where each uncertain parameter attains its average. This is searching for a deterministic optimal solution, so we call this method *Det*. The other two variants use descriptive sampling to obtain a set of scenarios. Scenario sets of sizes 10 and 100 are used and the methods are denoted as *Des10* and *Des100*. Size 10 has shown to be effective for some problems in literature [4,28], while size 100 is assumed to be large enough to be representative for the input distribution.

Two methods are proposed to compare to these benchmarks. The first method is sequential *difference* sampling (*SeqDif*) based on [2] as shown in Algorithm 2. A random difference is sampled between the two solutions that are compared. However, we do not assume Gaussian noise with known variance, so we estimate the variance by updating it throughout the search:

$$\hat{\sigma}_i^2 \leftarrow (\hat{\sigma}_{i-1}\sqrt{i-1} + c_{n_i}^i)^2/i,$$

where i is the iteration and $c_{n_i}^i$ is the cumulative difference as defined in Eq. 2 when an accept or reject decision is made after n_i samples for iteration i.

The second method is our proposed sequential *predictive* sampling approach (*SeqPre*), as shown in Algorithm 3. Hyper-parameters as defined in Sect. 4 are set at $k = 50$ and $r = 0.9$ (chosen based on empirical results). Both sequential approaches use as decision rule the steps 9–12 in Algorithm 2, and a descriptive set of 100 scenarios as Ω. The final obtained solutions are evaluated on a test set of 1000 randomly drawn scenarios with a unique seed.

5.3   Description of Example Problems

Surgery Scheduling in Flexible Operating Rooms under Uncertainty (SSFORU) has as goal to plan as many elective surgeries as possible in a given set of

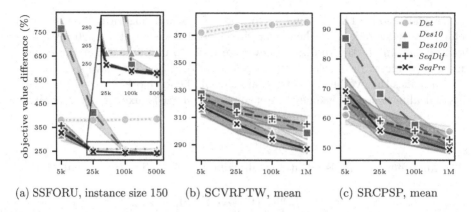

(a) SSFORU, instance size 150   (b) SCVRPTW, mean   (c) SRCPSP, mean

Fig. 1. Objectives values and 95% bounds in % difference to the optimal deterministic solution obj. value based on the test set and relative to budgets b_0.

operating rooms, while minimizing idle and overtime for the staff and waiting time for the patients. At any time during the day an emergency surgery can come in that needs to be performed as soon as any operating room is available. This problem has been studied in literature in several modifications [26] and was proposed in this form by [34], with instance sizes of $n \in \{70, 100, 150, 200\}$.

We use the most generally used costs configuration ($c_{\text{over}} = 26$, $c_{\text{idle}} = \frac{2}{3}c_{\text{over}}$, $c_{\text{waiting}} = \frac{1}{5}c_{\text{over}}$, $c_{\text{not-scheduling}} = 30 \times \frac{3}{4}c_{\text{over}}$) in literature [23,24,33]. Emergency inter-arrival times are modelled as exponentially distributed with a rate of 4 per day and surgery duration times are modelled as log-normal distributed fitted by type. In the realization of a schedule, emergency surgeries are assigned to an operating room based on the lowest expected duration of the remaining planned elective surgeries. In the SA procedure moves are defined as removing or adding elective surgeries and shifting the times of already planned surgeries.

Stochastic Resource Constrained Project Scheduling Problem (SRCPSP) is a well-studied problem, both the deterministic variant and different stochastic adaptations. The goal is to minimize the total duration of a project that consists of activities. Some activities can only start when certain others are finished (precedence relationships), while they also have resource requirements and there are limited resources available. We define a solution as a priority list of activities as in Example 1, while the evaluation of a schedule consists of simulating the realized schedule and obtaining the makespan. We test on instances 1.1–1.5 from the PSPLIB [19] with 30 activities. The duration of the activities is uncertain and modelled as exponentially distributed with as mean the given duration in the instance. In the SA procedure we define a move as a swap between two consecutive activities in the priority list. These are done randomly, limited to swaps that adhere to the precedence relationships.

Stochastic Capacitated Vehicle Routing Problem with Time Windows (SCVRPTW) has a given set of the same vehicles stationed at a depot, with the goal to serve a given set of customers. The vehicles have some capacity, while each customer has demand and a time window in which delivery is allowed. A solution is a route for reach vehicle. We test on instances R101–R105 introduced by [35].

Both demand and travel times are uncertain and modelled as exponentially distributed with as mean the given demand and the Euclidean distance respectively. Similarly to [22], costs are determined by distance travelled. The evaluation of a solution and scenario consists of a simulation in which vehicles travel to the customers. When a vehicle does not have enough product or arrives after the end of the time window, a cost penalty is incurred equal to a single trip back and forth to the depot. When a vehicle arrives before the start of a time window it waits. In the SA procedure a move is defined as for each route: 1) randomly adding or removing a customer, or doing nothing; 2) randomly swapping two consecutive customers.

5.4   Experimental Results

Tables 1 and 2 show the experimental results. Baseline *Det* performs poorly, due to over-fitting on the single scenario it uses. We see similar behaviour in lower magnitudes for *Des10*. Figure 1a illustrates that this over-fitting causes solutions to not improve for larger budgets. On the other hand, *Des100* does improve with larger budgets, but performs poorly on smaller budgets.

In general the aim of the sequential sampling methods is to reach similar objectives as *Des100* in less time. We see that for SSFORU, *SeqDif* and *SeqPre* perform similarly and better than the benchmarks. For SRCPSP and SCVRPTW however, *SeqPre* performs significantly better than *SeqDif*. This could be caused by the fact that differences of individual scenarios are not good enough estimates for the actual differences, or because the estimated variance is not accurate enough. Another difference between problem results is that *Des10* performs close to the best methods for SRCPSP and SCVRPTW. Specifically for SRCPSP we can state that this is caused by the fact that uncertainty does not change solutions that much, as the obtained solution is still competitive despite over-fitting. This confirms the findings in [3].

Table 1. Results SSFORU per instance size n for different evaluation budgets b_0: average test ($\pm$ SE) and train objectives in % difference to the optimal deterministic objective based on 20 random seeds, and average time s in seconds. $*$ shows significant difference between the 2 best methods (paired t-test, $\alpha = 0.05$).

n	Method	$b_0 = 5k$			$b_0 = 25k$			$b_0 = 100k$			$b_0 = 500k$		
		Test	Train	s	Test	Train	s	Test	Train	s	Test	Train	s
70	Det	298.1±4.2	32	9	344.9±5.5	29	47	327.2±6.2	29	189	322.1±6.0	29	972
	Des10	**188.9**±2.3	120	8	179.0±2.0	103	40	170.7±1.7	94	151	165.8±1.1	89	748
	Des100	394.7±15.7	388	6	204.5±9.0	194	44	163.1±1.2	152	190	158.2±0.6	144	943
	SeqDif	199.3±3.4	190	10	**165.6**±1.3	155	59	*159.4±0.6	146	233	**156.4**±0.9	142	1172
	SeqPre	190.9±2.4	181	11	169.3±0.8	158	61	165.8±0.9	152	243	158.5±0.6	143	1309
100	Det	809.4±7.0	60	11	819.0±6.9	48	58	837.6±7.9	46	227	836.4±6.4	46	1156
	Des10	**594.5**±13.5	455	12	547.6±3.6	384	56	540.8±3.7	365	241	539.0±3.3	355	1202
	Des100	1194.8±41.8	1177	8	669.6±35.5	647	60	514.6±1.6	483	238	505.4±1.4	466	1290
	SeqDif	655.3±18.6	629	14	526.9±2.1	501	78	**507.7**±1.5	475	354	*501.7±1.1	462	1747
	SeqPre	**594.5**±5.3	571	14	*521.1±1.7	494	83	510.9±1.8	478	349	505.5±1.4	469	1807
150	Det	381.5±3.5	32	17	380.9±3.0	24	66	382.9±2.5	20	367	385.8±2.1	18	1796
	Des10	317.9±12.3	257	17	259.1±1.0	188	71	258.9±0.9	178	288	258.7±0.9	175	1323
	Des100	764.7±21.5	764	13	412.7±21.6	407	76	248.9±3.2	238	291	242.2±0.3	228	1412
	SeqDif	357.0±12.5	353	19	250.2±0.5	242	77	**243.7**±0.4	232	327	*241.3±0.3	227	1400
	SeqPre	327.4±11.2	323	17	**249.1**±0.7	239	93	244.1±0.3	233	341	242.2±0.4	229	1753
200	Det	**210.7**±0.7	26	13	209.2±0.5	20	60	209.3±0.4	20	228	208.6±0.3	18	1156
	Des10	219.8±10.8	188	22	140.6±0.7	106	69	138.9±0.7	102	252	140.0±0.8	100	1030
	Des100	631.7±16.7	624	31	335.1±17.3	329	144	138.7±2.4	135	384	**130.7**±0.1	126	1485
	SeqDif	274.2±18.0	269	21	135.9±0.4	132	87	**131.8**±0.2	128	321	**130.7**±0.2	125	1354
	SeqPre	257.2±10.6	252	32	**134.7**±0.3	132	90	132.1±0.2	128	306	131.3±0.1	126	1672

Table 2. Results SRCPSP & SCVRPTW for different evaluation budgets b_0: test ($\pm$ SE) and train objectives in % difference to the optimal deterministic objective averaged over 10 random seeds and 5 instances, and average time s in seconds. $*$ shows significant difference between the 2 best methods (paired t-test, $\alpha = 0.05$).

	Method	$b_0 = 5k$			$b_0 = 25k$			$b_0 = 100k$			$b_0 = 1M$		
		Test	Train	s	Test	Train	s	Test	Train	s	Test	Train	s
SRCPSP	Det	**61.1**±1.6	9	2	58.8±1.2	5	8	56.6±1.3	3	31	55.5±0.9	1	317
	Des10	64.0±2.4	52	2	56.4±2.1	43	8	**52.6**±1.1	39	35	50.1±0.7	35	351
	Des100	86.9±3.2	85	2	68.2±2.8	66	8	57.7±2.0	55	33	50.7±0.9	48	348
	SeqDif	65.7±2.3	64	2	59.1±1.8	57	9	55.7±1.7	54	39	52.8±1.1	51	397
	SeqPre	69.2±2.2	66	3	**55.8**±1.5	53	17	**52.6**±1.1	50	67	**49.4**±0.7	47	677
SCVRPTW	Det	372.1±1.7	175	6	376.2±1.3	163	32	377.8±1.2	155	127	379.6±1.0	145	1278
	Des10	**316.1**±2.7	313	1	305.3±2.6	302	6	299.5±2.6	297	22	287.2±1.9	285	217
	Des100	327.5±2.0	327	1	318.2±2.5	318	4	308.5±2.1	308	15	298.7±2.4	298	129
	SeqDif	324.2±2.2	324	2	313.6±2.5	313	10	309.0±2.7	308	49	305.2±2.8	305	551
	SeqPre	317.8±2.7	317	4	**305.2**±3.3	305	19	*294.0±2.4	294	81	**287.0**±1.2	287	844

6   Conclusions and Future Work

The central idea of this paper is that metaheuristic efficiency can be improved by reducing the number evaluations per solution. We propose a general methodology based on a sequential sampling procedure and a predictive model that utilizes earlier obtained information memory-efficiently. This is done by modelling linear relationships between both solutions and scenarios. We applied the approach to SA on three diverse problems and found consistently strong performance.

Future work includes applying this method on metaheuristics in addition to SA: For single solution metaheuristics the method is readily applicable, while for population-based metaheuristcs the procedure of accepting/rejecting solutions needs to be adjusted. A second avenue is exploring potential further improvements: modelling more complex relationships between evaluations; and including solution characteristics in the predictive model, as currently it only uses solution evaluations. Finally, more sophisticated machine learning methods could be applied to estimate a general scenario distribution or predict objectives directly from all obtained information. It can require more computing power to learn a strong predictive model than to perform the search, but this could still be useful when offline computation is cheaper.

Acknowledgements. We thank the reviewers for their suggestions. This work was partially supported by Epistemic AI and by TAILOR, both funded by EU Horizon 2020 under grants 964505 and 952215 respectively.

References

1. Amaran, S., Sahinidis, N.V., Sharda, B., Bury, S.J.: Simulation optimization: a review of algorithms and applications. Ann. Oper. Res. **240**, 351–380 (2016). https://doi.org/10.1007/s10479-015-2019-x
2. Ball, R.C., Branke, J., Meisel, S.: Optimal sampling for simulated annealing under noise. INFORMS J. Comput. **30**, 200–215 (2018). https://doi.org/10.1287/ijoc.2017.0774
3. Ballestín, F.: When it is worthwhile to work with the stochastic RCPSP? J. Sched. **10**, 153–166 (2007). https://doi.org/10.1007/s10951-007-0012-1
4. Ballestín, F., Leus, R.: Resource-constrained project scheduling for timely project completion with stochastic activity durations. Prod. Oper. Manag. **18**, 459–474 (2009). https://doi.org/10.1111/j.1937-5956.2009.01023.x
5. Bartz-Beielstein, T., Blum, D., Branke, J.: Particle swarm optimization and sequential sampling in noisy environments. In: Doerner, K.F., Gendreau, M., Greistorfer, P., Gutjahr, W., Hartl, R.F., Reimann, M. (eds.) Metaheuristics. ORSIS, vol. 39, pp. 261–273. Springer, Boston, MA (2007). https://doi.org/10.1007/978-0-387-71921-4_14
6. Bellman, R.: Dynamic programming. Science **153**(3731), 34–37 (1966). https://doi.org/10.1126/science.153.3731.34
7. Bianchi, L., Dorigo, M., Gambardella, L.M., Gutjahr, W.J.: A survey on metaheuristics for stochastic combinatorial optimization. Nat. Comput. **8**, 239–287 (2009). https://doi.org/10.1007/s11047-008-9098-4

8. Birge, J.R., Louveaux, F.: Introduction to Stochastic Programming. Springer, New York (2011). https://doi.org/10.1007/978-1-4614-0237-4
9. Bouneffouf, D., Rish, I., Aggarwal, C.: Survey on applications of multi-armed and contextual bandits. In: 2020 IEEE Congress on Evolutionary Computation (CEC), pp. 1–8. (2020). https://doi.org/10.1109/CEC48606.2020.9185782
10. Bulgak, A.A., Sanders, J.L.: Integrating a modified simulated annealing algorithm with the simulation of a manufacturing system to optimize buffer sizes in automatic assembly systems. In: 1988 Winter Simulation Conference Proceedings, pp. 684–690. (1988). https://doi.org/10.1109/WSC.1988.716241
11. Chen, Z., Demeulemeester, E., Bai, S., Guo, Y.: Efficient priority rules for the stochastic resource-constrained project scheduling problem. Eur. J. Oper. Res. **270**, 957–967 (2018). https://doi.org/10.1016/j.ejor.2018.04.025
12. Dumouchelle, J., Julien, E., Kurtz, J., Khalil, E.B.: Neur2ro: neural two-stage robust optimization. arXiv preprint (2023). https://doi.org/10.48550/ARXIV.2310.04345
13. Gelman, A., Carlin, J.B., Stern, H.S., Rubin, D.B.: Bayesian Data Analysis. Chapman and Hall/CRC, Boca Raton (1995). https://doi.org/10.1201/9780429258411
14. Groves, M., Branke, J.: Sequential sampling for noisy optimisation with CMA-ES. In: Proceedings of the 2018 Genetic and Evolutionary Computation Conference, pp. 1023–1030. Association for Computing Machinery, Inc., (2018). https://doi.org/10.1145/3205455.3205559
15. Juan, A.A., Faulin, J., Grasman, S.E., Rabe, M., Figueira, G.: A review of simheuristics: extending metaheuristics to deal with stochastic combinatorial optimization problems. Oper. Res. Perspect. **2**, 62–72 (2015). https://doi.org/10.1016/j.orp.2015.03.001
16. Juan, A.A., et al.: A review of the role of heuristics in stochastic optimisation: from metaheuristics to learnheuristics. Ann. Oper. Res. (2021). https://doi.org/10.1007/s10479-021-04142-9
17. Kirkpatrick, S., Gelatt, C.D., Vecchi, M.P.: Optimization by simulated annealing. Science **220**, 671–680 (1983). https://doi.org/10.1126/science.220.4598.671
18. Kleywegt, A.J., Shapiro, A., Homem-de-mello, T.: The sample average approximation method for stochastic discrete optimization. Soc. Ind. Appl. Math. **12**, 479–502 (2001). https://doi.org/10.1137/S1052623499363220
19. Kolisch, S.: Psplib a project scheduling problem library. Eur. J. Oper. Res. **96**, 205–216 (1996). https://doi.org/10.1016/S0377-2217(96)00170-1
20. Kolisch, R., Hartmann, S.: Heuristic algorithms for the resource-constrained project scheduling problem: classification and computational analysis. In: Weglarz, J. (eds.) Project Scheduling. International Series in Operations Research & Management Science, LNCS, vol. 14, pp. 147–178. Springer, Boston, MA (1999). https://doi.org/10.1007/978-1-4615-5533-9_7
21. Lattimore, T., Szepesvári, C.: Bandit Algorithms. Cambridge University Press, Cambridge (2020). https://doi.org/10.1017/9781108571401
22. Lei, H., Laporte, G., Guo, B.: The capacitated vehicle routing problem with stochastic demands and time windows. Comput. Oper. Res. **38**, 1775–1783 (2011). https://doi.org/10.1016/j.cor.2011.02.007
23. Liu, N., Truong, V.A., Wang, X., Anderson, B.R.: Integrated scheduling and capacity planning with considerations for patients' length-of-stays. Prod. Oper. Manag. **28**, 1735–1756 (2019). https://doi.org/10.1111/poms.13012
24. Min, D., Yih, Y.: Scheduling elective surgery under uncertainty and downstream capacity constraints. Eur. J. Oper. Res. **206**, 642–652 (2010). https://doi.org/10.1016/j.ejor.2010.03.014

25. Prudius, A.A., Andradóttir, S.: Averaging frameworks for simulation optimization with applications to simulated annealing. Nav. Res. Logist. **59**, 411–429 (2012). https://doi.org/10.1002/nav.21496

26. Rahimi, I., Gandomi, A.H.: A comprehensive review and analysis of operating room and surgery scheduling. Arch. Comput. Methods Eng. **28**, 1667–1688 (2021). https://doi.org/10.1007/s11831-020-09432-2

27. Ritzinger, U., Puchinger, J., Hartl, R.F.: A survey on dynamic and stochastic vehicle routing problems. Int. J. Prod. Res. **54**, 215–231 (2016). https://doi.org/10.1080/00207543.2015.1043403

28. Rostami, S., Creemers, S., Leus, R.: New strategies for stochastic resource-constrained project scheduling. J. Sched. **21**, 349–365 (2018). https://doi.org/10.1007/s10951-016-0505-x

29. Saliby, E.: Descriptive sampling: a better approach to Monte Carlo simulation. Source J. Oper. Res. Soc. **41**, 1133–1142 (1990). https://doi.org/10.2307/2583110

30. Schutte, N.: Codebase experiments sequential predictive sampling. https://github.com/NoahJSchutte/sequential-predictive-sampling. Accessed 21 Mar 2024

31. Schutte, N., van den Houten, K., Eigbe, E.: Dynamic scenario reduction for simulation based optimization under uncertainty, working notes of the data science meets optimisation workshop at IJCAI 2022. https://drive.google.com/file/d/1kxzgO8ZhW2bjXo1vwVskK4_5LNRbp5vj/view?usp=sharing. Accessed 21 Mar 2024

32. Seyyedabbasi, A.: A reinforcement learning-based metaheuristic algorithm for solving global optimization problems. Adv. Eng. Softw. **178**, 103411 (2023). https://doi.org/10.1016/j.advengsoft.2023.103411

33. Shehadeh, K.S.: Data-driven distributionally robust surgery planning in flexible operating rooms over a wasserstein ambiguity. Comput. Oper. Res. **146** (2022). https://doi.org/10.1016/j.cor.2022.105927

34. Shehadeh, K.S., Zuluaga, L.F.: 14th AIMMS-MOPTA optimization modeling competition 2022: surgery scheduling in flexible operating rooms under uncertainty. https://iccopt2022.lehigh.edu/competition-and-prizes/aimms-mopta-competition/. Accessed 21 Mar 2024

35. Solomon, M.M.: Algorithms for the vehicle routing and scheduling problems with time window constraints. Oper. Res. **35**, 254–265 (1987). https://doi.org/10.1287/opre.35.2.254

36. Vajda, S.: Mathematical Programming. Courier Corporation, Chelmsford (2009)

37. Zakaria, A., Ismail, F.B., Lipu, M.S., Hannan, M.A.: Uncertainty models for stochastic optimization in renewable energy applications. Renew. Energy **145**, 1543–1571 (2020). https://doi.org/10.1016/j.renene.2019.07.081

38. Zhu, S., Fan, W., Yang, S., Pei, J., Pardalos, P.M.: Operating room planning and surgical case scheduling: a review of literature. J. Comb. Optim. **37**, 757–805 (2019). https://doi.org/10.1007/s10878-018-0322-6

SMT-Based Repair of Disjunctive Temporal Networks with Uncertainty: Strong and Weak Controllability

Ajdin Sumic[1]([✉]), Alessandro Cimatti[2], Andrea Micheli[2], and Thierry Vidal[1]

[1] LGP/ENIT, Technical University of Tarbes, Tarbes, France
aidin.sumic@doctorant.uttop.fr
[2] Fondazione Bruno Kessler, Trento, Italy

Abstract. Temporal Networks with Uncertainty are a powerful and widely used formalism for representing and reasoning over temporal constraints in the presence of uncertainty. Since their introduction, they have been used in planning and scheduling applications to model situations where some activity durations or event timings are not under the control of the scheduling agent. Moreover, a wide variety of classes of temporal networks (depending on the types of constraints) have been defined, and many algorithms for dealing with these emerged.

We are interested in the repair problem, consisting of reconsidering the bounds of the uncertain durations when the network is not *controllable*, i.e., when no strategy for scheduling exists. This problem is important, as it allows for the explanation and negotiation of the problematic uncertainties. In this paper, we address the repair problem for a very expressive class of temporal networks, namely the Disjunctive Temporal Networks with Uncertainty. We use the Satisfiability Modulo Theory framework to formally encode and solve the problem, and we devise a uniform solution encompassing different "levels" of controllability, namely strong and weak. Moreover, we provide specialized encodings for important sub-classes of the problem, and we experimentally evaluate our approaches.

Keywords: Disjunctive Temporal Network with Uncertainty · Plan Repair · Satisfiability Modulo Theory · Strong and Weak Controllability

1 Introduction

Since their introduction in [11], Temporal Networks have been recognized as a fundamental tool to represent and reason about temporal dependencies among tasks and events. Several variants have been studied in depth: e.g. Simple, Disjunctive [4,21]. The important extension to temporal networks with uncertainty [25] allows to model tasks of uncontrollable duration. This yields the problem of controllability, i.e., devising a strategy that will work regardless of the uncertainty, that comes in different forms depending on the observability

B. Dilkina (Ed.): CPAIOR 2024, LNCS 14743, pp. 176–192, 2024.
https://doi.org/10.1007/978-3-031-60599-4_11

assumptions, i.e., Weak, Strong, and Dynamic controllability. Various algorithms have been proposed to *check* controllability, also based on constraints propagation [17, 25], Satisfiability Modulo Theories [3], and Timed Games [8].

Despite the widespread use of these formalisms in the literature, little attention has been devoted to the case in which a temporal network is deemed non-controllable.

In this case, the source of non-controllability could be unfeasible requirements (e.g., even with a less uncertain environment, it would still be impossible to schedule the controllable time points). Still, the assumptions about the environment could also be too loose given the constraints (i.e., if only an uncontrollable activity had stricter bounds, we could recover controllability). In this paper, we tackle the problem of *repair* of a Disjunctive Temporal Network with Uncertainty (DTNU) [24]. Starting from an uncontrollable network $\mathcal{D}$, we define the repair problem as finding a variant $\mathcal{D}'$ that is controllable and that is "sufficiently" close to $\mathcal{D}$. In particular, we consider the case where the $\mathcal{D}'$ variant is obtained from $\mathcal{D}$ by restricting the bounds on uncontrollable durations.

This problem is of theoretical importance but also has a strong practical relevance: consider the case of a multi-agent system, in which the uncontrollability bounds can be negotiated, for example, when the duration of an activity is considered to be uncontrollable for the scheduling agent, but it is, in fact, controllable for another agent. In a collaborative scenario, the agent can try to repair its own network by asking for stronger assumptions on the controlling agent instead of relaxing its objectives by creating a new controllable network. In other words, one agent could ask the controlling agent to reduce its flexibility in order to help in recovering controllability. Another application of the repair problem concerns explainability: by providing a (minimal) repair to an uncontrollable network, we suggest a strategy to change the constraints to recover controllability: this is an immediate counterfactual explanation [22] for the non-controllability and, even if the constraint is not negotiable, explanation allows to focus on where/what to replan.

We tackle the repair problem in the setting of Satisfiability Modulo Theories, proposing a solution for the cases of Weak and Strong controllability. We first define a general approach for DTNUs that uniformly works for both controllability levels: we define a quantified formula encoding all valid repairs for a given DTNU. Any model of such formula is a valid relaxation of the contingent bounds that recovers the controllability of the network. Furthermore, we prove that if the formula is unsatisfiable, then the problem admits no repair. Then, we use Optimization Modulo Theory [20] to select, among the possible repairs the one that sacrifices the least amount of flexibility. Second, we devise a specialized encoding for the specific but important case of Simple Temporal Networks with Uncertainty (STNU) (that is, when disjunctions are not allowed). By exploiting the convexity of the problem formulation, we obtain a much more efficient algorithm for synthesizing a repaired STNU.

We implemented the proposed approaches within the pySMT framework [12], using the Z3 solver as backend [10], and we experimentally evaluated them on a large set of non-controllable DTNU and STNU benchmarks. The empirical evaluation shows that the general case of DTNU repair may incur scalability issues, heavily dependent on the number of uncontrollables. On the other hand, the specialized algorithm leads to substantial increases in efficiency by leveraging the features of STNUs, hence reducing the cost of quantified reasoning.

Structure of the Paper. The next Section discusses the related work. In Sect. 3, we present the needed background and in Sect. 4 we define the problem of DTNU repair. In Sect. 5, we propose a general repair encoding for the general case of DTNU, while Sect. 6 tackles the special case of STNU. In Sect. 7, we empirically evaluate our approaches and in Sect. 8 we draw our conclusions and present directions for future work.

2   Related Work

The problem of repair for Disjunctive Temporal Networks under Uncertainty has never been tackled before. Most of the literature considered problems of checking controllability [13,15,16] and flexibility in execution [19]. The closest related works typically focus on STNUs and diagnosis, i.e., pinpointing reasons for non-controllability. For example, in [14], a diagnosis approach for the temporal checker is proposed for dynamic controllability. This approach is able to give the set of constraints involved in the non-DC of an STNU. Then, the *designer*(planner) is in charge of solving the problem. In the case where multiple contingents are involved, the repair will help the designer in deciding what to do by providing the best way to shrink the contingent to restore controllability.

The notion of repair arises in [6]. The authors propose a MILP approach to decouple an STNU into sub-networks (STNU), one per agent, in a multi-agent setting. The authors state that it should be possible to modify the encoding of the MILP approach so that it reduces the bounds of the contingents so that all sub-networks are DC. Then, in [2], the authors compute the volume space of an STNU to assess just how far from being controllable an uncontrollable STNU is by defining some metrics for SC and DC. Later, they propose an incomplete LP approach to repair a non-DC STNU by repairing the negative cycles [1]. However, it is incomplete because it doesn't consider inter-dependence between negative cycles.

In this paper, we formally define the repair problem as tightening the contingent constraints and propose a series of SMT encodings for synthesizing valid repairs. Unlike the literature reported above, we consider the very general case of DTNUs (instead of limiting ourselves to STNUs) and uniformly tackle both strong and weak controllability. Moreover, we provide specialized encodings for the STNU case.

3   Background

3.1   Satisfiability and Optimization Modulo Theory

The Satisfiability Modulo Theory (SMT) [3] is the problem of deciding whether there exists a model that satisfies a first-order formula ϕ (i.e., an assignment to the free variables in ϕ such that ϕ is satisfied. Given the formula $\phi_1 = (x <= y) \wedge (x + y = 10)$ a valid model of ϕ_1 is $\{x := 4, y := 6\}$, where $x, y \in \mathbb{R}$.

SMT solvers can support multiple *Theories* such as Linear Real Arithmetic (LRA). In LRA a formula is a Boolean combination, or a universal and existential quantification (respectively $\forall$, $\exists$) of atoms, which are linear constraints in the form: $\sum_i a_i x_i \bowtie c$ where $\bowtie \in \{>, <, \leqslant, \geqslant, \neq, =\}$, every x_i is a real variable and every a_i, c are real constants. QF_LRA denotes the quantifier-free fragment of LRA.

Optimization Modulo Theory (OMT) generalizes SMT with optimization procedures to find a model of ϕ that is optimal for an objective function f (or a combination of multiple objective functions) under all models of a formula ϕ. The objective function f can be expressed as a term in different theories, but this paper focuses only on objective functions expressed as QF_LRA terms. In the literature, several OMT solvers exist, such as OptiMathSAT [20] and Z3 [5].

3.2 Temporal Networks (With Uncertainty)

A Temporal Network (TN) is a formalism that is used to represent temporal constraints over time-valued variables called time points. Two families of TN exist in the literature: TN that does not consider uncertainty, to which the scheduler freely assigns all time points [11,23], and TN with Uncertainty (TNU), where only a subset of time points are assigned by the scheduler, while another exogenous entity decides the others [24,25]. This paper focuses on the Disjunctive Temporal Network with Uncertainty (DTNU) and Simple Temporal Network with Uncertainty (STNU), which is a restricted case of DTNU.

Definition 1. *A **DTNU** is a tuple $\langle \mathcal{V}, \mathcal{E}, \mathcal{C} \rangle$ where:*

- *$\mathcal{V}$ is a set of time points, partitioned into controllable ($\mathcal{V}_c$) and uncontrollable ($\mathcal{V}_u$);*
- *$\mathcal{E}$ is a set of free constraints (requirement constraints): each constraint e_i is of the form, $\bigvee_{j=1}^{D_i} v_{1j} - v_{2j} \in [L_{ij}, U_{ij}]$, for some $v_{1j}, v_{2j} \in \mathcal{V}$ and $L_{ij}, U_{ij} \in \mathbb{R} \cup \{+\infty, -\infty\}$;*
- *$\mathcal{C}$ is a set of contingent constraints: each $c_i \in \mathcal{C}$ is of the form, $\langle b_i, \mathcal{B}_i, d_i \rangle$, where $b_i \in \mathcal{V}_c$, $d_i \in \mathcal{V}_u$, and $\mathcal{B}_i$ is a set of pairs $\langle L_{ij}, U_{ij} \rangle$ such that $0 \leqslant L_{ij} \leqslant U_{ij} < \infty$, $j \in [1, |\mathcal{B}_i|]$; and for any distinct pairs, $\langle L_{ij}, U_{ij} \rangle$ and $\langle L_{ik}, U_{ik} \rangle \in \mathcal{B}_i$, either $L_{ij} > U_{ik}$ or $U_{ij} < L_{ik}$.*

In a DTNU, free constraints are disjunctions of intervals, each between any time points, possibly overlapping, while a contingent is a disjunction of non-overlapping intervals restricted to a single pair of time points.

Intuitively, controllable time points ($\mathcal{V}_c$) are moments in time to be decided by the scheduling agent, which is trying to satisfy all the free constraints ($\mathcal{E}$) under any possible instantiation of the uncontrollable time points ($\mathcal{V}_u$) respecting the contingent constraints ($\mathcal{C}$).

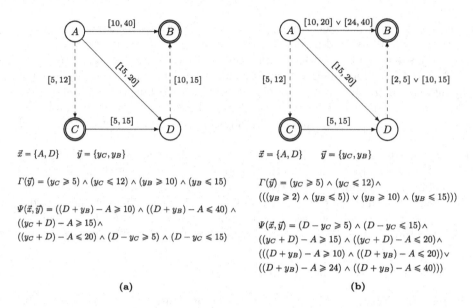

$\vec{x} = \{A, D\}$ $\vec{y} = \{y_C, y_B\}$

$\Gamma(\vec{y}) = (y_C \geqslant 5) \wedge (y_C \leqslant 12) \wedge (y_B \geqslant 10) \wedge (y_B \leqslant 15)$

$\Psi(\vec{x}, \vec{y}) = ((D + y_B) - A \geqslant 10) \wedge ((D + y_B) - A \leqslant 40) \wedge$
$((y_C + D) - A \geqslant 15) \wedge$
$((y_C + D) - A \leqslant 20) \wedge (D - y_C \geqslant 5) \wedge (D - y_C \leqslant 15)$

(a)

$\vec{x} = \{A, D\}$ $\vec{y} = \{y_C, y_B\}$

$\Gamma(\vec{y}) = (y_C \geqslant 5) \wedge (y_C \leqslant 12) \wedge$
$(((y_B \geqslant 2) \wedge (y_B \leqslant 5)) \vee (y_B \geqslant 10) \wedge (y_B \leqslant 15)))$

$\Psi(\vec{x}, \vec{y}) = (D - y_C \geqslant 5) \wedge (D - y_C \leqslant 15) \wedge$
$((y_C + D) - A \geqslant 15) \wedge ((y_C + D) - A \leqslant 20) \wedge$
$(((D + y_B) - A \geqslant 10) \wedge ((D + y_B) - A \leqslant 20)) \vee$
$((D + y_B) - A \geqslant 24) \wedge ((D + y_B) - A \leqslant 40)))$

(b)

Fig. 1. Graph representations of the STNU (a) and DTNU (b) examples, together with their SMT encodings. Nodes represent time points, doubly circled nodes are uncontrollable; solid edges are free constraints while dashed edges are contingent constraints.

Definition 2. *An **STNU** is a DTNU without disjunctions. Hence, an STNU is a DTNU $\langle V, \mathcal{E}, \mathcal{C} \rangle$ where $D_i = 1$ for all free constraints and $|\mathcal{B}_i| = 1$ for all contingent constraints.*

Example 1. Bob is working in a production line and has 40 min to build a product. This product is done in 3 phases: first Bob needs to wait for a machine to transform raw materials into a component of the final product. This task performed by the machine has an uncontrollable duration that lasts between 5 to 12 min. Then, Bob needs to compare the quality of the component with some regulation (size, weight, etc.) and he can decide to take 5 to 15 min to do the checking. However, Bob knows that to be efficient the two tasks need to be performed between 15 to 20 min. Then, he must give the component to another machine that will create the final product. This task is an uncontrollable event that lasts between 10 to 15 min. An STNU representing the example is shown in Fig. 1a. For the sake of the example, we suppose that the product needs at least some minimal amount of time to be produced; therefore we put a minimum duration of 10. A DTNU version is shown in Fig. 1b, where Bob can produce two products using two different machines: the first product that needs to be produced in 20 min with the first robot that can produce it in 2 to 5 min, the second needs to be produced between 24 to 40 min with the second machine that can produce it in 10 to 15 min. We will explain the encoding of the two examples in Sect. 3.3.

The basic computational problem arising given a DTNU is controllability [25]: we want to check whether there exists a strategy for scheduling the controllable time points that will respect all the free constraints under any possible realization of the uncontrollable time points respecting the contingent constraints. However, depending on the assumptions on the observability of uncontrollable time points, different *levels* of controllability have been defined. In this paper, we focus on strong and weak controllability, corresponding to the cases of no-observation and clairvoyant observation, respectively. In the following, we report the basic definitions for these two cases.

Definition 3. *A **controllable assignment** $\delta : \mathcal{V}_c \to \mathbb{R}$ of a DTNU $\langle \mathcal{V}, \mathcal{E}, \mathcal{C} \rangle$ is a mapping from all the controllable time points to real values.*

Definition 4. *A **DTN** $\mathcal{D} = \langle \mathcal{V}, \mathcal{E}, \varnothing \rangle$ is a DTNU with no contingent constraints (and no uncontrollable time points: $\mathcal{V}_c = \mathcal{V}$, $\mathcal{V}_u = \varnothing$). A **consistent schedule** for a DTN $\mathcal{D}$ is a controllable assignment δ that satisfies each free constraint e_i in $\mathcal{E}$:*

$$\bigvee_{j=1}^{D_i} \delta(v_{1j}) - \delta(v_{2j}) \in [L_{ij}, U_{ij}].$$

Definition 5. *given a DTNU $\mathcal{D} = \langle \mathcal{V}, \mathcal{E}, \mathcal{C} \rangle$, let $m = |\mathcal{V}_u|$. The **situations** of $\mathcal{D}$ is a set of tuples $\Omega_\mathcal{D}$ defined as the cartesian product of:*

$$\underset{\langle b_i, \mathcal{B}_i, d_i \rangle \in \mathcal{C}}{\times} \left(\underset{\langle L_{ij}, U_{ij} \rangle \in \mathcal{B}_i}{\bigcup} [L_{ij}, U_{ij}] \right).$$

*A **situation** is an element ω of $\Omega_\mathcal{D}$ and we write $\omega(c)$ with $c \in \mathcal{C}$ to indicate the element in ω associated with c in the cross product.*

Intuitively, the set of situations defines the region of uncertainty, all the allowed durations of contingent constraints. A network will be deemed controllable if (with the provided observations) it is possible to schedule the controllable time points so that all free constraints are satisfied in any possible situation.

Definition 6. *Given a DTNU $\mathcal{D} = \langle \mathcal{V}, \mathcal{E}, \mathcal{C} \rangle$ and a situation ω, the **projection** $\mathcal{D}_\omega$ is the DTN $\langle \mathcal{V}, \mathcal{E} \cup \mathcal{C}_\omega, \varnothing \rangle$ where $\mathcal{C}_\omega = \{d_i - b_i \in [\omega(c), \omega(c)] \mid c \in \mathcal{C} \text{ and } c = \langle b_i, \mathcal{B}_i, d_i \rangle\}$.*

Intuitively, the projection $\mathcal{D}_\omega$ substitutes all the contingent links with requirements, forcing the duration of a contingent link to the value indicated in ω.

Definition 7. *A DTNU $\mathcal{D}$ is **weakly controllable** iff for all $\omega \in \Omega$, there exists a consistent schedule δ for $\mathcal{D}_\omega$.*

In weak controllability, we assume that the uncontrollable durations are decided before execution, and therefore, the scheduling agent can condition its decisions on the situation.

Definition 8. *A DTNU $\mathcal{D}$ is **strongly controllable** iff there exists a consistent schedule δ, for all $\mathcal{D}_\omega$ where $\omega \in \Omega$.*

This level of controllability implies that the controllable schedule is so robust that it's possible to fix the execution of the controllable time points in time so that it carries any execution of the uncontrollable time point observed at execution time.

A third level of controllability exists, dynamic controllability, but it's not in the scope of this paper. However, many works in the literature tackled this level of controllability, which is of great practical importance [13,16,17].

3.3   Controllability Checking Using SMT

In this section, we present the basic SMT encodings devised in [7] and [9], targeting strong and weak controllability, respectively. We limit ourselves to the basic formulations, as these are the only requirements for our paper. Still, the authors provide different variations of these encodings that turn out to be more efficient than the basic formulation for the sake of checking controllability.

Definition 9. *Given a DTNU $\mathcal{D} = \langle \mathcal{V}, \mathcal{E}, \mathcal{C} \rangle$, we define the following sets of SMT variables:*

- *the **uncontrollable SMT durations** $\vec{y}$ is a set of real SMT variables, one for each contingent constraint $c_i \in C$ (we write y_i to indicate the variable corresponding to c_i);*
- *the **controllable SMT timepoints** $\vec{x}$ is a set of real SMT variables, one for each controllable time point (we write x_v to indicate the variable corresponding to $v \in \mathcal{V}_c$).*

We also define the following basic SMT formulae:

- *$\Gamma(\vec{y})$ representing the **SMT contingent constraints**:*

$$\Gamma(\vec{y}) \doteq \bigwedge_{c_i \doteq \langle b_i, \mathcal{B}_i, d_i \rangle \in C} \left(\bigvee_{\langle L_{ij}, U_{ij} \rangle \in \mathcal{B}_i} (y_i \geq L_{ij} \wedge y_i \leq U_{ij}) \right) \tag{1}$$

- *$\Psi(\vec{x}, \vec{y})$ representing the **SMT free constraints** (requirements):*

$$\Psi(\vec{x}, \vec{y}) \doteq \bigwedge_{e_i \in \mathcal{E}} (((e_i[(b_1 + y_1)/d_1])[(b_2 + y_2)/d_2]) \dots [(b_m + y_m)/d_m]) \tag{2}$$

where $m = |\mathcal{C}|$ and the logical notation $\phi[a/b]$ indicates the substitution of the term b with term a in the formula ϕ.

In this approach, each uncontrollable time point $d_i \in \mathcal{V}_u$ is encoded by its difference with the starting time point $b_i \in \mathcal{V}_c$ represented by a universally quantified variable $y_i \in \mathbb{R}$ such that: $d_i - b_i \in [L_{ij}, U_{ij}]$, and $y_i \in [L_{ij}, U_{ij}]$. Thus, y_i represents the duration of the interval $[b_i, d_i]$ and the execution of d_i is $(b_i + y_i)$. We define $\vec{x}$, $\vec{y}$ as the sets of controllable time points and uncontrollable durations, respectively. Therefore, the rewriting of the contingent constraints depends only on $\vec{y}$. Next, $\Gamma(\vec{y})$ is the formula representing the conjunction of all the contingent constraints over the uncontrollable durations, and $\Phi(\vec{x}, \vec{y})$ is the formula representing the conjunction of all free constraints over $\vec{x}$ and $\vec{y}$. We show in Figs. 1a and 1b this encoding for the running examples.

Given the formulae presented in Definition 9, the authors of [7] prove that a DTNU $\mathcal{D}$ is weakly controllable if the following formula Φ_{weak} is valid.

$$\Phi_{weak} \doteq \forall \vec{y}.\exists \vec{x}.\Gamma(\vec{y}) \rightarrow \Psi(\vec{x}, \vec{y}) \tag{3}$$

Moreover, in [9], they prove that $\mathcal{D}$ is strongly controllable if the following formula ϕ_{strong} is valid.

$$\Phi_{strong} \doteq \exists \vec{x}.\forall \vec{y}.\Gamma(\vec{y}) \rightarrow \Psi(\vec{x}, \vec{y}) \tag{4}$$

We will use these encodings as the basis for our repair problem formulation.

4   DTNU Repair: Problem Definition

The concept of repair in DTNU arises when the network is not controllable; in these situations, we want to find a tightening (if it exists) of the bounds of the contingent constraints such that controllability is recovered. This is useful as a counterfactual explanation for non-controllability and in multi-agent applications where contingents are controlled by other agents and can be negotiated. In the following, we formalize the repair problem.

Definition 10. Let $\mathcal{D} = \langle \mathcal{V}, \mathcal{E}, \mathcal{C} \rangle$ be a not τ-controllable DTNU, with $\tau = \{weak, strong\}$. The τ-**repair problem** consists in finding a τ-controllable DTNU $\mathcal{D}' = \langle \mathcal{V}, \mathcal{E}, \mathcal{C}' \rangle$ such that $\mathcal{C}' = \{\langle b_i, \mathcal{B}'_i, d_i \rangle \mid \langle b_i, \mathcal{B}_i, d_i \rangle \in \mathcal{C}\}$ and $\mathcal{B}'_i = \{\langle L'_{ij}, U'_{ij} \rangle \mid \langle L_{ij}, U_{ij} \rangle \in \mathcal{B}_i$ and $L_{ij} \leqslant L'_{ij} \leqslant U'_{ij} \leqslant U_{ij}\}$.

Intuitively, given a DTNU $\mathcal{D}$ which is not weakly (resp. strongly) controllable, we want to find a DTNU $\mathcal{D}'$ which is weakly (resp. strongly) controllable. $\mathcal{D}'$ must be identical to $\mathcal{D}$ except for the bounds of the contingent constraints, which can be restricted (but not enlarged).

For example, consider a DTNU $\mathcal{D} = \langle \mathcal{V}, \mathcal{E}, \mathcal{C} \rangle$ identical to the DTNU depicted in Fig. 1b. This DTNU is not controllable, because the DTN projection $\mathcal{D}_\omega$ where $\omega(c_1) = 12$ and $\omega(c_2) = 4$ (c_1 being the contingent AC and c_2 the contingent DB) does not admit a consistent schedule. A solution to the weak-repair problem for $\mathcal{D}$ is a DTNU $\mathcal{D}' = \langle \mathcal{V}, \mathcal{E}, \mathcal{C}' \rangle$ such that c_1 is replaced by $c'_1 = (A, \{\langle 5, 8 \rangle\}, C)$. $\mathcal{D}'$ is one possible solution to the weak-repair problem.

We are interested in repair solutions that minimize the reduction in the size of the contingent constraints' bounds. This intuitively corresponds to minimizing the flexibility for scheduling uncontrollable time points that are removed by the repair.

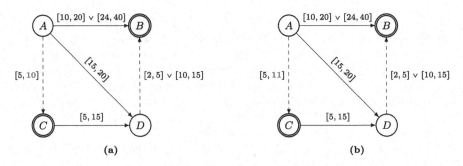

Fig. 2. Graph representations of the DTNU (a) an optimal solution to the strong-repair problem of $\mathcal{D}$, and DTNU (b) for the weak-repair problem

Definition 11. *Let* $\mathcal{D} = \langle \mathcal{V}, \mathcal{E}, \mathcal{C} \rangle$, *be a non* τ-*controllable DTNU and let* $\mathcal{R}_\mathcal{D}$ *be the set of all the solutions to the* τ-*repair problem for* $\mathcal{D}$. *An **optimal** τ-**repair** for D is defined as:*

$$\operatorname*{argmin}_{\mathcal{D}' \in \mathcal{R}_\mathcal{D}} \left(\sum_{\langle b_i, \mathcal{B}_i, d_i \rangle \in \mathcal{C}} \sum_{\langle L_{ij}, U_{ij} \rangle \in \mathcal{B}_i} ((L'_{ij} - L_{ij}) + (U_{ij} - U'_{ij})) \right)$$

We show in Fig. 2, the optimal repair of $\mathcal{D}$ for strong and weak controllability.

5   DTNU Repair: SMT Encoding

The base encoding presented in Sect. 2 encodes a DTNU into an SMT model to check weak and strong controllability: in fact, all the encodings presented in [7,9] are focused only on the problem of checking if a given DTNU is τ-controllable. To tackle the repair problem, we need to synthesize the new uncontrollable bounds. Hence, we need to extend the encoding to have the uncontrollable bounds as free variables so that the solver can assign concrete values to the uncontrollable bounds, as detailed below.

Compared to the basic encoding for controllability checking, we add two types of variables. We denote the original bounds of the i-th contingent constraints with L_{ij} and U_{ij} as in Definition 1. Then, for each such pair of contingent bounds, we introduce two free variables l_{ij} and u_{ij}; these represent the repaired bounds of the i-th contingent and will be constrained such that $L_{ij} \leqslant l_{ij} \leqslant u_{ij} \leqslant U_{ij}$. To simplify the notation, we indicate with $\vec{l}, \vec{u}$ the set of free variables for the lower and upper bounds, respectively. We redefine the formula for the contingent constraints in Definition 9 (indicated as $\Gamma'(\vec{y}, \vec{l}, \vec{u})$) which now depend not only on $\vec{y}$ but also on $\vec{l}$ and $\vec{i}$; instead, Ψ remains the same. With these new formulae, we will define the encodings for tackling the strong and weak repair problems for DTNU. In the following definition, we formalize these basic components of the encoding.

Definition 12. *Given a DTNU $\mathcal{D} = \langle \mathcal{V}, \mathcal{E}, \mathcal{C} \rangle$, we define the following sets of SMT variables:*

- *$\vec{y}$ and $\vec{x}$ as in Definition 9.*
- *the **uncontrollable SMT lower bounds** $\vec{l}$ is a set of real SMT variables, one for each contingent constraint bound $L_{ij} \in \mathcal{B}_i$. We denote l_{ij}, the variable associated with L_{ij}.*
- *the **uncontrollable SMT upper bounds** $\vec{u}$ is a set of real SMT variables, one for each contingent constraint bound $U_{ij} \in \mathcal{B}_i$. We denote u_{ij}, the variable associated with U_{ij}.*

We also define the following SMT formulae:

- *$\Gamma'(\vec{y}, \vec{l}, \vec{u})$ representing the **SMT contingent constraints**:*

$$\Gamma'(\vec{y}, \vec{l}, \vec{u}) \doteq \bigwedge_{c_i \doteq \langle b_i, \mathcal{B}_i, d_i \rangle \in \mathcal{C}} \left(\bigvee_{\langle L_{ij}, U_{ij} \rangle \in \mathcal{B}_i} (y_i \geq l_{ij} \wedge y_i \leq u_{ij}) \right) \tag{5}$$

- *$\Psi(\vec{x}, \vec{y})$ as in Definition 9.*
- *$\Xi(\vec{l}, \vec{u})$ representing the **valid repair SMT constraints**:*

$$\Xi(\vec{l}, \vec{u}) \doteq \bigwedge_{\langle b_i, \mathcal{B}_i, d_i \rangle \in \mathcal{C}} \bigwedge_{\langle L_{ij}, U_{ij} \rangle \in \mathcal{B}_i} L_{ij} \leq l_{ij} \leq u_{ij} \leq U_{ij} \tag{6}$$

Considering the DTNU example in Fig. 1b, we report the new variables and formulae below.

$$\vec{l} = \{l_{C1}, l_{B1}, l_{B2}\}$$
$$\vec{u} = \{u_{C1}, u_{B1}, l_{B2}\}$$

$$\Gamma'(\vec{y}, \vec{l}, \vec{u}) = (y_C \geq l_{C1}) \wedge (y_C \leq u_{C1}) \wedge$$
$$(((y_B \geq l_{B1}) \wedge (y_B \leq u_{B1})) \vee ((y_B \geq l_{B2}) \wedge (y_B \leq u_{B2})))$$

$$\Xi(\vec{l}, \vec{u}) = (5 \leq l_{C1} \leq u_{C1} \leq 10) \wedge (2 \leq l_{B1} \leq u_{B1} \leq 5) \wedge (10 \leq l_{B2} \leq u_{B2} \leq 15)$$

The new variable l_{C1} indicates the repaired lower bound of the contingent constraint $\langle A, \{\langle 5, 12 \rangle\}, C \rangle$; for the other contingent constraint in the example, we need two variables l_{B1} and l_{B2} because the constraint is disjunctive. Similarly for the upper bounds.

With these basic formulae, we can now define two encodings (called "quantified encodings"), one for the weak repair problem (Θ_{weak}) and one for the strong one (Θ_{strong}).

$$\Theta_{weak}(\vec{l}, \vec{u}) \doteq (\forall \vec{y}. \exists \vec{x}. \Gamma'(\vec{y}, \vec{l}, \vec{u}) \rightarrow \Psi(\vec{x}, \vec{y})) \wedge \Xi(\vec{l}, \vec{u}) \tag{7}$$

$$\Theta_{strong}(\vec{x}, \vec{l}, \vec{u}) \doteq \left(\forall \vec{y}. \Gamma'(\vec{y}, \vec{l}, \vec{u}) \rightarrow \Psi(\vec{x}, \vec{y})\right) \wedge \Xi(\vec{l}, \vec{u}) \tag{8}$$

Note that for strong controllability, we do not need to quantify the controllable time points because by solving the SMT (or OMT) problem, we implicitly perform an existential quantification. Both encodings will have the property that all the models represent valid repairs for the input DTNU network, encoded in the values of the $\vec{l}$ and $\vec{u}$ variables. Suppose μ is a model for Θ_τ, then the DTNU with the new contingent constraints $\mathcal{C}'$ defined below is a valid τ-repair.

$$\mathcal{C}' \doteq \{\langle b_i, \{\langle \mu(l_{i,1}), \mu(u_{i,1})\rangle, \langle \mu(l_{i,2}), \mu(u_{i,2})\rangle, \ldots\}, d_i\rangle \mid \langle b_i, \mathcal{B}_i, d_i\rangle \in \mathcal{C}\}$$

Intuitively, we simply consider the value of l_{ij} in the model μ as the repaired value for L_{ij} and equivalently for the upper bounds.

Proposition 1. *Let $\mathcal{D} = \langle \mathcal{V}, \mathcal{E}, \mathcal{C}\rangle$ be a DTNU such that $\mathcal{D}$ is not τ-controllable. Any model of Θ_τ yields a solution to the τ-repair problem and if the formula is unsatisfiable, the τ-repair problem admits no solution.*

It is easy to see why Proposition 1 holds: both the encodings are direct derivations from the ones for checking strong and weak controllability in [9] and [7], where we changed the constant bounds into SMT variables. Therefore, Proposition 1 follows from the correctness of these basic encodings.

Thanks to the guarantees of Proposition 1, we can use Optimization Modulo Theory to solve the optimal τ-repair problem by simply imposing an optimization objective over the τ-repair encodings.

$$\text{minimize} \sum_{\langle b_i, \mathcal{B}_i, d_i\rangle \in \mathcal{C}} \sum_{\langle L_{ij}, U_{ij}\rangle \in \mathcal{B}_i} ((l_{ij} - L_{ij}) + (U_{ij} - u_{ij})) \tag{9}$$

$$\text{s.t. } \Theta_\tau$$

This problem formulation can be solved by any OMT solver capable of dealing with LRA formulations such as Z3 [10]. Alternatively, it is possible to apply quantifier-elimination techniques to construct a quantifier-free formula equivalent to Ω_τ to be used by any OMT solver for QF_LRA.

6   STNU Repair: SMT Encoding

In this Section, we tackle the problem of temporal network repair in the special case of STNU. In fact, the general encoding for DTNU requires the use of quantifiers that are decidable in LRA but bring a very high computational cost. To optimize the encoding for STNU, we removed the quantification of the variables in $\vec{y}$ exploiting the convexity of the problem, as proved in [25], considering all combinations of the lower and upper bounds of the contingents is enough to check the controllability of an STNU. This allows us to fix the duration of contingent constraints to either their lower bound (in $\vec{l}$) or upper bound (in $\vec{u}$) and disregard the values within the interval because the problem convexity guarantees that the problem is controllable if it is controllable "on the bounds." In the following, we formalize the basic components of the encoding for STNU.

Definition 13. *Given an STNU $\mathcal{D} = \langle \mathcal{V}, \mathcal{E}, \mathcal{C} \rangle$, we define the following sets of SMT variables:*

- *$\vec{x}$ as in Definition 9.*
- *$\vec{l}$ and $\vec{u}$ as in Definition 12*
- *the **set of symbolic boundary projections** $\mathcal{P}(\vec{l}, \vec{u})$ is a set of vectors of SMT variables, one for each possible combination of lower and upper bound for each contingent constraint (hence a total of $2^{\mathcal{C}}$ vectors):*

$$\mathcal{P}(\vec{l}, \vec{u}) = \{\vec{v} \mid v_i \in \{l_{i1}, u_{i1}\}, c_i \in \mathcal{C}\}$$

- *a set of "fresh" real SMT variables[1], one for each element $\vec{v}$ of $\mathcal{P}(\vec{l}, \vec{u})$, indicated as $\vec{x}^{\vec{v}}$. We indicate the set of all these free variables as $\vec{x}^P$.*

By exploiting the convexity of the STNU fragment, we can re-formulate the encodings of the strong and weak repair as follows.

$$\Theta_{weak}(\vec{x}^P, \vec{l}, \vec{u}) = \left(\bigwedge_{\vec{v} \in \mathcal{P}(\vec{l}, \vec{u})} \Psi(\vec{x}^{\vec{v}}, \vec{v}) \right) \wedge \Xi(\vec{l}, \vec{u}) \tag{10}$$

$$\Theta_{strong}(\vec{x}, \vec{l}, \vec{u}) = \left(\bigwedge_{\vec{v} \in \mathcal{P}(\vec{l}, \vec{u})} \Psi(\vec{x}, \vec{v}) \right) \wedge \Xi(\vec{l}, \vec{u}) \tag{11}$$

We call this the "on-bounds" encoding. The key intuition here is that for the STNU fragment, we can ensure the controllability of the network by checking that a strategy for scheduling controllable time points exists for every possible combination of the bounds. The difference between strong and weak controllability here becomes evident: in Θ_{strong} we have one variable for each controllable time point ($\vec{x}$), and we need a single assignment that works for all the combinations of lower and upper bounds; in weak controllability we have one vector of controllable time points variables ($\vec{x}^{\vec{v}}$) for every combination $\vec{v}$ of lower and upper bounds, and therefore the solver can assign different values to the controllable for every bound combination.

We remark that this encoding only works in the STNU fragment but retains all the properties of Proposition 1 and can be used for solving the optimal repair problem analogously to the DTNU case.

Finally, we remark that another strength of our proposal is that both the quantified and the on-bounds encodings can compute a schedule for the controllable time points in the strong controllability case.

---

[1] In SMT jargon, a fresh variable is a new variable with a name that is unused in the rest of the encoding and with no additional constraints.

7   Experiments

In this Section, we empirically evaluate the effectiveness of the proposed approaches. We implemented all the encodings in Python using the pySMT framework [12]. For our experiments, we use the Z3 solver as the backend.

We experimented on a large set of DTNU and STNU benchmarks. For the DTNU case, we consider the benchmark set of random networks used by Osanlou et al. in [18], which is composed of 1500 DTNU, 500 of them of the size between 10 to 20, 500 between 20 to 25, and 500 between 25 to 30 time points.

For the STNU case, we implemented a custom random generator: we create an STNU in the form of a complete directed acyclic graph (DAG); then, we randomly and safely remove some of the edges according to the following parameters:

- the number of time points n;
- the number of uncontrollable time points set by a range $m = [min_m, max_m]$, i.e., it creates an STNU with $|\mathcal{V}_u| = k$ with $k \in [min_m, max_m]$;
- the contingency rate per edge, i.e., the probability for a constraint to become a contingent. We keep this rate low to scatter the contingent in the graph (less than 10%);
- the rate of removal q, i.e., the probability for a constraint to be removed representing the sparsity of the graph.

We generated STNUs of different sizes: $n \in \{5, 10, 20, 50, 100\}$ and with different degrees of uncontrollability depending on the size: $m = \{[1, 2], [2, 3], [3, 4], [4, 5], [5, 6], [8, 10]\}$. The combinations are as follows: n = 5 with $\{[1,2], [2,3]\}$, n = 10 with $\{[2,3], [3,4]\}$, n = 20 with $\{[3,4], [4,5]\}$, n = 50 with $\{[5,6], [8,10]\}$, and n = 100 with $\{[8,10]\}$. Finally, for each relevant combination of n and m, we change the density of the graph with $q = \{0\%, 30\%, 50\%, 75\%, 90\%\}$ to get a benchmark set of 1100 non-controllable STNUs.

We performed a total of 7400 tests: 2 * 1500 for weak and strong repairs on DTNUs and 4 * 1100 on STNUs for both the quantified and on-bounds encodings and both controllability levels. We ran all the experiments on an Xeon E5-2620 2.10 GHz with 3600 s/10 GB time/memory limits.

Figure 3 shows the results of the encodings for DTNUs (in logarithmic scale) with the number of instances solved on the x-axis and the time on the y-axis. The plot shows that the strong-repair problem is much easier to solve for our encodings than the weak-repair: we could solve the strong-repair for all the instances, while the weak one approximately solves the easiest 500 instances. This is not surprising, as in Θ_{strong}, we only have one existential quantifier, while in Θ_{weak}, we have a quantifier alternation.

The cactus plot for STNU instances is shown in Fig. 4. We compare the quantified and on-bounds encodings for strong (left) and weak (right) repair. In general, the on-bounds encoding for STNU can solve many more instances than the quantified one. In Fig. 5, we also report scatter plots for strong and weak repair that highlight dominance of the on-bounds encoding for this class of problems. Moreover, as shown in the scatter plots, we never hit the memory limit

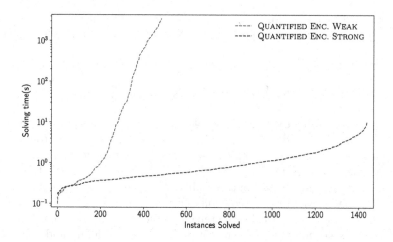

Fig. 3. DTNU cactus plot of weak and strong repair: for each encoding, we plot the solving time for the instances sorted from the easiest to the hardest.

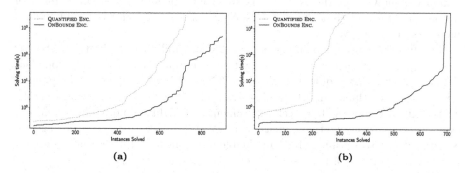

Fig. 4. STNU cactus plots for the strong (a) and weak (b) repair problems.

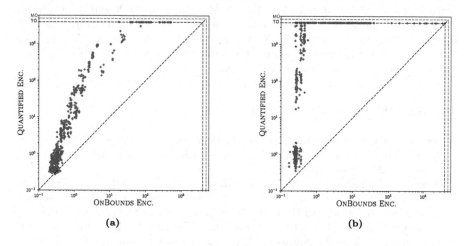

Fig. 5. STNU scatter plots for the strong (a) and weak (b) repair problems.

with our encodings, the limiting dimension is time, while memory consumption is not an issue (by manual inspection, we report that we never exceed 2 GB of memory usage).

8   Conclusions and Future Work

In this paper, we tackled the repair problem in temporal networks under uncertainty. The problem is defined as finding a variant of an uncontrollable network that is obtained by reducing the bounds of the contingent constraints and is controllable. We presented two SMT-based algorithms, tackling the specific problems of Weak and Strong controllability. The first one is a general encoding for DTNUs that works for any controllability that can be encoded into SMT. The second one efficiently tackles the repair problem for the weak controllability of STNU by exploiting the problem convexity. We also define and solve using OMT the optimal repair problem, that is finding the repair that removes the least amount of flexibility from the original temporal network. The experimental evaluation shows that the general case is heavily dependent on the number of uncontrollables and that the STNU specialized algorithm leads to substantial increases in efficiency.

For future work, we plan to explore the possibility of tackling the STNU repair problem using constraint propagation techniques. Using propagation algorithms, the checking problem for SC and DC on STNU has proven to be polynomial. Such algorithms can infer a negative cycle if the STNU is not controllable. Thus, fixing all the negative cycles incrementally might be more efficient. Moreover, we plan to apply these ideas to a multi-agent setting where an agent's contingent constraint might be controllable decisions for another, and the repair problem can serve as a means for negotiation among agents.

Acknowledgements. Andrea Micheli and Alessandro Cimatti have been partly supported by the PNRR project FAIR - Future AI Research (PE00000013) and by the Interconnected Nord-Est Innovation Ecosystem (iNEST) (ECS00000043), under the NRRP MUR program funded by the NextGenerationEU. This manuscript reflects only the authors' views and opinions, neither the European Union nor the European Commission can be considered responsible for them.

References

1. Akmal, S., Ammons, S., Li, H., Gao, M., Popowski, L., Boerkoel Jr, J.C.: Quantifying controllability in temporal networks with uncertainty. Artif. Intell. **289**, 103384 (2020). ISSN 0004-3702
2. Akmal, S., Ammons, S., Li, H., Boerkoel Jr, J.C.: Quantifying degrees of controllability in temporal networks with uncertainty. In: Proceedings of the International Conference on Automated Planning and Scheduling, vol. 29, pp. 22–30 (2019)
3. Barrett, C., Tinelli, C.: Satisfiability Modulo Theories. In: Handbook of Model Checking, pp. 305–343. Springer, Cham (2018). https://doi.org/10.1007/978-3-319-10575-8_11

4. Bettini, C., Wang, X.S., Jajodia, S.: Solving multi-granularity temporal constraint networks. Artif. Intell. **140**(1–2), 107–152 (2002)
5. Bjørner, N., Phan, A.-D., Fleckenstein, L.: νZ - an optimizing SMT solver. In: Baier, C., Tinelli, C. (eds.) TACAS 2015. LNCS, vol. 9035, pp. 194–199. Springer, Heidelberg (2015). https://doi.org/10.1007/978-3-662-46681-0_14
6. Casanova, G., Pralet, C., Lesire, C., Vidal, T.: Solving dynamic controllability problem of multi-agent plans with uncertainty using mixed integer linear programming. In: Gal, A. et al., (eds.), ECAI 2016 - 22nd European Conference on Artificial Intelligence, 29 August–2 September 2016, The Hague, The Netherlands - Including Prestigious Applications of Artificial Intelligence (PAIS 2016), volume 285 of Frontiers in Artificial Intelligence and Applications, pp. 930–938. IOS Press (2016)
7. Cimatti, A., Micheli, A., Roveri, M.: An SMT-based approach to weak controllability for disjunctive temporal problems with uncertainty. Artif. Intell. **224**, 1–27 (2015)
8. Cimatti, A., Micheli, A., Roveri, M.: Dynamic controllability of disjunctive temporal networks: validation and synthesis of executable strategies. In: Proceedings of the AAAI Conference on Artificial Intelligence, vol. 30 (2016)
9. Cimatti, A., Micheli, A., Roveri, M.: Solving strong controllability of temporal problems with uncertainty using SMT. Constraints **20**, 1–29 (2015)
10. de Moura, L., Bjørner, N.: Z3: an efficient SMT solver. In: Ramakrishnan, C.R., Rehof, J. (eds.) TACAS 2008. LNCS, vol. 4963, pp. 337–340. Springer, Heidelberg (2008). https://doi.org/10.1007/978-3-540-78800-3_24
11. Dechter, R., Meiri, I., Pearl, J.: Temporal constraint networks. Artif. Intell. **49**(1–3), 61–95 (1991)
12. Gario, M., Micheli, A.: PySMT: a solver-agnostic library for fast prototyping of SMT-based algorithms. In: SMT Workshop (2015)
13. Hunsberger, L., Posenato, R.: Speeding up the RUL dynamic-controllability-checking algorithm for simple temporal networks with uncertainty. In: Proceedings of the AAAI Conference on Artificial Intelligence, vol. 36, pp. 9776–9785 (2022)
14. Lubas, J., Franceschetti, M., Eder, J.: Resolving conflicts in process models with temporal constraints. In: Proceedings of the ER Forum and PhD Symposium (2022)
15. Micheli, A.: Disjunctive temporal networks with uncertainty via SMT: recent results and directions. Intell. Artif. **11**(2), 155–178 (2017)
16. Morris, P.H., Muscettola, N.: Temporal dynamic controllability revisited. In: Veloso, M.M., Kambhampati, S. (eds.), Proceedings, The Twentieth National Conference on Artificial Intelligence and the Seventeenth Innovative Applications of Artificial Intelligence Conference, 9–13 July 2005, Pittsburgh, Pennsylvania, USA, pp. 1193–1198. AAAI Press/The MIT Press (2005). http://www.aaai.org/Library/AAAI/2005/aaai05-189.php
17. Morris, P.H., Muscettola, N., Vidal, T.: Dynamic control of plans with temporal uncertainty. In: Nebel, B. (ed.), Proceedings of the Seventeenth International Joint Conference on Artificial Intelligence, IJCAI 2001, Seattle, Washington, USA, 4–10 August 2001, pp. 494–502. Morgan Kaufmann (2001)
18. Osanlou, K., et al.: Solving disjunctive temporal networks with uncertainty under restricted time-based controllability using tree search and graph neural networks. In: Proceedings of the AAAI Conference on Artificial Intelligence, vol. 36, pp. 9877–9885 (2022)
19. Posenato, R., Combi, C.: Adding flexibility to uncertainty: flexible simple temporal networks with uncertainty (FTNU). Inf. Sci. **584**, 784–807 (2022)

20. Sebastiani, R., Tomasi, S.: Optimization modulo theories with linear rational costs. ACM Trans. Comput. Log. **16**(2), 12:1–12:43 (2015)
21. Shah, J.A., Williams, B.C.: Fast dynamic scheduling of disjunctive temporal constraint networks through incremental compilation. In: ICAPS, pp. 322–329 (2008)
22. Stepin, I., Alonso, J.M., Catala, A., Pereira-Farina, M.: A survey of contrastive and counterfactual explanation generation methods for explainable artificial intelligence. IEEE Access **9**, 11974–12001 (2021)
23. Stergiou, K., Koubarakis, M.: Backtracking algorithms for disjunctions of temporal constraints. Artif. Intell. **120**(1), 81–117 (2000)
24. Venable, K.B., Yorke-Smith, N.: Disjunctive temporal planning with uncertainty. In: IJCAI-05, Proceedings of the Nineteenth International Joint Conference on Artificial Intelligence, Edinburgh, Scotland, UK, July 30–August 5, 2005, pp. 1721–1722. Professional Book Center (2005)
25. Vidal, T., Fargier, H.: Handling contingency in temporal constraint networks: from consistency to controllabilities. J. Exp. Theor. Artif. Intell. **11**(1), 23–45 (1999)

CaVE: A Cone-Aligned Approach for Fast Predict-then-optimize with Binary Linear Programs

Bo Tang(ID) and Elias B. Khalil(✉)(ID)

SCALE AI Research Chair in Data-Driven Algorithms for Modern Supply Chains,
Department of Mechanical and Industrial Engineering, University of Toronto,
Toronto, Canada
{botang,khalil}@mie.utoronto.ca

Abstract. The *end-to-end predict-then-optimize* framework, also known as *decision-focused learning*, has gained popularity for its ability to integrate optimization into the training procedure of machine learning models that predict the unknown cost (objective function) coefficients of optimization problems from contextual instance information. Naturally, most of the problems of interest in this space can be cast as integer linear programs. In this work, we focus on binary linear programs (BLPs) and propose a new end-to-end training method to predict-then-optimize. Our method, Cone-aligned Vector Estimation (CaVE), aligns the predicted cost vectors with the normal cone corresponding to the *true* optimal solution of a training instance. When the predicted cost vector lies inside the cone, the optimal solution to the linear relaxation of the binary problem is optimal. This alignment not only produces decision-aware learning models, but also dramatically reduces training time as it circumvents the need to solve BLPs to compute a loss function with its gradients. Experiments across multiple datasets show that our method exhibits a favorable trade-off between training time and solution quality, particularly with large-scale optimization problems such as vehicle routing, a hard BLP that has yet to benefit from predict-then-optimize methods in the literature due to its difficulty.

Keywords: Integer programming · predict-then-optimize · data-driven optimization · machine learning

1 Introduction

Theoretical and experimental results reported over the past few years, starting with Elmachtoub and Grigas [7] and Ban and Rudin [2], have demonstrated the need for end-to-end training of Machine Learning (ML) models that predict the cost coefficients of optimization problems. This contrasts with the more traditional two-stage approach, where an ML model is first trained to minimize regression loss for prediction, and then its predictions are applied to new

B. Dilkina (Ed.): CPAIOR 2024, LNCS 14743, pp. 193–210, 2024.
https://doi.org/10.1007/978-3-031-60599-4_12

test instances for decision. This conventional approach often leads to substantial *regret* in terms of the quality of the solutions obtained, especially when the training set is small. Given that such predict-then-optimize settings are commonly encountered in many applications (e.g., predicting product demand to manage inventory or travel time on a road network to route trucks), researchers in ML and optimization have proposed a wide range of end-to-end training methods, many of which have been recently surveyed and compared [16,24,28].

With a few exceptions [6,11], these methods largely follow the now prevalent mini-batch stochastic gradient descent training algorithm. This process involves the following steps in each iteration: (0) a small batch of training instances is fed into the ML model; (1) the model predicts the cost coefficients; (2) a loss function, incorporating the concept of "decision error" such as *regret*, is calculated; (3) the gradients of this loss w.r.t. the parameters of the ML model are computed using backpropagation; and (4) a gradient descent step is employed to update the model parameters. A common feature of many of these methods is the necessity to solve the optimization problem for each training instance at least once in the forward pass (Steps (1, 2)). Since these repeated calls to the (integer) optimization solver represent a significant computational bottleneck, there have been attempts to improve efficiency by replacing them with much cheaper linear optimization calls [17], solution caching [14,20], and function approximation [9, 26,27]. However, these measures often come at a sacrifice in solution quality, as we empirically demonstrate in this paper.

In this work, we are interested in efficient end-to-end training of ML models that predict cost coefficients of challenging binary linear optimization problems, a prime example of which is the Capacitated Vehicle Routing Problem (CVRP) [29]. CVRP is defined on a graph comprising customers and a depot, where binary variables represent the edges of the graph and indicate whether or not a vehicle traverses that edge. The number of vehicles, their capacities, and customer demands are fixed and known. Linear constraints capture the requirements of valid tours for all vehicles that start and end at the depot, comply with vehicle capacity limits, and visit each customer exactly once. In the predict-then-optimize setting, each CVRP instance is associated with a "feature vector" (e.g., weather conditions, time-of-day, whether it is a holiday or not, etc.) that is predictive of the (unknown) travel times on the graph's edges, which are also the cost coefficients of the objective function, namely the total travel time. CVRP is notoriously hard to solve, even for tens of customers, making end-to-end training of ML models extremely time-consuming.

A distinguishing feature of our CaVE loss functions is that they do not require solving the original optimization problem during training. Instead, they rely on easier projection problems, continuous and quadratic, which are significantly faster to solve. Our key insight is as follows: By ensuring that the predicted cost vector falls inside a specific cone, namely, one that corresponds to the optimal solution under the true cost vector, we are able to recover this optimal solution. The binding (or active) constraints at the optimal solution define this critical cone. To align the ML model's predicted cost vector with the cone, we need to minimize the angle between the prediction and the cone; this is done through projection onto the cone, our main optimization routine. We show that CaVE trains ML models in a fraction of the time required by state-of-the-art methods such as SPO+ [7] and PFYL [3] while yielding equally effective cost predictions as measured by regret in unseen test instances.

2 Related Work

In the field of operations research, the integration of ML methodologies has emerged as a crucial area of research, significantly reshaping traditional approaches. End-to-end predict-then-optimize, also known as decision-focused learning, effectively utilizes data to tackle optimization problems involving unknown (cost) coefficients.

KKT-Based Methods. A notable advancement in this area is the KKT-based method: Amos and Kolter [1] obtain both optimal solutions and gradients by solving a linear system derived from KKT conditions. Wilder et al. [31] adapted the method for linear programs (LPs) by incorporating a small quadratic term into the objective function, while Mandi and Guns [15] introduced a logarithmic barrier term. Furthermore, Ferber et al. [8] employ the cutting-plane method, which allows for integer variables. These KKT-based implicit differentiation methods require the use of specialized solving algorithms, which inherently limit their flexibility and often compromise the efficiency of solving problems such as linear programming. Furthermore, aside from the time-intensive cutting-plane method, KKT-based approaches generally struggle to tackle discrete models.

Black-Box Methods. In contrast, other methodologies approach the optimization solver as a black box, functioning independently of the solver and algorithm. Elmachtoub and Grigas [7] propose a convex surrogate of *regret* for linear objective functions, which has nonzero subgradients. In contrast to this convex and theoretically sound loss function, Pogančić et al. [23] present a linear interpolation, transforming optimization into a piecewise linear function. As a specific case of the interpolation method, Sahoo et al. [25] adopt a more direct straight-through estimator by using the negative identity matrix as the surrogate optimization gradient. Niepert et al. [21] extend the interpolated method with perturbations with random noise. In the context of linear objective functions, more perturbation approaches [3,4] involve adding a perturbation to the predicted cost coefficients to smooth the optimization function and further construct a loss function based on duality.

Since the training process involves solving the optimization problem in each iteration, many solutions naturally accumulate as samples. Under the assumption that the feasible region remains fixed, these solutions are all feasible. Therefore, Mulamba et al. [20] proposes a contrastive loss designed to maximize the distinction between suboptimal solutions in the sample and the optimal solution. Additionally, it utilizes these accumulated solutions as a cache, effectively reducing computational cost. Inspired by the contrastive approach, Mandi et al. [14] employed "learning-to-rank" [13] by ranking the objective value of the cached solutions.

Function Approximation Methods. Due to the computational inefficiency often encountered in solving optimization problems, function approximation methods that do not require constraint optimization are appealing. The critical component of function approximation is a learnable surrogate function, which is learned to mimic the original objective or loss function. Shah et al. [26,27] samples datasets and employs an additional neural network model. This model is trained to approximate the actual loss of a decision. In doing so, they effectively deploy this approximate loss in end-to-end training. With the approximation method, the complexity of directly dealing with the original function is significantly reduced, allowing for more efficient learning. However, the accuracy of the approximate loss function directly impacts the final model performance. In practice, training a model to learn and approximate a specific function effectively can be a challenging endeavour.

3   Problem Statement and Preliminaries

3.1   Definitions and Notation

For the sake of clarity, we can define a binary linear program as follows: There are binary decision variables $w \in \{0,1\}^d$. The cost coefficients associated with these decision variables are represented by $c \in \mathbb{R}^d$; the constraints are $Aw \leq b$, where $A \in \mathbb{R}^{k \times d}$ and $b \in \mathbb{R}^k$:

$$\min_{w} \quad c^\mathsf{T} w$$
$$\text{s.t.} \quad Aw \leq b, \tag{1}$$
$$w \in \{0,1\}^d.$$

Let Ω represent the feasible region of the problem; then the optimal solution is expressed as $w^*(c) = \arg\min_{w \in \Omega} c^\mathsf{T} w$. Given its computational complexity, the process of determining $w^*(c)$ can be extremely time-consuming, especially when the number of decision variables and constraints is large.

In the predict-then-optimize setting, the coefficients c are unknown and linked to a feature vector $x \in \mathbb{R}^p$. This relationship facilitates the use of an ML model $g(x, \theta)$ to estimate the predicted coefficients $\hat{c}$. In this model, θ represents the learnable parameters, which are adjusted to minimize a decision loss $\mathcal{L}(\cdot)$, a metric that quantifies the discrepancy between the true optimal solution $w^*(c)$ and the solution derived from the prediction $\hat{c}$.

3.2   Metric

To measure the quality of the decision, *regret* is introduced by Elmachtoub and Grigas [7], which is defined as the absolute gap between the objective value of the solution $w^*(\hat{c})$ obtained using the predicted coefficients $\hat{c}$, and the optimal value of the solution $w^*(c)$ obtained using the actual coefficients c. It is expressed in the following equation:

$$\mathcal{L}_{\text{Regret}}(\hat{c}, c) = c^\mathsf{T}\big(w^*(\hat{c}) - w^*(c)\big). \tag{2}$$

In line with Elmachtoub and Grigas [7], we adopt *normalized regret*, which serves as an adjusted metric that takes into account the scale of the problem, providing a more standardized and comparable measure with n samples:

$$\frac{\sum_{i=1}^{n} \mathcal{L}_{\text{Regret}}\left(\hat{c}_i, c_i\right)}{\sum_{i=1}^{n} |c_i^\mathsf{T} w_i^*(c_i)|}. \tag{3}$$

The regret is best understood as follows: If a method records a test regret of 0.07 for example, it means that it produces solutions that are 7% worse than the true optimal solutions under the true but unknown cost vectors.

4 Methodology

4.1 Optimal Cones and Subcones

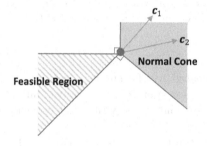

Fig. 1. Illustration of a normal cone: the cost vectors c_1 and c_2 produce the same optimal solution if and only if they lie within this cone.

In (continuous) LP, one can associate a normal cone as *optimal cone* with a given cost vector $c \in \mathbb{R}^d$. Within this cone, all cost coefficients yield the same optimal solution $w^*(c)$. As depicted on the right of Fig. 1, the construction of the normal cone leverages the conical combination of the binding constraints $\widetilde{A}(c)$ at $w^*(c)$: each vector within the cone can be represented as a non-negative combination of the binding constraints coefficients $a_j \in \widetilde{A}(c)$, which can be written as

$$\sum_{a_j \in \widetilde{A}(c)} \lambda_i a_j, \forall \lambda \geq 0,$$

Now consider the case of a binary linear program. If linear cuts $A'(c)$ correspond to the convex hull of integer points of the BLP, the same logic to obtain the optimal cone of an integer solution would have been applied. This parallel is illustrated on the right side of Fig. 2 for a BLP optimal cone, defined as

$$C^*(c) = \left\{ v \in \mathbb{R}^d : v = \sum_{a_j \in A'(c)} \lambda_i a_i, \forall \lambda \geq 0 \right\}.$$

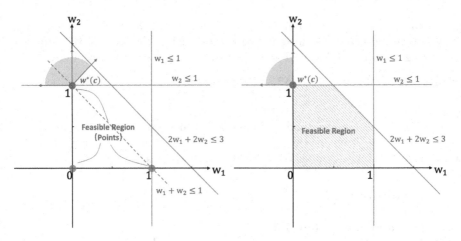

Fig. 2. Illustration of the optimal cone and optimal subcone: On the left, the green cone is the optimal cone of a BLP. On the right, the green cone is a subset of the left cone and the optimal cone of the LP relaxation of the BLP on the left. (Color figure online)

Here, the cone is delineated by the cuts $w_1 \geq 0$, $w_2 \leq 1$ and $w_1 + w_2 \leq 1$. However, we typically do not operate on the convex hull of a BLP, necessitating an alternate definition for the cones with which we will attempt to align the predictions.

To this end, we keep the normal cone of a BLP as the *optimal subcone*:

$$\mathcal{SC}^*(c) = \left\{ v \in \mathbb{R}^d : v = \sum_{a_j \in \tilde{A}(c)} \lambda_i a_i, \forall \lambda \geq 0 \right\}.$$

For a BLP with an optimal solution $w^*(c)$, the optimal subcone is the optimal cone of the same solution, but with the LP relaxation of the BLP instead. In Fig. 2, the cone on the left side is the optimal cone for the BLP; the cone on the right side is the optimal subcone. Since $\mathcal{SC}^*(c) \subset \mathcal{C}^*(c)$, all rays within the subcone also belong to the optimal cone. Although the LP relaxation leads to an expanded feasible region, all binary feasible solutions of the original BLP remain feasible vertices in the LP. As such, recovering an optimal subcone for a given solution $w^*(c)$ is trivial: for a BLP problem defined as in Eq. 1, its optimal subcone is the set of vectors that can be expressed as nonnegative combinations of the coefficient vectors $a_j \in \tilde{A}(c)$ that satisfy $a_j w^*(c) = b_j$, including the variable bounds for which we have either $w_i^*(c) = 0$ or $w_i^*(c) = 1$.

4.2   Cone-Aligned Vector Estimation: Three Variants

Cone-aligned Vector Estimation (CaVE) is an approach specially designed for end-to-end training in BLP. The core idea is to leverage the optimal subcone defined in Sect. 4.1. This approach aims to train an ML model so that the predicted cost coefficients reside within this optimal subcone.

To drive the predicted vector $\hat{c}$ into the subcone, the ML model is trained to reduce the *angle* ϕ between the cost prediction and the subcone. Accordingly, the loss $\mathcal{L}_{\text{CaVE}}(\cdot)$ to be minimized can be defined as the negative cosine similarity between the prediction and its projection onto the subcone, namely:

$$\mathcal{L}_{\text{CaVE}}(\hat{c}, \widetilde{A}(c)) = -\text{cosine\_similarity}(\hat{c}, p_{\hat{c}}) = -\frac{\hat{c}^{\mathsf{T}} p_{\hat{c}}}{\|\hat{c}\| \|p_{\hat{c}}\|}, \tag{4}$$

where, assuming there are m binding constraints, i.e., $\widetilde{A}(c) \in \mathbb{R}^{m \times d}$, the projection writes:

$$p_{\hat{c}} = \widetilde{A}(c)^{\mathsf{T}} \lambda^*, \quad \lambda^* = \arg\min_{\lambda \geq 0} \|\widetilde{A}(c)^{\mathsf{T}} \lambda - \hat{c}\|^2. \tag{5}$$

Algorithm 1. Cone-aligned Vector Estimation (CaVE)

Require: Pairs of feature vectors and binding constraints $\{(x^i, \widetilde{A}^i)\}_{i=1}^n$ for n training
    instances; learning rate $\alpha > 0$
1: Initialize model parameters θ
2: **for** each training epoch **do**
3:    **for** each batch of training samples $(x, \widetilde{A})$ **do**
4:       Predict cost coefficient $\hat{c} \leftarrow g(x, \theta)$
5:       Compute projection $p_{\hat{c}}$ with quadratic program (5)
6:       Compute cosine similarity loss $\mathcal{L}_{\text{CaVE}}(\hat{c}, \widetilde{A})$ (4)
7:       Compute the gradient $\nabla_\theta \mathcal{L}_{\text{CaVE}}(\hat{c}, \widetilde{A})$ with backpropagation
8:       Update ML model parameters $\theta \leftarrow \theta - \alpha \nabla_\theta \mathcal{L}_{\text{CaVE}}(\hat{c}, \widetilde{A})$
9:    **end for**
10: **end for**
11: **return** $g(\cdot, \theta)$

When the alignment is precise, i.e., the predicted cost vector falls within the correct optimal subcone, the CaVE loss achieves its minimum value of -1, indicating an optimal decision. Although our method still requires a quadratic program (QP) to compute the projection of the prediction values during each training iteration, it effectively circumvents the need to solve the more challenging binary linear program. Algorithm 1 presents a detailed, step-by-step description of the CaVE training process.

Figure 3 illustrates the three types of projection that can be employed in the CaVE framework. The *exact projection* projects the cost coefficients $\hat{c}$ directly onto the surface of the optimal subcone; this is the approach that we have just laid out in Algorithm 1. The *inner projection* ensures that the projected vector

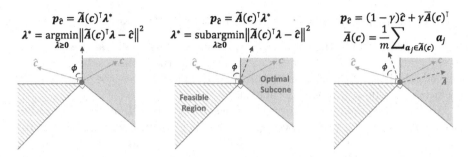

Fig. 3. Illustration of the three projections: Exact projection on the left, inner projection in the middle, and heuristic projection on the right.

lies strictly within the subcone. The *heuristic projection* is an approximation of the true projection onto the optimal subcone, used to reduce the computational cost. We will detail these three variants next.

CaVE-E with Exact Projection. CaVE-E performs exact projection, wherein the optimal solution of the quadratic programming problem (5) is computed to locate the projection on the surface of the cone, as illustrated on the left in Fig. 3. Nevertheless, the CaVE-E method encounters a significant drawback due to its projection onto the face of the cone and the use of cosine similarity as the loss function. This approach results in the vanishing of gradients as the predicted cost vector nears the surface of the optimal subcone but is yet to enter it. Experimental evidence in Sect. 6 corroborates this issue – the regret of CaVE-E is typically higher than existing end-to-end methods, necessitating a modification.

CaVE+ with Inner Projection. Due to the issue of vanishing gradients associated with CaVE-E, CaVE+ replaces the exact projection with what we refer to as "inner projection". The goal is to obtain a projection of the predicted cost vector that lies *inside the subcone*. As all optimal projections will lie on a face of the subcone, we thus require a suboptimal solution to the projection problem. This is readily achieved by simply limiting the number of iterations in the quadratic programming solver and thus terminating prematurely. Since the solver uses the primal-dual interior point method, the feasibility is guaranteed at each iteration. In our experiments, the maximum number of iterations of the QP solver used by CaVE+ was set to 3. A suboptimal projection will lie inside the optimal subcone, resulting in nonzero loss and a strong gradient signal that will push the ML model parameters to produce predictions that move toward the inside of the subcone. Compared to exact projection, this approach is more computationally efficient with fewer iterations of the interior point method. The inner projection is illustrated in the middle of Fig. 3.

CaVE-H with a Mix of Inner and Heuristic Projections. To alleviate the computational burden of repeated QP solving in both CaVE-E and CaVE+, a hybrid strategy is employed in CaVE-H. We interleave inner projections (obtained with a QP just as in CaVE+) with much cheaper heuristic projections. Unlike exact and inner projections, the heuristic projection does not necessitate solving a quadratic program and instead requires a simple convex combination of $\overline{A}(c)$ and $\hat{c}$ with weight $\gamma \in [0, 1]$:

$$p_{\hat{c}} = (1 - \gamma)\hat{c} + \gamma\overline{A}(c)^\mathsf{T},$$

where $\overline{A}(c) = \frac{1}{m}\sum_{a_j \in \tilde{A}(c)} a_j$ is the average of all binding constraints. As illustrated on the right in Fig. 3, it is crucial to note that the heuristic projection is not guaranteed to be in the optimal subcone, but it still ensures that the cost coefficient vector is pushed in the direction of the optimal subcone. With probability $\beta < 0.5$, i.e., in the minority of the iterations of Algorithm 1, CaVE-H performs an inner projection via QP. With probability $(1 - \beta)$, the heuristic projection is used instead, without any optimization required. In our experiments, γ and β are set to 0.2 and 0.3, respectively, and were not tuned any further for performance.

Table 1. Comparison of state-of-the-art predict-then-optimize methods SPO+, PFYL, and NCE with the three CaVE variants w.r.t. "Iteration cost" (per training instance), i.e., the frequency of time-consuming solver calls required to compute the loss of a method during gradient descent. The methods are sorted in decreasing order of their iteration costs. SPO+ requires solving the BLP with predicted costs and PFYL requires solving the K (typically 1–5) BLPs, whereas NCE with solution cache needs to solve only a small fraction ($100 \times \beta\%$) of the BLPs. CaVE methods require solving a single QP, or partially solving a single QP in ($100 \times \beta\%$) of the iterations.

Method	Iteration cost
PFYL	$K \times$ BLP
SPO+	$1 \times$ BLP
NCE	$\beta \times$ BLP
CaVE-E	$1 \times$ QP
CaVE+	$1 \times$ QP (partial)
CaVE-H	$\beta \times$ QP (partial)

Comparison with Existing Methods. Table 1 summarizes the computational cost of training predict-then-optimize models using SPO+ [7], PFYL [3], NCE [20] and the three CaVE variants. We choose SPO+ and PFYL as they represent the state-of-the-art based on an extensive evaluation carried out in [28], though our experiments will include NCE with solution cache, a fast method, as well as accelerated variants of SPO+ and PFYL. The table shows that CaVE-H and NCE are the fastest to train, whereas PFYL and SPO+ exhibit similar costs, particularly when PFYL's number of random perturbations of the predicted cost vector is set to $K = 1$. CaVE+ sits between CaVE-E and CaVE-H in terms of training cost. We will empirically examine these theoretical complexities in the next section.

5   Benchmark Datasets

We utilized synthetic datasets [28] for our experiments. The synthetic dataset, denoted as $\mathcal{D}$, comprises feature vectors $\boldsymbol{x}$ and the corresponding cost coefficients $\boldsymbol{c}$. Each feature vector $\boldsymbol{x}$ adheres to a standard Gaussian distribution, and the associated cost vector $\boldsymbol{c}$ is derived from a polynomial function of $\boldsymbol{x}$, with added random noise.

Shortest Path. The shortest path problem is a fundamental task in graph theory and network analysis, where the objective is to find a minimum weight path from a fixed source node to a fixed target node. Our instances are based on 5×5 grid networks (SP5) where the source is the node in the northwest corner of the grid and the target is the node in the southeast corner. The cost coefficient c_{ij} comes from

$$\left[\frac{1}{3.5^{\deg}} \left(\frac{1}{\sqrt{5}} (\mathcal{B}\boldsymbol{x}_i)_j + 3 \right)^{\deg} + 1 \right] \cdot \epsilon_{ij},$$

where feature size is 5, $\mathcal{B}$ follows a Bernoulli distribution, ϵ_{ij} is random noise uniformly distributed between 0.5 and 1.5, and deg is the polynomial degree of feature mapping. This is a standard and easy task introduced by Elmachtoub and Grigas [7].

Traveling Salesperson and Vehicle Routing. We utilized the traveling salesperson problem (TSP) dataset. TSP, known for its NP-hardness, is a classic problem in combinatorial optimization where the goal is to find the shortest tour that visits a set of locations exactly once and returns to the original starting point. In our study, we do not only test TSP instances with 20 and 50 nodes, but also explore the same graphs under the much more challenging CVRP problem. In this dataset, the cost coefficient c_{ij} comes from two parts: The first part consists of the Euclidean distances among the nodes, and the second part resembles the shortest path problem, formulated as

$$\left(\frac{1}{\sqrt{10}} (\mathcal{B}\boldsymbol{x}_i)_j + 3 \right)^{\deg} \cdot \epsilon_{ij},$$

where the feature vector $\boldsymbol{x} \in \mathbb{R}^{10}$. For CVRP, the capacity of each vehicle is set at 30, and each customer's demand follows a uniform distribution from 0 to 10. To our knowledge, our work is the first to use a CVRP in a predict-then-optimize setting. As will become apparent in the next section, this is due to the time required to solve the BLP representing the CVRP, making SPO+ and PFYL extremely time-consuming as per Table 1.

6   Experimental Results

6.1   Experimental Setup

The experiments were designed to evaluate both the training time and normalized regret, with the size of both the training and test sets being 1,000 instances.

To account for randomness in data generation and stochastic model training, 5 or 10 random seeds are used to generate and train training/test sets and the corresponding ML models. Our comparative analysis encompassed several methods, including three variants of CaVE, a two-stage approach (least-squares regression on cost vectors), SPO+, PFYL, and NCE. A linear model was used as the cost prediction model class for all the aforementioned approaches; this is a standard choice in the literature starting with [7]. In the case of PFYL, the size of random sampling is fixed at $K = 1$; for NCE, the solving ratio is set at $\beta = 5\%$.

The Adam optimizer was used for gradient descent for all our experiments. For SP5, a learning rate of 0.01 was used with all methods for 10 epochs, except for the two-stage and NCE method which allotted 20 epochs at the same learning rate. For both TSP and CVRP, the only modification was to increase the learning rate to 0.05, maintaining the number of epochs at 10. In selecting the number of epochs, careful consideration was given to the convergence of each model, ensuring that our results are not skewed by over- or under-training. The loss curves in Fig. 4 show that all methods exhibit a convergent behavior, with CaVE methods doing so earlier than the baselines. Note that the 2-Stage models are trained essentially to its optimality as they are convex optimization problems.

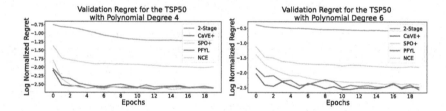

Fig. 4. Validation regeret curves for TSP50 with a polynomial of degree 4 (left) and 6 (right). The vertical axis represents the average normalized regret values of each of the five methods for the validation dataset with 1000 instances.

Our numerical experiments were conducted using Python v3.9.6 on a system with 8 Intel E5-2683 v4 CPUs and 32 GB memory. We utilized SciPy [30] v1.11.2 and Clarabel 0.6.0 for QP solving, Gurobi [10] 10.0.3 for BLP, and PyTorch [22] v2.0.1 with PyEPO [28] v0.3.6 for end-to-end training, where PyEPO provided implementations for SPO+, PFYL and NCE. Our code is available at https://github. com/khalil-research/CaVE.

6.2 Results

Tables 2, 3, 4, 5, and 6 summarize the results. The tables correspond to the Shortest Path problem on 5×5 grid (SP5), TSP with 20 nodes (TSP20), TSP with 50 nodes (TSP50), CVRP with 20 customers (CVRP20), and CVRP with 30 customers (CVRP30), respectively. Each table contains two sub-tables: (a) reports the normalized regret metric defined in Sect. 3.2 on 1,000 test instances and (b) reports the training time in seconds; both quantities are averages over 5 random seeds for CVRP20 and 10 random seeds for others with standard deviation, while the experiments for CVRP30 is not repeated due to running time. Each sub-table has two rows, one corresponding to datasets that use a degree-4 polynomial in the (unknown) mapping from instance features to cost coefficients, and the other for a degree-6 polynomial; we refer to Sect. 5 for details on the role of the polynomial in the function that we are attempting to learn, but note that the higher the degree the more difficult the learning task. In each sub-table, we bold the best-performing method, excluding the 2-Stage method as it is often fast to train, but has much worse regret than end-to-end methods.

Shortest Path. This is the easiest BLP (in fact, LP) we will look at. It serves as a sanity check for any new method in this space. Table 2 shows that SPO+ and PFYL achieve the lowest test regrets, closely followed by CaVE-H and CaVE+. The latter two trains roughly 10 times faster than SPO+ and PFYL, already substantiating our claim that QP solving is faster in solving the optimization problem itself, even for small-scale polynomial-time solvable shortest path problems. In addition, although NCE is fast, it has a higher regret than others.

Table 2. Experimental Results for SP5

(a) Average Test Normalized Regret (%) with Standard Deviation

Methods	2-Stage	CaVE-E	CaVE+	CaVE-H	SPO+	PFYL	NCE
Deg 4	8.82 ± 1.15	10.73 ± 1.54	8.39 ± 0.95	8.35 ± 0.88	7.79 ± 1.00	$\mathbf{7.68 \pm 0.99}$	11.34 ± 1.11
Deg 6	12.58 ± 2.14	11.30 ± 1.30	8.89 ± 0.90	8.84 ± 1.00	$\mathbf{7.72 \pm 1.11}$	7.86 ± 0.96	13.78 ± 1.58

(b) Average Training Time (Sec) with Standard Deviation

Methods	2-Stage	CaVE-E	CaVE+	CaVE-H	SPO+	PFYL	NCE
Deg 4	1.52 ± 0.14	4.64 ± 0.09	4.89 ± 0.12	$\mathbf{2.57 \pm 0.19}$	17.64 ± 0.12	18.52 ± 0.31	4.50 ± 0.48
Deg 6	1.38 ± 0.13	3.52 ± 0.11	3.72 ± 0.14	$\mathbf{2.39 \pm 0.19}$	18.68 ± 0.40	17.78 ± 0.13	4.38 ± 0.42

TSP. For the TSP, SPO+, PFYL and NCE employ the Dantzig-Fulkerson-Johnson (DFJ) formulation [5] to solve this BLP efficiently. For both TSP20 and TSP50, CaVE+ achieves the best time-regret trade-off across all methods: its regret is the best or second-best across all methods and its training time is the second-best after CaVE-H. As shown in part (b) of Tables 3 and 4, CaVE+ trains in roughly half the time of SPO+ and PFYL, achieving the same or even better test regret. Additionally, we also compared our methods with SPO+ Rel and PFYL Rel, which employ a linear relaxation of the BLP during training, as detailed in the Appendix A.1. While they do reduce the training time of vanilla SPO+/ PFYL by a bit, this typically comes at an increase in regret. Similarly, NCE has a prohibitively high regret.

Table 3. Experimental Results for TSP20

(a) Average Test Normalized Regret (%) with Standard Deviation

Methods	2-Stage	CaVE-E	CaVE+	CaVE-H	SPO+	PFYL	NCE
Deg 4	12.12 ± 0.89	7.35 ± 0.40	6.20 ± 0.24	7.69 ± 0.33	$\mathbf{5.95 \pm 0.16}$	6.56 ± 0.21	12.21 ± 0.88
Deg 6	21.32 ± 1.81	8.01 ± 0.45	$\mathbf{6.97 \pm 0.37}$	9.52 ± 0.64	7.48 ± 0.36	7.41 ± 0.37	14.31 ± 0.40

(b) Average Training Time (Sec) with Standard Deviation

Methods	2-Stage	CaVE-E	CaVE+	CaVE-H	SPO+	PFYL	NCE
Deg 4	1.52 ± 0.10	113.56 ± 3.16	107.15 ± 3.80	27.06 ± 2.17	175.23 ± 4.95	220.21 ± 24.20	$\mathbf{25.92 \pm 4.23}$
Deg 6	1.53 ± 0.19	158.66 ± 9.65	102.19 ± 10.38	30.17 ± 2.62	185.13 ± 7.44	185.02 ± 5.09	$\mathbf{25.48 \pm 3.66}$

Table 4. Experimental Results for TSP50

(a) Average Test Normalized Regret (%) with Standard Deviation

Methods	2-Stage	CaVE-E	CaVE+	CaVE-H	SPO+	PFYL	NCE
Deg 4	28.16 ± 1.08	15.19 ± 0.65	7.69 ± 0.22	9.59 ± 0.44	$\mathbf{7.57 \pm 0.20}$	8.03 ± 0.23	14.31 ± 0.40
Deg 6	52.61 ± 2.36	23.25 ± 2.41	$\mathbf{8.57 \pm 0.38}$	11.28 ± 0.80	10.26 ± 0.46	9.00 ± 0.52	17.12 ± 0.48

(b) Average Training Time (Sec) with Standard Deviation

Methods	2-Stage	CaVE-E	CaVE+	CaVE-H	SPO+	PFYL	NCE
Deg 4	1.55 ± 0.18	611.47 ± 23.52	518.07 ± 51.89	196.96 ± 35.92	1220.68 ± 85.39	1328.99 ± 28.87	$\mathbf{151.80 \pm 24.21}$
Deg 6	1.16 ± 0.13	502.71 ± 16.03	573.87 ± 20.19	253.93 ± 27.67	1191.29 ± 42.63	1456.21 ± 34.18	$\mathbf{155.95 \pm 24.46}$

CVRP. As mentioned earlier, we are the first to tackle a CVRP in an end-to-end predict-then-optimize setting. To solve CVRP as a BLP, we formulated the problem with k-path cuts [12] and solved it using Gurobi. For CVRP20, CaVE+, SPO+ and PFYL compete for the lowest regret, with CaVE-H running closely behind. However, CaVE+ completes its 10 training epochs in roughly 2–3 min, whereas SPO+ and PFYL require more than 1 h each, on average. The regret of NCE is high and the training time has increased a lot. Similar as TSP, we also employ a linear relaxation of the BLP during training for SPO+ and PFYL (see Appendix A.2), which results in a significant decrease in training time at the cost of regret.

Table 5. Experimental Results for CVRP20

(a) Average Test Normalized Regret (%) with Standard Deviation

Methods	2-Stage	CaVE-E	CaVE+	CaVE-H	SPO+	PFYL	NCE
Deg 4	10.10 ± 0.64	9.26 ± 1.56	6.44 ± 0.24	7.92 ± 0.52	$\mathbf{5.94 \pm 0.25}$	6.32 ± 0.28	15.77 ± 0.96
Deg 6	19.50 ± 1.22	11.64 ± 0.25	$\mathbf{7.94 \pm 0.54}$	11.44 ± 1.14	8.75 ± 0.28	8.09 ± 0.57	18.96 ± 1.01

(b) Average Training Time (Sec) with Standard Deviation

Methods	2-Stage	CaVE-E	CaVE+	CaVE-H	SPO+	PFYL	NCE
Deg 4	1.65 ± 0.48	213.56 ± 42.36	153.56 ± 11.08	$\mathbf{44.52 \pm 6.27}$	7020.11 ± 1043.05	3773.31 ± 288.84	583.56 ± 170.67
Deg 6	1.54 ± 0.25	208.95 ± 12.90	127.94 ± 13.84	$\mathbf{51.83 \pm 8.78}$	2204.83 ± 99.86	6197.84 ± 288.63	470.20 ± 84.46

Table 6. Experimental Results for CVRP30

(a) Test Normalized Regret

Methods	2-Stage	CaVE-E	CaVE+	CaVE-H	SPO+	PFYL	NCE
Deg 4	0.1972	0.1254	**0.0913**	0.0999	N/A		0.1828

(b) Training Time (Sec)

Methods	2-Stage	CaVE-E	CaVE+	CaVE-H	SPO+	PFYL	NCE
Deg 4	9.27	331.73	287.77	**132.62**	$\geq$ 100h		884.95

This performance gap is even more pronounced for CVRP30 in Table 6. It takes roughly 20 s on average to solve a single CVRP30 instance, thus requiring 8 or so hours to traverse the entire dataset of 1,000 training instances once for SPO+ and PFYL. This makes end-to-end training with these methods impractical for real-world applications. In contrast, CaVE demonstrates its ability to handle such a challenging problem efficiently. Note that due to the scale of the problem, our experimental evaluation was not repeated with random seeds and we used a smaller test set comprising only 10 instances. All CaVE variants achieve test regrets of 9–12% compared to the 2-Stage method's far higher 20%, while requiring only 2–6 min of training each. In addition, NCE achieves 18% with 15 min of training and is thus worse than all variants CaVE in both metrics. To our knowledge, this is the hardest optimization problem ever targeted in the predict-then-optimize literature, a feat that is only possible due to the computational efficiency brought about by CaVE.

7   Conclusion

CaVE reframes the end-to-end training problem for predict-then-optimize as a regression task. Unlike the traditional two-stage approach, which regresses on the cost vectors, our framework instead regresses on cones that correspond to optimal solutions under the true costs. CaVE can be seen as an attempt at obtaining the best of both worlds: fast training with a regression loss that does not require solving hard integer optimization problems in every iteration of gradient descent, and a loss function that penalizes cost predictions that point in the wrong direction relative to the optimal decision. We proposed three versions of our method with varying performance trade-offs.

The best of the three appears to be CaVE+, which regresses on an inner vector of the optimal subcone of a training instance, resulting in stable and efficient training, as well as test regret results that compare to state-of-the-art methods that require training 30 times longer on CVRP20, or do not even complete a single training epoch in 8 h for CVRP30. We note that if we were to use early termination during training, CaVE methods would record even smaller training times as they do converge in fewer gradient descent iterations than competing methods as per Fig. 4. We hope that our framework will enable the adoption of end-to-end predict-then-optimize in a wider range of applications and have made our implementation available with plans to make CaVE one of the standard methods within the PyEPO package following the publication of this work.

CaVE is limited to binary problems. In practice, this is not too problematic, as bounded integer variables can be represented using a set of binary variables. Another limitation of our method is the lack of theoretical guarantees. In particular, we currently do not know whether the loss function in Eq. (4) or a modification thereof could be proven to be a valid upper bound on regret, as does the SPO+ loss of Elmachtoub and Grigas [7] for example. This direction merits further investigation. An interesting connection may be established between CaVE and recent ML methods such as [19] to predict the active constraints of a family of similar optimization problems for which optima are known. Rather than predict-then-optimize, the goal of Misra et al. [19] is to accurately predict the active set to solve a reduced optimization problem over only that set.

A More Experiments

A.1 TSP Relaxation

To enhance training efficiency, it has been proposed in [17] that a linear relaxation can be used as a substitute for the original BLP during training. Because the DFJ formulation utilizes constraint generation to handle subtour elimiation constraints, it is challenging to achieve linear relaxation due to the exponential number of constraints. In this study, we used the LP relaxation of the Miller-Tucker-Zemlin (MTZ) formulation [18] of the TSP, and trained SPO+ Rel and PFYL Rel on the same TSP instances with 20 and 50 nodes. Although relaxation methods are more efficient than CaVE in TSP20, they result in higher regret. As the size of the model increases, the relaxation approach also loses its efficiency advantage (Tables 7 and 8).

Table 7. Experimental Results for TSP20 Relaxation

(a) Average Test Normalized Regret (%) with Standard Deviation

Methods	SPO+ Rel	PFYL Rel
Deg 4	7.75 ± 0.32	9.15 ± 0.50
Deg 6	9.83 ± 0.57	11.28 ± 0.90

(b) Average Training Time (Sec) with Standard Deviation

Methods	SPO+ Rel	PFYL Rel
Deg 4	71.92 ± 2.05	77.25 ± 0.60
Deg 6	67.99 ± 0.60	53.32 ± 3.74

A.2 CVRP Relaxation

Similarly, the CVRP formulation with k-path cuts also struggles to obtain a linear relaxation. Thus, the MTZ formulation can be modified with constraints

$$u_j - u_i \geq Q(x_{ij} - 1) + q_j \quad \forall i \neq j, i \neq 0, j \neq 0$$

Table 8. Experimental Results for TSP50 Relaxation

(a) Average Test Normalized Regret (%) with Standard Deviation

Methods	SPO+ Rel	PFYL Rel
Deg 4	10.17 ± 0.23	11.11 ± 0.33
Deg 6	13.14 ± 0.46	13.38 ± 0.58

(b) Average Training Time (Sec) with Standard Deviation

Methods	SPO+ Rel	PFYL Rel
Deg 4	386.06 ± 9.69	536.67 ± 4.94
Deg 6	636.99 ± 3.04	510.37 ± 3.46

for subtour elimilation and customer demand, where Q is the capacity and q_i is the demand of customer i. The `SPO+ Rel` and `PFYL Rel` are trained for the same VRP instances with 20 customers. While the linear relaxation approach demonstrates high efficiency in these scenarios, it does not guarantee a low regret compared to `CaVE` (Table 9).

Table 9. Experimental Results for CVRP20 Relaxation

(a) Average Test Normalized Regret (%) with Standard Deviation

Methods	SPO+ Rel	PFYL Rel
Deg 4	8.03 ± 0.38	17.07 ± 0.63
Deg 6	15.73 ± 0.39	19.19 ± 1.66

(b) Average Training Time (Sec) with Standard Deviation

Methods	SPO+ Rel	PFYL Rel
Deg 4	78.95 ± 0.73	78.80 ± 1.19
Deg 6	78.74 ± 3.82	81.80 ± 0.86

References

1. Amos, B., Kolter, J.Z.: Optnet: differentiable optimization as a layer in neural networks. In: International Conference on Machine Learning, pp. 136–145. PMLR (2017)
2. Ban, G.Y., Rudin, C.: The big data newsvendor: practical insights from machine learning. Oper. Res. **67**(1), 90–108 (2019)
3. Berthet, Q., Blondel, M., Teboul, O., Cuturi, M., Vert, J.P., Bach, F.: Learning with differentiable perturbed optimizers. arXiv preprint arXiv:2002.08676 (2020)
4. Dalle, G., Baty, L., Bouvier, L., Parmentier, A.: Learning with combinatorial optimization layers: a probabilistic approach. arXiv preprint arXiv:2207.13513 (2022)
5. Dantzig, G., Fulkerson, R., Johnson, S.: Solution of a large-scale traveling-salesman problem. J. Oper. Res. Soc. Am. **2**(4), 393–410 (1954)
6. Elmachtoub, A., Liang, J.C.N., McNellis, R.: Decision trees for decision-making under the predict-then-optimize framework. In: International Conference on Machine Learning, vol. 119, pp. 2858–2867. PMLR (2020)
7. Elmachtoub, A.N., Grigas, P.: Smart predict, then optimize. Manag. Sci. (2021)
8. Ferber, A., Wilder, B., Dilkina, B., Tambe, M.: Mipaal: mixed integer program as a layer. In: Proceedings of the AAAI Conference on Artificial Intelligence, vol. 34, pp. 1504–1511 (2020)

9. Ferber, A.M., et al.: Surco: learning linear surrogates for combinatorial nonlinear optimization problems. In: International Conference on Machine Learning, pp. 10034–10052. PMLR (2023)

10. Gurobi Optimization, LLC: Gurobi Optimizer Reference Manual (2021). https://www.gurobi.com

11. Jeong, J., Jaggi, P., Butler, A., Sanner, S.: An exact symbolic reduction of linear smart predict+ optimize to mixed integer linear programming. In: International Conference on Machine Learning, pp. 10053–10067. PMLR (2022)

12. Kohl, N., Desrosiers, J., Madsen, O.B., Solomon, M.M., Soumis, F.: 2-path cuts for the vehicle routing problem with time windows. Transp. Sci. **33**(1), 101–116 (1999)

13. Liu, T.Y., et al.: Learning to rank for information retrieval. Found. Trends® Inf. Retr. **3**(3), 225–331 (2009)

14. Mandi, J., Bucarey, V., Tchomba, M.M.K., Guns, T.: Decision-focused learning: through the lens of learning to rank. In: International Conference on Machine Learning, pp. 14935–14947. PMLR (2022)

15. Mandi, J., Guns, T.: Interior point solving for lp-based prediction+optimisation. In: Larochelle, H., Ranzato, M., Hadsell, R., Balcan, M.F., Lin, H. (eds.) Advances in Neural Information Processing Systems, vol. 33, pp. 7272–7282, Curran Associates, Inc. (2020)

16. Mandi, J., et al.: Decision-focused learning: foundations, state of the art, benchmark and future opportunities. arXiv preprint arXiv:2307.13565 (2023)

17. Mandi, J., Stuckey, P.J., Guns, T., et al.: Smart predict-and-optimize for hard combinatorial optimization problems. In: Proceedings of the AAAI Conference on Artificial Intelligence, vol. 34, pp. 1603–1610 (2020). https://doi.org/10.1609/aaai.v34i02.5521

18. Miller, C.E., Tucker, A.W., Zemlin, R.A.: Integer programming formulation of traveling salesman problems. J. ACM (JACM) **7**(4), 326–329 (1960)

19. Misra, S., Roald, L., Ng, Y.: Learning for constrained optimization: identifying optimal active constraint sets. INFORMS J. Comput. **34**(1), 463–480 (2022)

20. Mulamba, M., Mandi, J., Diligenti, M., Lombardi, M., Bucarey, V., Guns, T.: Contrastive losses and solution caching for predict-and-optimize. arXiv preprint arXiv:2011.05354 (2020)

21. Niepert, M., Minervini, P., Franceschi, L.: Implicit mle: backpropagating through discrete exponential family distributions. Adv. Neural Inf. Process. Syst. **34**, 14567–14579 (2021)

22. Paszke, A., et al.: Pytorch: an imperative style, high-performance deep learning library. Adv. Neural Inf. Process. Syst. **32** (2019)

23. Pogančić, M.V., Paulus, A., Musil, V., Martius, G., Rolinek, M.: Differentiation of blackbox combinatorial solvers. In: International Conference on Learning Representations (2019)

24. Sadana, U., Chenreddy, A., Delage, E., Forel, A., Frejinger, E., Vidal, T.: A survey of contextual optimization methods for decision making under uncertainty. arXiv preprint arXiv:2306.10374 (2023)

25. Sahoo, S.S., Paulus, A., Vlastelica, M., Musil, V., Kuleshov, V., Martius, G.: Backpropagation through combinatorial algorithms: identity with projection works. arXiv preprint arXiv:2205.15213 (2022)

26. Shah, S., Perrault, A., Wilder, B., Tambe, M.: Leaving the nest: going beyond local loss functions for predict-then-optimize. arXiv preprint arXiv:2305.16830 (2023)

27. Shah, S., Wilder, B., Perrault, A., Tambe, M.: Learning (local) surrogate loss functions for predict-then-optimize problems. arXiv e-prints pp. arXiv–2203 (2022)

28. Tang, B., Khalil, E.B.: PyEPO: a pytorch-based end-to-end predict-then-optimize library for linear and integer programming. arXiv preprint arXiv:2206.14234 (2022)
29. Toth, P., Vigo, D.: Vehicle routing: problems, methods, and applications. SIAM (2014)
30. Virtanen, P., et al.: Scipy 1.0: fundamental algorithms for scientific computing in python. Nat. Methods **17**(3), 261–272 (2020)
31. Wilder, B., Dilkina, B., Tambe, M.: Melding the data-decisions pipeline: decision-focused learning for combinatorial optimization. In: Proceedings of the AAAI Conference on Artificial Intelligence, vol. 33, pp. 1658–1665 (2019)

A Constraint Programming Approach for Aircraft Disassembly Scheduling

Charles Thomas[✉][iD] and Pierre Schaus[iD]

UCLouvain, Ottignies-Louvain-la-Neuve, Belgium
{charles.thomas,pierre.schaus}@uclouvain.be
https://uclouvain.be

Abstract. The dismantling and recycling of aircrafts is one of the future challenges for the air transport industry in terms of sustainability. This problem is hard to solve and optimize as planning operations are highly constrained. Indeed, extracting each part requires technicians with the necessary qualifications and equipment. The parts to be extracted are constrained by precedence relations and the number of simultaneous technicians on specific zones is restricted. It is also essential to avoid unbalancing the aircraft during disassembly. Cost is a significant factor, influenced by the duration of ground mobilization and the choice of technicians for each operation. This paper presents a first constraint programming model for this problem using optional interval variables. This model is used to solve variations of a large instance involving up to 1500 tasks, based on real-life data provided by our industrial partner. The results show that the model can find feasible solutions for all variations of the instance and compares the solutions obtained to lower bounds.

Keywords: Aircraft Dismantling · Scheduling · RCPSP · Constraint Programming · Application · Industrial Problem

1 Introduction

As environmental concerns are more and more present, finding ways to reduce the impact of industries is pressing. In addition to carbon emissions, another sustainability concern in the air transport industry is the disposal of aircrafts retired from service [19]. During this process, parts and materials can be collected in order to be reused or recycled [1,10,25]. This limits the amount of material discarded but also indirectly decreases the impacts of the construction and maintenance of new aircrafts as less raw materials are needed. While the recycling process may yield parts and materials, it is costly in itself. Thus, finding ways to increase the amount and value of what is recouped and decrease the costs of the recycling is crucial in order to incentivize companies towards sustainable disposal of their aircrafts.

This research is done as part of the Planum project which consists in studying and developing technologies and tools in order to facilitate the recycling of end-of-life aircrafts. The problem studied in this paper concerns the scheduling of the

B. Dilkina (Ed.): CPAIOR 2024, LNCS 14743, pp. 211–220, 2024.
https://doi.org/10.1007/978-3-031-60599-4_13

operations taking part in the dismantling phase of the recycling process, from the reception of the plane to the sectional cutting and shredding of the carcass. These operations mostly consist in parts removal but also include inspection and pollutant disposal tasks. In addition to the scheduling aspect, workers and other resources must be assigned in order to complete the operations. In this paper, a CP approach is proposed to tackle this problem and evaluated on large-scale instances derived from real data by an industrial partner.

2   Problem

The problem consists in ordering the different tasks to perform from the reception of the plane to the sectional cutting of its carcass. Several considerations must be taken into account: Different operations may involve a different number of technicians and may require specific certification levels for some technicians. Some technicians may not be available during the whole planning horizon. There are precedences between some operations.

The different parts of the plane may have space restrictions that limit the number of technicians working at the same time there. Thus, the plane is divided into locations that each have an occupancy limit corresponding to the maximum number of technicians allowed to work there at the same time. Finally, the plane must be kept balanced during the whole disassembly process by ensuring that the difference of mass between its extremities does not overstep given thresholds.

The main objective is to minimize the total time taken by the whole extraction process. This is modelled with a *makespan* value that corresponds to the time step at which the last operation finishes. A secondary objective is to minimize the dismantling cost by limiting the use of more costly resources.

The problem is formally defined as such: The set of all operations to perform is denoted $\mathcal{O}$. With each operation $i \in \mathcal{O}$ is associated the duration needed to perform the operation d_i, a location l_i where the operation takes place, an occupancy τ_i, a mass removed m_i, a set of precedences $\mathcal{P}_i$ referencing operations that must be finished before the start of the operation and a set of requirements needed to perform the operation $\mathcal{Q}_i$. Each element $q \in \mathcal{Q}_i$ of this set is a tuple $(\mathcal{C}_{i,q}, n_{i,q})$ where $\mathcal{C}_{i,q}$ is a set of categories of the resource needed and $n_{i,q}$ indicates the amount of this resource needed.

All the available resources are part of the set $\mathcal{R}$. Each resource $j \in \mathcal{R}$ is associated with a category c_j, a set of unavailabilities $\mathcal{U}_j$ consisting of time windows when the resource is not available and a cost f_j which corresponds to the cost per time step to use this resource.

A set of locations $\mathcal{L}$ contains all the locations where operations can take place. Each location $l \in \mathcal{L}$ is associated to a capacity k_l that indicates the maximum number of technicians that can work simultaneously in this location and optionally a zone z_l which corresponds to one of the balance zones of the aircraft. There are four balance zones in total: Aft and Fwd which correspond to the rear and front of the aircraft and Left and Right which correspond to the wings.

A global planning horizon H is given. Two global parameters: B_{af} and B_{lr} indicate the maximum difference of mass allowed at any point in the planning between the Aft and Fwd zones and the Left and Right zones respectively.

The objective is to minimize first the makespan under the following constraints: (1) The makespan must be lower or equal to the global planning horizon H; (2) Precedences between tasks must be respected; (3) A resource cannot be allocated to different operations at the same time; (4) The difference of mass between the Aft and Fwd zones cannot overstep the balance parameter B_{af} at any time during the planning; (5) The difference of mass between the Left and Right zones cannot overstep the balance parameter B_{lr} at any time during the planning; (6) The capacity k_l of a location must not be overloaded at any time; (7) A resource may not be used during its unavailabilities; (8) All the resources needed for a task must be allocated during its whole duration.

Once the optimal makespan has been reached or after some limit, a secondary objective is to minimize the cost of the planning under the same constraints.

3   State of the Art

The Aircraft Dismantling Scheduling problem presented in the previous section is a variation of the Resource Constrained Project Scheduling Problem (RCPSP) [4,28] which consists in scheduling a series of tasks consuming several resources under precedence constraints. The objective is to find a feasible schedule that minimizes the makespan of the tasks. This problem is NP-complete [8]. Several variants of the problem exist [9]. The closest one to our current problem is probably the Multi-Skill Project Scheduling Problem (MCPSP) introduced in [2]. It consists in scheduling tasks and assigning workers with different skill levels to them. It is essentially a relaxed version of the Aircraft Disassembly Scheduling problem without the capacity and balance constraints. In [24], the authors use a CP model to solve several instances of the MSPSP with up to 60 tasks, 19 workers and 15 different skills.

Other publications are related to the problem studied in this paper: In [20], the authors propose a genetic algorithm to solve an aircraft assembly RCPSP. The authors of [17] propose an integer programming approach to schedule aircraft engine assembly lines which also involves workers with several skills on up to 100 tasks. In [18] the authors propose an approach to schedule technicians on short-term aviation maintenance processes (up to 48 h). In [5,6,21,26] different approaches are studied to solve problems linked to aircraft disassembly by finding optimal sequences to access specific components based on spatial and geometrical data. Several CP approaches have also been proposed for problems linked to disassembly scheduling: In [16], a disassembly problem with capacity constraints is studied. The stochastic aspects of disassembly processes are studied in [3] and [22]. In [7,11,27] several MILP and CP models are proposed to solve disassembly problems but are only able to solve instances up to 150 tasks. To our knowledge, this work is the first one to propose a CP model able to solve large-scale RCPSP industrial instances with up to 1500 tasks.

4   Model

The CP model proposed relies on conditional time-intervals [13,15] implemented in CP Optimizer [14]. This modeling approach operates under a paradigm where each interval can be present or not. Resource constraints within this framework are represented as cumulative functions, which are applied over the time intervals that can be constrained within a predefined range. A detailed description of the complete model follows.

$$\text{minimize } \max_{a_i \in \mathcal{A}}(e_i) \tag{1}$$

$$\text{minimize } \sum_{r_{j,i,q} \in R} (x_{j,i,q} \times d_i \times f_j) \tag{2}$$

subject to

$$S_j = sequence(\{\omega_{j,i,q} \forall i \in \mathcal{O}, q \in \mathcal{Q}_i\} \cup \{v_{j,u} \forall u \in \mathcal{U}_j\}) \qquad \forall j \in \mathcal{R} \tag{3}$$

$$noOverlap(S_j) \qquad \forall j \in \mathcal{R} \tag{4}$$

$$b_{af} = step(0, B_{af}) + \sum_{a_i \in \mathcal{A}|z_{l_i}=\texttt{Aft}} stepAtStart(a_i, m_i) + \\ \sum_{a_i \in \mathcal{A}|z_{l_i}=\texttt{Fwd}} stepAtStart(a_i, -m_i) \tag{5}$$

$$b_{lr} = step(0, B_{lr}) + \sum_{a_i \in \mathcal{A}|z_{l_i}=\texttt{Left}} stepAtStart(a_i, m_i) + \\ \sum_{a_i \in \mathcal{A}|z_{l_i}=\texttt{Right}} stepAtStart(a_i, -m_i) \tag{6}$$

$$0 \le b_{af} \le B_{af} \times 2 \tag{7}$$

$$0 \le b_{lr} \le B_{lr} \times 2 \tag{8}$$

$$o_l = \sum_{a_i \in \mathcal{A}|l_i=l} pulse(a_i, \tau_i) \qquad \forall l \in \mathcal{L} \tag{9}$$

$$0 \le o_l \le k_l \qquad \forall l \in \mathcal{L} \tag{10}$$

$$alternative(a_i, \{\omega_{j,i,q} \forall j \in \mathcal{R}|c_j \in \mathcal{C}_{i,q}\}, n_{i,q}) \qquad \forall i \in \mathcal{O}, q \in \mathcal{Q}_i \tag{11}$$

$$e_p \le s_i \qquad \forall i \in \mathcal{O}, p \in \mathcal{P}_i \tag{12}$$

Variables. Interval variables represent the operations to perform. Each operation $i \in \mathcal{O}$ is thus modelled with an interval variable $a_i \in \mathcal{A}$ characterized by a start s_i and an end e_i initialized to $[0, H - d_i]$ and $[d_i, H]$ respectively. These interval variables are always present and their duration is fixed to the duration of the corresponding operation: d_i. The assignment of resources to operations is also represented by interval variables. For each requirement $q \in \mathcal{Q}_i$ of each operation $i \in \mathcal{O}$, all the compatible required resources ($j \in \mathcal{R}|c_j \in \mathcal{C}_{i,q}$) are associated to a corresponding optional interval variable $\omega_{j,i,q} \in \Omega$ which presence $x_{j,i,q}$ indicates whether the resource is assigned to the operation. The initial domain of the interval variables corresponds to the whole planning horizon ($[0, H]$). Operation variables are always set to present while assignment variables are optional.

The unavailabilities of the resources are also modelled as interval variables $v_{j,u}$ which are set to the time windows corresponding to the unavailabilities. All the optional assignment and unavailability interval variables of a same resource are added to a sequence variable S_j (3).

Constraints. Each sequence variable is subject to a *noOverlap* constraint (4). This constraint ensures that a resource is never assigned to more than one operation simultaneously and is not assigned when unavailable.

Balance and occupancy constraints are modelled using cumulative functions. There are two cumulative functions used for the balance constraints: The cumulative function b_{af} (5) represents the difference of mass between the Aft and Fwd zones of the aircraft. The cumulative function b_{lr} (6) does the same for the Left and Right zones. When weight is removed in a balance zone as part of an operation, it is either added to or subtracted from the relevant cumulative function. For example, if an operation removes a weight of 50 in the tail of the aircraft, this amount will be added to the cumulative function b_{af} while an operation that removes weight in the cockpit will have this weight subtracted from the b_{af} function. In order to avoid having to deal with negative cumulative functions, these are shifted by the amount of tolerated mass difference (B_{af} or B_{lr}). Thus, the cumulative function starts at the tolerated mass difference and must at all time be comprised between 0 and twice this amount (7, 8).

Occupancy constraints also use cumulative functions: For each location in the airplane $l \in \mathcal{L}$, a cumulative function o_l (9) models the number of technicians working in this location. This cumulative function is linked to the operation activities taking place at this location and must not overstep the capacity of the location k_l (10).

An *Alternative* constraint is used to link the operation activity $a_i \in \mathcal{A}$ with the optional assignment activities $\{\omega_{j,i,q} \forall j \in \mathcal{R} | c_j \in \mathcal{C}_{i,q}\}$ for each requirement $q \in \mathcal{Q}_i$ of each operation $i \in \mathcal{O}$ (11). Note that its third parameter is its cardinality which is set to the amount of the resource required $n_{i,q}$ so that the constraint selects exactly the required number of resources among the optional activities. Finally, precedence constraints ensure that preceding activities are finished when an activity starts (12).

Objectives. The main objective of the problem is to minimize the makespan which is modelled as the maximum of the ends of the operation activities (1). The secondary objective is the total operating cost which corresponds to the sum of the costs of each assignment. For each assignment activity $\omega_{j,i,q}$, its cost is computed as the duration of the activity multiplied by the cost of the corresponding resource. This cost is then multiplied by the boolean attribute corresponding to the presence of the assignment activity $x_{j,i,q}$ (the attribute *presenceOf* of an interval variable in CP Optimizer is considered as a value of 1 if true and 0 otherwise in an expression). Thus, only present activities contribute towards the global cost. The cost objective is the sum of all these costs (2).

The two objectives of the problem are solved using a lexicographical search: First, the makespan objective is solved to optimality or until a given limit is reached. Second, the cost objective is minimized subject to an additional constraint that prevents the makespan objective to regress.

5 Experiments

Data. The instances used in the experiments are based on data provided by an industrial partner from the Planum research project. It was collected during the full dismantling of a Boeing 737-600 aircraft. It consists in a list of 1459 operations that are performed as part of the aircraft disassembly. Each operation details the section of the plane where the task takes place; the estimated time and the man power needed to perform the task. Note that this data is not enough to make complete instances of the problem as several items are currently missing and in the process of being collected by the industrial partner. The following data had to be completed with arbitrary values: the level of certification needed for each operation as well as the mass removed; all the data relative to the technicians and some of the precedences between operations.

The instances used in the experiments were created based on this dataset. Each instance uses the same set of technicians with 21 technicians available. Some unavailability periods are randomly assigned to some of the technicians. Four different certification categories are considered: uncertified (11 technicians), B1 (6), B2 (2) and B1 and B2 (2). A subset of operations chosen randomly requires either a B1 or a B2 certification. The cost of each technician is expressed as a value per time period that varies between 750 and 1250 depending on the certification level of the technician. A mass value between 0 and 50 is assigned to operations of the four balance zones. The maximum difference of mass allowed is 100 on the aft - forward axis and 50 on the left - right axis. The instance B737-600-Full corresponds to the whole set of operations. All the other instances are subsets of this instance where some of the operations were randomly removed. An anonymized version of these instances is made available at https://github.com/cftmthomas/AircraftDisassemblyScheduling as well as the model and results.

Experimental Protocol. The model is implemented in the java API of CP Optimizer 22.1.1 [14]. Experiments were run on a laptop with a 2.6 GHz Intel i5 processor and a memory of 16 GB. The model was run on each instance with a lexicographical search of 1 h with 40 min allocated to the first objective (makespan) and 20 min allocated to the secondary objective (cost). If an optimal solution is reached for the first objective before the end of its allocated time, the remaining time is added to the allowed search time for the secondary objective. The automatic search of CP Optimizer was used. It consists in an adaptive large neighbourhood search [12] that automatically switches to a failure directed search [23] if stagnation is detected in order to improve the lower bound of the objective and prove the solution optimal.

Table 1. Lexicographical search results

Instance	# ops.	Makespan				Cost			
		1st sol.		best sol.		obj. switch		best sol.	
		gap	time	gap	time	gap	time	gap	time
01	16	0	0.02	0	0.02	0.38	0.23	0	0.26
02	29	3.61	0.07	0	0.07	0.08	0.49	0.01	1.48
03	41	0.53	0.14	0	0.15	0.15	0.81	0.01	19.59
04	54	0.34	0.43	0	0.45	0.22	1.51	0.02	153.99
05	63	0.38	0.46	0	0.51	0.16	1.08	0.02	67.18
06	105	0.73	0.55	0	1.05	0.20	42.37	0.11	1739.50
07	104	0.78	0.66	0	1.38	0.16	1.94	0.11	25.54
08	111	0.32	0.56	0	1.08	0.17	2.86	0.07	16.58
09	145	0.59	0.62	0.05	7.85	0.17	2401.06	0.15	2428.92
10	126	1.28	0.57	0.04	2.60	0.19	2401.04	0.12	2424.98
20	294	2.03	1.48	0.05	21.67	0.18	2401.99	0.08	2543.88
30	442	0.16	1.61	0.04	113.61	0.15	2402.62	0.10	2875.37
40	588	1.83	5.49	0.03	547.91	0.16	2402.96	0.10	3362.61
50	729	1.38	7.42	0.09	104.45	0.13	2401.92	0.10	3118.61
60	890	0.54	25.16	0.10	477.57	0.16	2403.34	0.11	3544.60
70	1028	0.62	30.61	0.19	711.38	0.16	2403.42	0.15	3583.29
80	1166	0.13	92.13	0.11	258.47	0.15	2403.65	0.13	3538.19
90	1310	0.17	97.57	0.16	739.67	0.17	2404.26	0.14	3521.01
Full	1459	0.20	178.82	0.17	2317.62	0.16	2403.99	0.15	3593.69

Search Results. Table 1 reports the experiment results for the makespan and the cost objective, which are reported as gap values, computed as $(obj - LB)/LB$ where LB is the lower bound found by the solver at the end of the search.

We can see that on large instances (> 800 tasks), even finding the first solution can take a lot of time which indicates that it is in itself a difficult problem. Interestingly, the quality of the first solutions found is already quite good as their objectives are relatively close to the best solutions obtained at the end of the search. The model is able to prove the optimality of the best solution found only on smaller instances (<120 tasks). For the other instances, the gap of the best solution obtained goes up to 17% of the lower bound. In subsequent experiments where the balance, capacity and certifications constraints were relaxed, the model was not able to find better solutions for most instances despite being noticeably faster to find a first solution and during the search. This might indicate that the solutions obtained by the full model are optimal.

Comparison of Objectives. In order to compare the impact of both objectives, the results presented above for the full instance are compared to a lex-

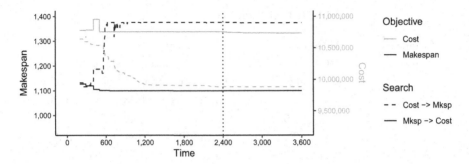

Fig. 1. Comparison of the objectives during a lexicographical search.

icographical search where the objectives are inverted: The cost is the primary objective and the makespan the secondary objective. Figure 1 shows the evolution of the two considered objectives during the search for both approaches.

The solid curves correspond to the lexicographical search on the makespan objective first. The dashed curves correspond to the inverted lexicographical search on the cost objective first. Blue curves correspond to the makespan objective while orange curves correspond to the cost objective. The vertical dotted line indicates the moment when the objective is changed at 40 min. For both graphs, the y value at the bottom corresponds to the lower bound computed for the objective (943 and 9206000 respectively).

We can see that both objectives are in conflict. Indeed, trying to improve one of them prevents the other to be improved or even degrades it. Furthermore, in both cases, once the switch of objective occurs, the secondary objective can only be marginally improved as the main one is constrained and limits the solution space.

6   Conclusion

This paper presents an aircraft disassembly scheduling problem. While it shares similarities with the Resource-Constrained Project Scheduling Problem (RCPSP), it incorporates several unique constraints specific to aircraft dismantling, such as capacity and balance limitations, as well as the need for specific certification levels of technicians to carry out certain tasks. We propose a Constraint Programming (CP) model that employs interval variables, sequence variables, and cumulative functions. The model is assessed using a set of scenarios comprising up to 1500 tasks, which are derived from real data provided by an industrial partner. Our experiments demonstrate that the model can effectively identify feasible solutions for all instances. However, proving optimality is only feasible for instances with a smaller scale.

Future Work. Several research opportunities remain open based on this work. One potential avenue is to compare the performance of the CP model with other optimization approaches. Another direction for research involves enhancing the performance of the current model, either by implementing more effective pruning techniques or by developing custom search heuristics. Additionally, considering the extensive number of tasks and the time horizon involved, investigating a rolling horizon search strategy for this problem could also prove worthwhile.

Acknowledgments. This work was funded by the Walloon Region (Belgium) as part of the Planum project. We thank Sabena Engineering for allowing the diffusion of an anonymized version of the dataset provided.

Disclosure of Interests. The authors have no competing interests to declare that are relevant to the content of this article.

References

1. Asmatulu, E., Overcash, M., Twomey, J.: Recycling of aircraft: state of the art in 2011. J. Ind. Eng. (2013)
2. Bellenguez, O., Néron, E.: Lower bounds for the multi-skill project scheduling problem with hierarchical levels of skills. In: Burke, E., Trick, M. (eds.) PATAT 2004. LNCS, vol. 3616, pp. 229–243. Springer, Heidelberg (2005). https://doi.org/10.1007/11593577_14
3. Bentaha, M.L., Battaïa, O., Dolgui, A.: Chance constrained programming model for stochastic profit–oriented disassembly line balancing in the presence of hazardous parts. In: Prabhu, V., Taisch, M., Kiritsis, D. (eds.) APMS 2013. IAICT, vol. 414, pp. 103–110. Springer, Heidelberg (2013). https://doi.org/10.1007/978-3-642-41266-0_13
4. Brucker, P., Drexl, A., Möhring, R., Neumann, K., Pesch, E.: Resource-constrained project scheduling: notation, classification, models, and methods. Eur. J. Oper. Res. **112**(1), 3–41 (1999). https://doi.org/10.1016/S0377-2217(98)00204-5
5. Camelot, A., Baptiste, P., Mascle, C.: Decision support tool for the disassembly of reusable parts on an end-of-life aircraft. In: Proceedings of 2013 International Conference on Industrial Engineering and Systems Management (IESM), pp. 1–8. IEEE (2013)
6. Dayi, O., Afsharzadeh, A., Mascle, C.: A lean based process planning for aircraft disassembly. IFAC-PapersOnLine **49**(2), 54–59 (2016)
7. Edis, E.B.: Constraint programming approaches to disassembly line balancing problem with sequencing decisions. Comput. Oper. Res. **126**, 105111 (2021)
8. Garey, M.R., Johnson, D.S.: Complexity results for multiprocessor scheduling under resource constraints. SIAM J. Comput. **4**(4), 397–411 (1975)
9. Hartmann, S., Briskorn, D.: An updated survey of variants and extensions of the resource-constrained project scheduling problem. Eur. J. Oper. Res. **297**(1), 1–14 (2022). https://doi.org/10.1016/j.ejor.2021.05.004
10. Khan, W., Soltani, S., Asmatulu, E., Asmatulu, R.: Aircraft recycling: a review of current issues and perspectives. In: International SAMPE Technical Conference (2013)
11. Kizilay, D.: A novel constraint programming and simulated annealing for disassembly line balancing problem with and/or precedence and sequence dependent setup times. Comput. Oper. Res. **146**, 105915 (2022)

12. Laborie, P., Godard, D.: Self-adapting large neighborhood search: application to single-mode scheduling problems. In: Proceedings MISTA-07, Paris **8** (2007)
13. Laborie, P., Rogerie, J.: Reasoning with conditional time-intervals. In: FLAIRS conference, pp. 555–560 (2008)
14. Laborie, P., Rogerie, J., Shaw, P., Vilím, P.: IBM ILOG CP optimizer for scheduling: 20+ years of scheduling with constraints at IBM/ILOG. Constraints **23**, 210–250 (2018)
15. Laborie, P., Rogerie, J., Shaw, P., Vilím, P., Katai, F.: Interval-based language for modeling scheduling problems: an extension to constraint programming. Algebraic Modeling Systems: Modeling and Solving Real World Optimization Problems, pp. 111–143 (2012)
16. Lee, D.H., Xirouchakis, P., Zust, R.: Disassembly scheduling with capacity constraints. CIRP Ann. **51**(1), 387–390 (2002)
17. da Matta Oliveira Borsato Pinhão, J., Ignacio, A.A.V., Coelho, O.: An integer programming mathematical model with line balancing and scheduling for standard work optimization: a realistic application to aircraft engines assembly lines. Comput. Ind. Eng. **173**, 108652 (2022). https://doi.org/10.1016/j.cie.2022.108652
18. Niu, B., Xue, B., Zhong, H., Qiu, H., Zhou, T.: Short-term aviation maintenance technician scheduling based on dynamic task disassembly mechanism. Inf. Sci. **629**, 816–835 (2023). https://doi.org/10.1016/j.ins.2023.01.137
19. Ribeiro, J.S., Gomes, J.D.O.: Proposed framework for end-of-life aircraft recycling. Procedia CIRP **26**, 311–316 (2015). https://doi.org/10.1016/j.procir.2014.07.048
20. Shan, S., Hu, Z., Liu, Z., Shi, J., Wang, L., Bi, Z.: An adaptive genetic algorithm for demand-driven and resource-constrained project scheduling in aircraft assembly. Inf. Technol. Manage. **18**, 41–53 (2017)
21. Srinivasan, H., Gadh, R.: Selective disassembly of components with geometric constraints. In: International Design Engineering Technical Conferences and Computers and Information in Engineering Conference, vol. 19746, pp. 571–579. American Society of Mechanical Engineers (1999)
22. Tian, G., Zhou, M., Chu, J.: A chance constrained programming approach to determine the optimal disassembly sequence. IEEE Trans. Autom. Sci. Eng. **10**(4), 1004–1013 (2013)
23. Vilím, P., Laborie, P., Shaw, P.: Failure-directed search for constraint-based scheduling. In: Michel, L. (ed.) CPAIOR 2015. LNCS, pp. 437–453. Springer, Cham (2015). https://doi.org/10.1007/978-3-319-18008-3_30
24. Young, K.D., Feydy, T., Schutt, A.: Constraint programming applied to the multi-skill project scheduling problem. In: Beck, J.C. (ed.) CP 2017. LNCS, vol. 10416, pp. 308–317. Springer, Cham (2017). https://doi.org/10.1007/978-3-319-66158-2_20
25. Zhao, D., Guo, Z., Xue, J.: Research on scrap recycling of retired civil aircraft. In: IOP Conference Series: Earth and Environmental Science, vol. 657, no.1, p. 012062 (2021). https://doi.org/10.1088/1755-1315/657/1/012062
26. Zhong, L., Youchao, S., Ekene Gabriel, O., Haiqiao, W.: Disassembly sequence planning for maintenance based on metaheuristic method. Aircr. Eng. Aerosp. Technol. **83**(3), 138–145 (2011)
27. Zwingmann, X., Ait-Kadi, D., Coulibaly, A., Mutel, B.: Optimal disassembly sequencing strategy using constraint programming approach. J. Qual. Maint. Eng. **14**(1), 46–58 (2008)
28. Özdamar, L., Ulusoy, G.: A survey on the resource-constrained project scheduling problem. IIE Trans. **27**(5), 574–586 (1995). https://doi.org/10.1080/07408179508936773

Optimization over Trained Neural Networks: Taking a Relaxing Walk

Jiatai Tong[1], Junyang Cai[1,2], and Thiago Serra[1(✉)]

[1] Bucknell University, Lewisburg, PA, USA
{jt037,jc092,thiago.serra}@bucknell.edu
[2] University of Southern California, Los Angeles, CA, USA
caijunya@usc.edu

Abstract. Besides training, mathematical optimization is also used in deep learning to model and solve formulations over trained neural networks for purposes such as verification, compression, and optimization with learned constraints. However, solving these formulations soon becomes difficult as the network size grows due to the weak linear relaxation and dense constraint matrix. We have seen improvements in recent years with cutting plane algorithms, reformulations, and an heuristic based on Mixed-Integer Linear Programming (MILP). In this work, we propose a more scalable heuristic based on exploring global and local linear relaxations of the neural network model. Our heuristic is competitive with a state-of-the-art MILP solver and the prior heuristic while producing better solutions with increases in input, depth, and number of neurons.

Keywords: Deep Learning · Mixed-Integer Linear Programming · Linear Regions · Neural Surrogate Models · Rectified Linear Units

1 Introduction

There is a natural role for mathematical optimization in machine learning with training, where discrete optimization has a "discreet" but growing presence in classification trees [2,3,12,13,21,28,37,45,81–83,89], decision diagrams [34,44], decision rules [1,51], and neural networks [7,11,14,47,50,64,68,77].

Now a new role has emerged with predictions from machine learning models being used as part of the formulation of optimization problems. For example, imagine that we train a neural network on historical data for approximating an objective function that we are not able to represent explicitly. Overall, we start with one or more trained machine learning models, and then we formulate an optimization model which—among other things—represents the relationship between decision variables for the inputs and outputs of those trained models. Since other discrete decision variables and constraints may be part of such optimization models, gradient descent is not as convenient here as it is for training.

These formulations are *neural surrogate models* if involving neural networks. Neural networks are essentially nonlinear, and thus challenging to model in mathematical optimization. However, we can use Mixed-Integer Linear Programming

(MILP) for popular activations such as the Rectified Linear Unit (ReLU) [36,39,53,61,67]. Neural networks with ReLUs represent piecewise linear functions [6,63], which we model in MILP with binary decision variables altering the slopes [72]. As with other activations [26,35,43], ReLU networks have been shown to be universal function approximators with one hidden layer but enough neurons [88] and with limited neurons per layer but enough layers [42,55,62].

Many frameworks to formulate neural surrogate models have emerged— JANOS [10], OMLT [22], OCL [32], OptiCL [56], and Gurobi Machine Learning [38]—in addition to stochastic and robust optimization variants [29,30,49]. The applications in machine learning include network verification [4,5,24,69,75], network pruning [31,73,74], counterfactual explanation [48], and constrained reinforcement learning [18,27]. In the broader line of work often denoted as *constraint learning*, these models have been used for scholarship allocation [10], patient survival in chemotherapy [56], power generation [60] and voltage regulation [23] in power grids, boiling point optimization in molecular design [57], and automated control of industrial operations in general [70,85,87].

However, these models can be difficult to solve as they grow in size. They have weak linear relaxations due to the dense constraint matrix within each layer and the big M constraints for each neuron, which sparked immediate and continued interest in calibrating big M coefficients [8,24,33,54,80] as well as in strengthening the formulation and generating cutting planes [4,5,80]. Other improvements include identifying stable neurons [78,86], exploiting the dependency among neural activations [16,71], and inducing sparser formulations by network pruning [19,70]. But at the rate of one binary variable per ReLU, typically-sized neural networks entail considerably large MILPs, hence limiting the applications where these models are solvable within reasonable time.

We may expect that improving scalability will require algorithms exploiting the model structure. For example, Fischetti and Jo [33] first observed that a feasible solution for the MILP mapping from inputs to outputs of a single neural network is immediate once a given input is chosen. This strategy has been shown effective at least twice [65,73], whereas finding a feasible solution for an MILP is generally NP-complete [25]. Another example of special structure comes from ReLU networks representing piecewise linear functions. Within each part of the domain mapped as a linear function, which is denoted as a *linear region*, there is a direction for locally improving the output. In fact, Perakis and Tsiourvas [65] developed a local search heuristic that moves along adjacent linear regions by solving restrictions of the MILP model with some binary variables fixed. However, the reliance on MILP eventually brings scalability issues back—although much later in comparison to solving the model without restrictions.

But is there hope for optimization over linear regions at scale? That is akin to thinking about MILPs as unions of polyhedra in disjunctive programming [9]. While earlier studies have shown that the number of linear regions may grow fast on model dimensions [6,59,63,66,76], later studies have shown that there are architectural tradeoffs limiting such growth [20,58,71,72]. Moreover, the networks with typical distributions of parameters have considerably fewer linear regions [40,41]; and gradients change little between adjacent linear regions [84].

Hence, we may conjecture that the search space is actually smaller and simpler than expected, and thus that a leaner algorithm may produce good results faster.

In this work, we propose an heuristic based on solving a Linear Programming (LP) model rather than an MILP model at each step of the local search, and we generate initial solutions with LP relaxations of the neural surrogate model. Confirming our intuition, this strategy is computationally better at scale, such as when neural networks have larger inputs, more neurons, or greater depths.

2   Notation and Conventions

In this paper, we consider feedforward networks with fully-connected layers of neurons having ReLU activation. Note that convolutional layers can be represented as fully-connected layers with a block-diagonal weight matrix. We also abstract that fully-connected layers are often followed by a softmax layer [17], since the largest input of softmax matches the largest output of softmax.

We assume that the neural network has an input $\boldsymbol{x} = [x_1 \ x_2 \ \ldots \ x_{n_0}]^\top$ from a bounded domain $\mathbb{X}$ and corresponding output $\boldsymbol{y} = [y_1 \ y_2 \ \ldots \ y_m]^\top$, and each layer $l \in \mathbb{L} = \{1, 2, \ldots, L\}$ has output $\boldsymbol{h}^l = [h_1^l \ h_2^l \ldots h_{n_l}^l]^\top$ from neurons indexed by $i \in \mathbb{N}_l = \{1, 2, \ldots, n_l\}$. Let $\boldsymbol{W}^l$ be the $n_l \times n_{l-1}$ matrix where each row corresponds to the weights of a neuron of layer l, $\boldsymbol{W}_i^l$ the i-th row of $\boldsymbol{W}^l$, and $\boldsymbol{b}^l$ the vector of biases associated with the units in layer l. With $\boldsymbol{h}^0$ for $\boldsymbol{x}$ and $\boldsymbol{h}^L$ for $\boldsymbol{y}$, the output of each unit i in layer l consists of an affine function $g_i^l = \boldsymbol{W}_i^l \boldsymbol{h}^{l-1} + b_i^l$ followed by the ReLU activation $h_i^l = \max\{0, g_i^l\}$. We denote the neuron *active* when $h_i^l = g_i^l > 0$ and *inactive* when $h_i^l = 0$ and $g_i^l < 0$. When $h_i^l = g_i^l = 0$, the state is given by the last nonzero value of g_i^l during local search.

In typical neural surrogate models, the parameters $\boldsymbol{W}^l$ and $\boldsymbol{b}^l$ of each layer $l \in \mathbb{L}$ are constant. The decision variables are the inputs of the network ($\boldsymbol{x} = \boldsymbol{h}^0 \in \mathbb{X}$) and, in each layer, the outputs before and after activation ($\boldsymbol{g}^l \in \mathbb{R}^{n_l}$ and $\boldsymbol{h}^l \in \mathbb{R}_+^{n_l}$ for $l \in \mathbb{L}$) as well as the activation states ($\boldsymbol{z}^l \in \{0,1\}^{n_l}$ for $l \in \mathbb{L}$). By linearly mapping these variables according to the parameters of the network, each possible combination of inputs, outputs, and activations become a solution of an MILP formulation. For each layer $l \in \mathbb{L}$ and neuron $i \in \mathbb{N}_l$, the following constraints associate its decision variables $\boldsymbol{h}^l$, g_i^l, h_i^l, and z_i^l:

$$\boldsymbol{W}_i^l \boldsymbol{h}^{l-1} + b_i^l = g_i^l \tag{1}$$

$$(z_i^l = 1) \to h_i^l = g_i^l \tag{2}$$

$$(z_i^l = 0) \to (g_i^l \leq 0 \wedge h_i^l = 0) \tag{3}$$

$$h_i^l \geq 0 \tag{4}$$

$$z_i^l \in \{0, 1\} \tag{5}$$

The indicator constraints (2)–(3) can be modeled with big M constraints [15].

We follow the convention of characterizing each linear region by the set of neurons that they activate [66]. For an input $\boldsymbol{x}$, let $\mathbb{S}^l(\boldsymbol{x}) \subseteq \{1, 2, \ldots, n_l\}$ denote the *activation set* of layer l. Hence, layer l defines an affine transformation of the form $\Omega^{\mathbb{S}^l(\boldsymbol{x})}(\boldsymbol{W}^l \boldsymbol{h}^{l-1} + \boldsymbol{b}^l)$, where $\Omega^{\mathbb{U}}$ is a diagonal $v \times v$

Algorithm 1. Local search to walk within and across linear regions.

1: **repeat**   ▷ Local search consists of an improvement loop
2:     $x^1 \leftarrow$ Optimal solution of **LP**(x^0) ▷ Finds best solution within linear region
3:     **if** $F(x^1) > F(x^0)$ **then**   ▷ Checks if there was an improvement
4:         $d \leftarrow x^1 - x^0$ ▷ Computes direction of improvement d
5:         $x^0 \leftarrow x^1 + \varepsilon d$ ▷ Leaves the linear region along direction d
6:         **for** $i \leftarrow 1, \ldots, n_0$ **do**   ▷ Loops over all input dimensions
7:             **if** $x^1 + \varepsilon d e^i \notin \mathbb{X}$ **then**   ▷ Checks if move is outside input space
8:                 $x_i^0 \leftarrow x_i^1$ ▷ Corrects move to be inside input space
9:             **end if**
10:        **end for**
11:    **end if**
12: **until** $F(x^1) \leq F(x^0)$ ▷ Stops when no improvement occurs
13: **return** x^0 ▷ Returns best solution found

---

matrix in which $\Omega_{ii}^{\mathbb{U}} = 1$ if $i \in \mathbb{U}$ and $\Omega_{ii}^{\mathbb{U}} = 0$ otherwise for a subset $\mathbb{U} \subseteq \mathbb{V} = \{1, 2, \ldots, v\}$. For the linear region containing $x = x^0$, the output of the neural network is the affine transformation $Tt + t$ for $T = \prod_{l=1}^{L} \Omega^{\mathbb{S}^l(x^0)} W^l$ and $t = \sum_{l'=1}^{L} \left(\prod_{l''=l'+1}^{L} \Omega^{\mathbb{S}^{l''}(x^0)} W^{l''} \right) \Omega^{\mathbb{S}^{l'}(x^0)} b^{l'}$, in comparison to which we note that the output h^ℓ of layer ℓ is obtained by replacing L with ℓ [46].

3   Walking Along Linear Regions

Let us consider a neural network representing the piecewise linear function $f(x)$, a linear objective function $F(x) = c^\top f(x)$ to be maximized, and an implicit set of linear constraints from assuming the input set $\mathbb{X}$ to be a polytope.

We can model our problem as an MILP on $(x, \{g\}_{i=1}^L, \{h\}_{i=0}^L, \{z\}_{i=1}^L, y)$:

$$\max \ c^\top y \tag{6}$$

$$\text{s.t. (1)-(5)} \qquad\qquad \forall l \in \mathbb{L}, i \in \{1, \ldots, n_l\} \tag{7}$$

$$x = h^0, y = h^L, x \in \mathbb{X} \tag{8}$$

For an input $x = x^0$, we can define an LP model by fixing the binary variables as $z_i^l = 1$ if $i \in \mathbb{S}^\ell(x^0)$ and $z_i^l = 0$ otherwise. Let us denote it as **LP**(x^0). By not fixing x, **LP**(x^0) finds an input maximizing $F(x)$ in a linear region with x^0.

We propose the local search outlined in Algorithm 1, which is a loop moving from an input x^0 to the input x^1 in the same linear region by solving LP(x^0). If we find an improvement, we continue moving along the same direction $d = x^1 - x^0$ to the next linear region with a step εd updating x^0. We expect that $F(x^1 + \varepsilon d) > F(x^1)$ since $\|\nabla F(x^1 + \varepsilon d) - \nabla F(x^1)\|$ is usually small [84]. Ideally, ε should be small enough to move only to the next linear region while being large enough to be numerically computed as in the relative interior of that next linear region. We also adjust the move along each dimension to ensure that $x^0 \in \mathbb{X}$.

Figure 1 illustrates three iterations of improvement with the local search algorithm, each characterized by a pair of points (x^0, x^1) denoting a direction of improvement: (A, B), (C, D), and (E, F). Among those, the second iteration shows that a larger step may skip a smaller linear region. Conversely, we could mistakenly conclude that no further improvement is possible if a smaller step d is numerically computed in such a way that $x^1 = H \approx x^1 + \varepsilon d$.

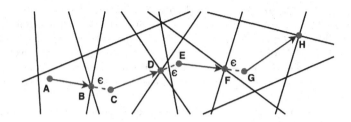

Fig. 1. From a starting point, our local search algorithm moves in a certain direction indicated by the blue arrow, and then takes a small step into the next linear region before moving again. We stop when the next linear region has no better solution. (Color figure online)

We embed the local search in a generator of initial solution outlined in Algorithm 2, which is based on solving variations of the linear relaxation of (6)-(8). We can compute an input $\tilde{x}$ that is somewhat aligned with maximizing function $F(x)$ by solving this relaxation. We denote this model as LR:

$$\max \ c^\top y \tag{9}$$
$$\text{s.t. (1)-(4)} \qquad\qquad \forall l \in \mathbb{L}, i \in \{1, \ldots, n_l\} \tag{10}$$
$$z_i^l \in [0, 1] \qquad\qquad \forall l \in \mathbb{L}, i \in \{1, \ldots, n_l\} \tag{11}$$
$$x = h^0, y = h^L, x \in \mathbb{X} \tag{12}$$

We then impose a random sequence of constraints on activation states, producing a solution $\tilde{x}$ in a different linear region after adding each constraint. The probability of fixing a neuron are calibrated to produce the most change to the linear relaxation. We start over from $\tilde{x}$ when no more activations can be fixed.

Related Work. We denote our approach as Relax-and-Walk (**RW**) and the local search in [65] as "Sample-and-MIP" (**SM**). SM is based on generating initial solutions by random sampling and then solving a restriction of the MILP (6)–(8) to identify the best solution among adjacent linear regions. SM may find the best solution locally, but it may take much longer to compute in networks with larger dimensions. Hence, SM may produce fewer solutions and less improvement.

4   Experiments

We benchmark our **RW** method with **SM** [65] and **Gurobi** 10.0.1. For local search, we use $\varepsilon = 0.01$ since by preliminary tests it was small enough to avoid skipping linear regions. We ran the code in [79] on 10 cores of a cluster with Intel(R) Xeon(R) Gold 6336Y CPU @ 2.40 GHz processors and 16 GB of RAM.

Algorithm 2. Generation of initial solutions and injection in local search

---

1: $(\widetilde{\boldsymbol{x}}, \{\widetilde{\boldsymbol{g}}\}_{i=1}^{L}, \{\widetilde{\boldsymbol{h}}\}_{i=0}^{L}, \{\widetilde{\boldsymbol{z}}\}_{i=1}^{L}, \widetilde{\boldsymbol{y}}) \leftarrow$ Optimal solution of LR
2: **DO LOCAL SEARCH FROM** $\widetilde{\boldsymbol{x}}$ ▷ First local search; and the only one at $\widetilde{\boldsymbol{x}}$
3: **loop**   ▷ Outer loop defines indefinite run until interrupted
4:     $(\bar{\boldsymbol{x}}, \bar{\boldsymbol{z}}) \leftarrow (\widetilde{\boldsymbol{x}}, \widetilde{\boldsymbol{z}})$ ▷ Uses first LR solution to decide where to go next
5:     **for** $\ell \leftarrow 1, \dots, L$ **do**   ▷ Loops sequentially over layers to fix neurons
6:         $\mathbb{N} \leftarrow \{1, 2, \dots, n_\ell\}$ ▷ Accounts for all neurons to fix in layer ℓ
7:         **while** $\mathbb{N} \neq \emptyset$ **do**   ▷ Loops to try fixing each neuron once
8:             **for** $i \in N$ **do**   ▷ Loops over unfixed neurons
9:                 **if** $i \in \mathbb{S}^\ell(\bar{\boldsymbol{x}})$ **then**   ▷ Checks if neuron i is active in last solution $\bar{\boldsymbol{x}}$
10:                    $\chi_i \leftarrow 1 - \bar{z}_i^\ell$ ▷ If so, measures distance of relaxed binary to 1
11:                **else**
12:                    $\chi_i \leftarrow \bar{z}_i^\ell$ ▷ Otherwise, measures distance of relaxed binary to 0
13:                **end if**
14:            **end for** ▷ Produces a shifted probability on χ values to pick a neuron
15:            $k \leftarrow$ Element $i \in \mathbb{N}$ with probability $\chi_i + \delta / \sum_{j \in \mathbb{N}} (\chi_j + \delta)$
16:            $\mathbb{N} \leftarrow \mathbb{N} \setminus \{k\}$ ▷ Records attempt to fix neuron $k \in \mathbb{N}$
17:            **if** $k \in \mathbb{S}^\ell(\bar{\boldsymbol{x}})$ **then**   ▷ Checks if neuron k is active in last solution $\bar{\boldsymbol{x}}$
18:                Add constraint $z_k^\ell = 0$ to LR ▷ If so, makes it inactive going forward
19:            **else**
20:                Add constraint $z_k^\ell = 1$ to LR   ▷ Otherwise, makes it active
21:            **end if**
22:            **if** LR is feasible **then**   ▷ Checks if new constraint keeps LR feasible
23:                $(\bar{\boldsymbol{x}}, \{\bar{\boldsymbol{g}}\}_{i=1}^{L}, \{\bar{\boldsymbol{h}}\}_{i=0}^{L}, \{\bar{\boldsymbol{z}}\}_{i=1}^{L}, \bar{\boldsymbol{y}}) \leftarrow$ Optimal solution of LR
24:                **DO LOCAL SEARCH FROM** $\bar{\boldsymbol{x}}$ ▷ Local search at new solution
25:            **else**   ▷ In case not, neuron can only have same activation as before
26:                Remove constraint on z_k^ℓ; revert $(\bar{\boldsymbol{x}}, \bar{\boldsymbol{z}})$ to last feasible solution of LR
27:            **end if**
28:        **end while**   ▷ Fixed the entire layer; moves on to the next
29:    **end for**   ▷ Fixed all layers; ready to drop constraints
30:    Remove all activation constraints from LR
31: **end loop**   ▷ Starts over from $(\widetilde{\boldsymbol{x}}, \widetilde{\boldsymbol{z}})$

4.1   Random ReLU Networks

Our first experiment replicates and extends the optimization of output value of randomly initialized neural networks in [65] to test scalability and solution quality. With a time limit of 1 h, we use 5 different networks for each choice of input sizes $n_0 \in \{10, 100, 1000\}$ and configurations of the form $L \times n_\ell$ for depth $L \in \{1, 2, 3\}$ and width $n_\ell \in \{100, 500\}$. We note that solving to optimality with Gurobi within 1 h is very unlikely, except for 1×100 with $n_0 = 10$.

RW vs. SM: Figure 2 shows the pair of values obtained for the same random network with RW and SM. RW outperforms SM for $n_0 \in \{100, 1000\}$ and $n_\ell = 500$. The performance is similar for $n_0 = 10$, except in the four cases where Gurobi fails to solve the linear relaxation. Those have minimum value in the plots. That happens more often when n_0 is the smallest while L is larger: the model is likely more sensitive to numerical issues as the linear regions get smaller.

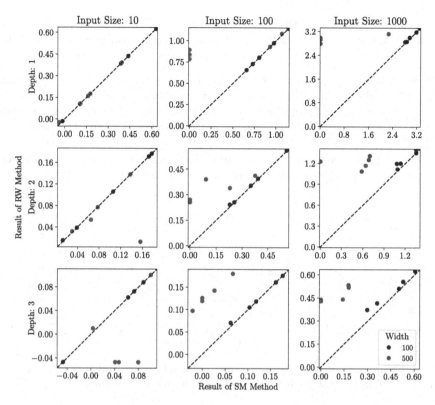

Fig. 2. Comparison of best objective values obtained by RW and SM in random networks. The points are favorable to RW above the line $Y = X$; and to SW below it. RW is at least 1% better in 64.4% of the cases, while SM is in 12.2%.

Table 1. Average number of final solutions produced by local search with each method.

Model Size	$n_0 = 10$		$n_0 = 100$		$n_0 = 1000$	
	RW	SM	RW	SM	RW	SM
1×100	2882.0	2562.2	1038.4	548.0	134.6	100.2
1×500	174.8	165.4	80.0	4.4	16.2	1.0
2×100	429.2	230.4	331.0	39.8	55.8	11.8
2×500	10.0	2.8	25.4	1.0	6.0	1.0
3×100	421.8	197.0	204.0	18.2	33.8	4.2
3×500	11.5	2.0	15.0	1.0	3.0	1.0

Moreover, we observe that **walking is cheaper than MIPing**: Table 1 shows that the walking algorithm RW converges more frequent to a local optimum by the time limit. Conversely, the average runtime of MILP restrictions in SM explodes very quickly when the network gets wider, consistent with Fig. 2.

This can be explained by the number of unfixed MILP variables growing with the network dimensions in the SM approach. Moreover, consecutive steps of the SM approach may reevaluate some neighboring linear regions again.

RW vs. Gurobi: Figure 3 shows a similar comparison between RW and Gurobi, but with depths combined for conciseness. Gurobi can handle a shallow network ($L = 1$) even with the largest input size $n_0 = 1000$. When the network is deeper and the structure of the linear regions more complex [72], directly solving an MILP is slower. When both width and depth are large, Gurobi cannot find a feasible solution—see points next to Y-axis. The same four cases with unbounded relaxation are also difficult for Gurobi—see points near the left bottom.

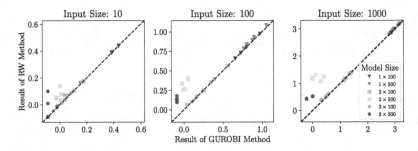

Fig. 3. Comparison of best objective values obtained by RW and Gurobi in random networks. RW is at least 1% better in 55.1% of the cases, while Gurobi is in 12.4%.

4.2   Optimal Adversary Experiment

Given an input $x = \hat{x}$, the output neuron for the predicted label c, and the output neuron for another likely label w, the optimal adversary problem aims to maximize $y_w - y_c$ for an input x sufficiently near $\hat{x}$. We solved this problem for $|x - \hat{x}|_1 < \Delta$ as in [80] by using a setup derived from the Gurobi Machine Learning repository [38]. Figure 4 shows the result from testing 50 images from the MNIST dataset [52] with $\Delta = 5$ for 1 hour, all of which on a 2×500 classifier with test accuracy 97.04%. In 10% of the cases, RW found an adversarial input (positive solution) and Gurobi did not. When RW does better, it does so by a wider margin.

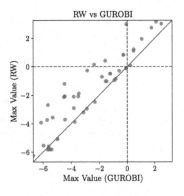

Fig. 4. Comparison of best objective values obtained by RW and Gurobi in optimal adversary models. RW is at least 1% better in 68% of the cases, while Gurobi is in 30%.

5   Conclusion

We introduced a local search algorithm for optimizing over trained neural networks. We designed our algorithm to leverage model structure based on what is known about linear regions in deep learning. Moreover, our algorithm scales more easily because it only solves LP models at every step. Last, but certainly not least, the solutions are usually better in comparison with other methods.

Acknowledgement. The authors were supported by the National Science Foundation (NSF) grant IIS 2104583, including Junyang Cai while at Bucknell.

References

1. Dash, S., Günlük, O., Wei, D.: Boolean decision rules via column generation. Neural Information Processing Systems (NeurIPS) (2018)
2. Aghaei, S., Gómez, A., Vayanos, P.: Strong optimal classification trees. arXiv:2103.15965 (2021)
3. Alston, B., Validi, H., Hicks, I.V.: Mixed integer linear optimization formulations for learning optimal binary classification trees. arXiv:2206.04857 (2022)
4. Anderson, R., Huchette, J., Tjandraatmadja, C., Vielma, J.: Strong mixed-integer programming formulations for trained neural networks. In: Integer Programming and Combinatorial Optimization (IPCO) (2019)
5. Anderson, R., Huchette, J., Ma, W., Tjandraatmadja, C., Vielma, J.P.: Strong mixed-integer programming formulations for trained neural networks. Math. Program. **183**, 3–39 (2020)
6. Arora, R., Basu, A., Mianjy, P., Mukherjee, A.: Understanding deep neural networks with rectified linear units. In: International Conference on Learning Representations (ICLR) (2018)
7. Aspman, J., Korpas, G., Marecek, J.: Taming binarized neural networks and mixed-integer programs. arXiv:2310.04469 (2023)
8. Badilla, F., Goycoolea, M., Muñoz, G., Serra, T.: Computational tradeoffs of optimization-based bound tightening in ReLU networks (2023)
9. Balas, E.: Disjunctive Programming. Springer, Cham (2018)
10. Bergman, D., Huang, T., Brooks, P., Lodi, A., Raghunathan, A.U.: JANOS: an integrated predictive and prescriptive modeling framework. INFORMS J. Comput. **34**, 807–816 (2022)
11. Bernardelli, A.M., Gualandi, S., Lau, H.C., Milanesi, S.: The BeMi stardust: a structured ensemble of binarized neural networks. In: Learning and Intelligent Optimization (LION) (2023)
12. Bertsimas, D., Dunn, J.: Optimal classification trees. Mach. Learn. **106**, 1039–1082 (2017)
13. Bertsimas, D., Shioda, R.: Classification and regression via integer optimization. Oper. Res. **55**, 252–271 (2007)
14. Bienstock, D., Muñoz, G., Pokutta, S.: Principled deep neural network training through linear programming. Discrete Optim. **49**, 100795 (2023)
15. Bonami, P., Lodi, A., Tramontani, A., Wiese, S.: On mathematical programming with indicator constraints. Math. Program. **151**, 191–223 (2015)

16. Botoeva, E., Kouvaros, P., Kronqvist, J., Lomuscio, A., Misener, R.: Efficient verification of relu-based neural networks via dependency analysis. In: AAAI Conference on Artificial Intelligence (AAAI) (2020)
17. Bridle, J.S.: Probabilistic interpretation of feedforward classification network outputs, with relationships to statistical pattern recognition. In: Soulié, F.F., Hérault, J. (eds.) Neurocomputing. NATO ASI Series, vol. 68, pp. 227–236. Springer, Heidelberg (1990). https://doi.org/10.1007/978-3-642-76153-9_28
18. Burtea, R.A., Tsay, C.: Safe deployment of reinforcement learning using deterministic optimization over neural networks. In: Computer Aided Chemical Engineering, vol. 52, pp. 1643–1648. Elsevier (2023)
19. Cacciola, M., Frangioni, A., Lodi, A.: Structured pruning of neural networks for constraints learning. arXiv:2307.07457 (2023)
20. Cai, J., et al.: Getting away with more network pruning: from sparsity to geometry and linear regions. In: International Conference on the Integration of Constraint Programming, Artificial Intelligence, and Operations Research (CPAIOR) (2023)
21. Carrizosa, E., Molero-Río, C., Morales, D.R.: Mathematical optimization in classification and regression trees. TOP **29**, 5–33 (2021)
22. Ceccon, F., et al.: Omlt: optimization & machine learning toolkit. J. Mach. Learn. Res. **23**(349), 1–8 (2022)
23. Chen, Y., Shi, Y., Zhang, B.: Data-driven optimal voltage regulation using input convex neural networks. Electr. Power Syst. Res. **189**, 106741 (2020)
24. Cheng, C.-H., Nührenberg, G., Ruess, H.: Maximum resilience of artificial neural networks. In: D'Souza, D., Narayan Kumar, K. (eds.) ATVA 2017. LNCS, vol. 10482, pp. 251–268. Springer, Cham (2017). https://doi.org/10.1007/978-3-319-68167-2_18
25. Conforti, M., Cornuéjols, G., Zambelli, G.: Valid inequalities for structured integer programs. In: Integer Programming. GTM, vol. 271, pp. 281–319. Springer, Cham (2014). https://doi.org/10.1007/978-3-319-11008-0_7
26. Cybenko, G.: Approximation by superpositions of a sigmoidal function. Math. Control, Signals Syst. **2**, 303–314 (1989)
27. Delarue, A., Anderson, R., Tjandraatmadja, C.: Reinforcement learning with combinatorial actions: an application to vehicle routing. In: NeurIPS (2020)
28. Demirović, E., et al.: MurTree: optimal decision trees via dynamic programming and search. J. Mach. Learn. Res. **23**, 1–47 (2022)
29. Dumouchelle, J., Julien, E., Kurtz, J., Khalil, E.B.: Neur2RO: neural two-stage robust optimization. arXiv:2310.04345 (2023)
30. Dumouchelle, J., Patel, R., Khalil, E.B., Bodur, M.: Neur2SP: neural two-stage stochastic programming. In: Neural Information Processing Systems (NeurIPS) (2022)
31. ElAraby, M., Wolf, G., Carvalho, M.: OAMIP: optimizing ANN architectures using mixed-integer programming. In: Cire, A.A. (ed.) Integration of Constraint Programming, Artificial Intelligence, and Operations Research (CPAIOR). LNCS, vol. 13884, pp. 219–237. Springer, Cham (2023). https://doi.org/10.1007/978-3-031-33271-5_15
32. Fajemisin, A., Maragno, D., den Hertog, D.: Optimization with constraint learning: a framework and survey. Eur. J. Oper. Res. **314**, 1–14 (2023)
33. Fischetti, M., Jo, J.: Deep neural networks and mixed integer linear optimization. Constraints **23**, 296–309 (2018)
34. Florio, A.M., Martins, P., Schiffer, M., Serra, T., Vidal, T.: Optimal decision diagrams for classification. In: AAAI Conference on Artificial Intelligence (AAAI) (2023)

35. Funahashi, K.I.: On the approximate realization of continuous mappings by neural networks. Neural Netw. **2**, 183–192 (1989)
36. Glorot, X., Bordes, A., Bengio, Y.: Deep sparse rectifier neural networks. In: International Conference on Artificial Intelligence and Statistics (AISTATS) (2011)
37. Günlük, O., Kalagnanam, J., Li, M., Menickelly, M., Scheinberg, K.: Optimal decision trees for categorical data via integer programming. J. Global Optim. **81**, 233–260 (2021)
38. Gurobi: Gurobi Machine Learning (2023). https://github.com/Gurobi/gurobi-machinelearning. Accessed 03 Dec 2023
39. Hahnloser, R., Sarpeshkar, R., Mahowald, M., Douglas, R., Seung, S.: Digital selection and analogue amplification coexist in a cortex-inspired silicon circuit. Nature **405**, 947–951 (2000)
40. Hanin, B., Rolnick, D.: Complexity of linear regions in deep networks. In: International Conference on Machine Learning (ICML) (2019)
41. Hanin, B., Rolnick, D.: Deep ReLU networks have surprisingly few activation patterns. In: Neural Information Processing Systems (NeurIPS), vol. 32 (2019)
42. Hanin, B., Sellke, M.: Approximating continuous functions by ReLU nets of minimal width. arXiv:1710.11278 (2017)
43. Hornik, K., Stinchcombe, M., White, H.: Multilayer feedforward networks are universal approximators. Neural Netw. **2**, 359–366 (1989)
44. Hu, H., Huguet, M.J., Siala, M.: Optimizing binary decision diagrams with maxsat for classification. In: AAAI Conference on Artificial Intelligence (AAAI) (2022)
45. Hu, X., Rudin, C., Seltzer, M.: Optimal sparse decision trees. In: Neural Information Processing Systems (NeurIPS) (2019)
46. Huchette, J., Muñoz, G., Serra, T., Tsay, C.: When deep learning meets polyhedral theory: a survey. arXiv:2305.00241 (2023)
47. Toro Icarte, R., Illanes, L., Castro, M.P., Cire, A.A., McIlraith, S.A., Beck, J.C.: Training binarized neural networks using MIP and CP. In: Schiex, T., de Givry, S. (eds.) CP 2019. LNCS, vol. 11802, pp. 401–417. Springer, Cham (2019). https://doi.org/10.1007/978-3-030-30048-7_24
48. Kanamori, K., Takagi, T., Kobayashi, K., Ike, Y., Uemura, K., Arimura, H.: Ordered counterfactual explanation by mixed-integer linear optimization. In: AAAI Conference on Artificial Intelligence (AAAI) (2021)
49. Kronqvist, J., Li, B., Rolfes, J., Zhao, S.: Alternating mixed-integer programming and neural network training for approximating stochastic two-stage problems. arXiv:2305.06785 (2023)
50. Kurtz, J., Bah, B.: Efficient and robust mixed-integer optimization methods for training binarized deep neural networks. arXiv:2110.11382 (2021)
51. Lawless, C., Dash, S., Günlük, O., Wei, D.: Interpretable and fair Boolean rule sets via column generation. J. Mach. Learn. Res. **24**(229), 1–50 (2023)
52. LeCun, Y., Bottou, L., Bengio, Y., Haffner, P.: Gradient-based learning applied to document recognition. In: Proceedings of the IEEE (1998)
53. LeCun, Y., Bengio, Y., Hinton, G.: Deep learning. Nature **521**, 436–444 (2015)
54. Liu, C., Arnon, T., Lazarus, C., Strong, C., Barrett, C., Kochenderfer, M.J., et al.: Algorithms for verifying deep neural networks. Found. Trends® Optim. **4**(3-4), 244–404 (2021)
55. Lu, Z., Pu, H., Wang, F., Hu, Z., Wang, L.: The expressive power of neural networks: a view from the width. In: Neural Information Processing Systems (NeurIPS) (2017)
56. Maragno, D., Wiberg, H., Bertsimas, D., Birbil, S.I., Hertog, D.d., Fajemisin, A.: Mixed-integer optimization with constraint learning. Oper. Res. (2023)

57. McDonald, T., Tsay, C., Schweidtmann, A.M., Yorke-Smith, N.: Mixed-integer optimisation of graph neural networks for computer-aided molecular design. arXiv:2312.01228 (2023)
58. Montúfar, G.: Notes on the number of linear regions of deep neural networks. In: Sampling Theory and Applications (SampTA) (2017)
59. Montúfar, G., Pascanu, R., Cho, K., Bengio, Y.: On the number of linear regions of deep neural networks. In: Neural Information Processing Systems (NeurIPS), vol. 27 (2014)
60. Murzakhanov, I., Venzke, A., Misyris, G.S., Chatzivasileiadis, S.: Neural networks for encoding dynamic security-constrained optimal power flow. In: Bulk Power Systems Dynamics and Control Symposium (2022)
61. Nair, V., Hinton, G.: Rectified linear units improve restricted Boltzmann machines. In: International Conference on Machine Learning (ICML) (2010)
62. Park, S., Yun, C., Lee, J., Shin, J.: Minimum width for universal approximation. In: International Conference on Learning Representations (ICLR) (2021)
63. Pascanu, R., Montúfar, G., Bengio, Y.: On the number of response regions of deep feedforward networks with piecewise linear activations. In: International Conference on Learning Representations (ICLR) (2014)
64. Patil, V., Mintz, Y.: A mixed-integer programming approach to training dense neural networks. arXiv:2201.00723 (2022)
65. Perakis, G., Tsiourvas, A.: Optimizing objective functions from trained ReLU neural networks via sampling. arXiv:2205.14189 (2022)
66. Raghu, M., Poole, B., Kleinberg, J., Ganguli, S., Dickstein, J.: On the expressive power of deep neural networks. In: International Conference on Machine Learning (ICML) (2017)
67. Ramachandran, P., Zoph, B., Le, Q.V.: Searching for activation functions. In: ICLR Workshop Track (2018)
68. Rosenhahn, B.: Mixed integer linear programming for optimizing a hopfield network. In: Amini, M.R., Canu, S., Fischer, A., Guns, T., Kralj Novak, P., Tsoumakas, G. (eds.) ECML PKDD. LNCS, vol. 13717, pp. 344–360. Springer, Cham (2022)
69. Rössig, A., Petkovic, M.: Advances in verification of ReLU neural networks. J. Global Optim. **81**, 109–152 (2021)
70. Say, B., Wu, G., Zhou, Y.Q., Sanner, S.: Nonlinear hybrid planning with deep net learned transition models and mixed-integer linear programming. In: International Joint Conference on Artificial Intelligence (IJCAI) (2017)
71. Serra, T., Ramalingam, S.: Empirical bounds on linear regions of deep rectifier networks. In: AAAI Conference on Artificial Intelligence (AAAI) (2020)
72. Serra, T., Tjandraatmadja, C., Ramalingam, S.: Bounding and counting linear regions of deep neural networks. In: International Conference on Machine Learning (ICML) (2018)
73. Serra, T., Yu, X., Kumar, A., Ramalingam, S.: Scaling up exact neural network compression by ReLU stability. In: Neural Information Processing Systems (NeurIPS) (2021)
74. Serra, T., Kumar, A., Ramalingam, S.: Lossless compression of deep neural networks. In: Hebrard, E., Musliu, N. (eds.) CPAIOR 2020. LNCS, vol. 12296, pp. 417–430. Springer, Cham (2020). https://doi.org/10.1007/978-3-030-58942-4_27
75. Strong, C.A., Wu, H., Zeljić, A., Julian, K.D., Katz, G., Barrett, C., Kochenderfer, M.J.: Global optimization of objective functions represented by ReLU networks. Mach. Learn. **112**, 3685–3712 (2021)

76. Telgarsky, M.: Representation benefits of deep feedforward networks. arXiv:1509.08101 (2015)
77. Thorbjarnarson, T., Yorke-Smith, N.: Optimal training of integer-valued neural networks with mixed integer programming. PLoS ONE **18**, e0261029 (2023)
78. Tjeng, V., Xiao, K., Tedrake, R.: Evaluating robustness of neural networks with mixed integer programming. In: International Conference on Learning Representations (ICLR) (2019)
79. Tong, J., Cai, J., Serra, T.: Relax-and-Walk Implementation (2024). https://github.com/JiataiTong/Optimization-Over-Trained-Neural-Networks-Taking-a-Relaxing-Walk. Accessed 28 Jan 2024
80. Tsay, C., Kronqvist, J., Thebelt, A., Misener, R.: Partition-based formulations for mixed-integer optimization of trained ReLU neural networks. In: Neural Information Processing Systems (NeurIPS), vol. 34 (2021)
81. Verhaeghe, H., Nijssen, S., Pesant, G., Quimper, C.G., Schaus, P.: Learning optimal decision trees using constraint programming. Constraints **25**, 226–250 (2020)
82. Verwer, S., Zhang, Y.: Learning decision trees with flexible constraints and objectives using integer optimization. In: International Conference on the Integration of Constraint Programming, Artificial Intelligence, and Operations Research (CPAIOR) (2017)
83. Verwer, S., Zhang, Y.: Learning optimal classification trees using a binary linear program formulation. In: AAAI Conference on Artificial Intelligence (AAAI) (2019)
84. Wang, Y.: Estimation and comparison of linear regions for ReLU networks. In: International Joint Conference on Artificial Intelligence (IJCAI) (2022)
85. Wu, G., Say, B., Sanner, S.: Scalable planning with deep neural network learned transition models. J. Artif. Intell. Res. **68**, 571–606 (2020)
86. Xiao, K.Y., Tjeng, V., Shafiullah, N.M., Madry, A.: Training for faster adversarial robustness verification via inducing ReLU stability. In: International Conference on Learning Representations (ICLR) (2019)
87. Yang, S., Bequette, B.W.: Optimization-based control using input convex neural networks. Comput. Chem. Eng. **144**, 107143 (2021)
88. Yarotsky, D.: Error bounds for approximations with deep ReLU networks. Neural Netw. **94**, 103–114 (2017)
89. Zhu, H., Murali, P., Phan, D., Nguyen, L., Kalagnanam, J.: A scalable MIP-based method for learning optimal multivariate decision trees. In: Neural Information Processing Systems (NeurIPS) (2020)

Learning from Scenarios for Repairable Stochastic Scheduling

Kim van den Houten[1]($\boxtimes$)(iD), David M. J. Tax[1](iD), Esteban Freydell[2](iD),
and Mathijs de Weerdt[1](iD)

[1] Delft University of Technology, Delft, The Netherlands
k.c.vandenhouten@tudelft.nl
[2] DSM-Firmenich, Delft, Netherlands

Abstract. When optimizing problems with uncertain parameter values in a linear objective, decision-focused learning enables end-to-end learning of these values. We are interested in a stochastic scheduling problem, in which processing times are uncertain, which brings uncertain values in the constraints, and thus repair of an initial schedule may be needed. Historical realizations of the stochastic processing times are available. We show how existing decision-focused learning techniques based on stochastic smoothing can be adapted to this scheduling problem. We include an extensive experimental evaluation to investigate in which situations decision-focused learning outperforms the state of the art, i.e., scenario-based stochastic optimization.

Keywords: Stochastic Scheduling · Repair · Decision-focused learning

1 Introduction

Decision-making can be challenging due to the stochastic nature of real-world processes. This complexity is evident in various contexts, such as manufacturing, where uncertain processing times make it challenging to meet strict customer deadlines. Formulating Constrained Optimization (CO) models for these problems is common, but unknown parameter values during decision-making add challenges, because wrong estimates of the parameters can lead to infeasibilities.

In practice, such infeasibilities are repaired when reality unfolds. For instance, in a manufacturing system, tasks may be postponed due to delays in earlier stages to maintain the factory's flow. Various repair policies and schedule definitions are used across different contexts.

Historical data, represented as scenarios of unknown parameters like task duration, are often available. Simple averaging of these scenarios is a common yet naive approach that ignores uncertainty. Stochastic programming [16] and robust optimization [1] offer alternatives, each with its challenges, such as scalability and too conservative solutions. Moreover, modeling realistic repair possibilities exactly is not always possible in such two-stage optimization approaches.

B. Dilkina (Ed.): CPAIOR 2024, LNCS 14743, pp. 234–242, 2024.
https://doi.org/10.1007/978-3-031-60599-4_15

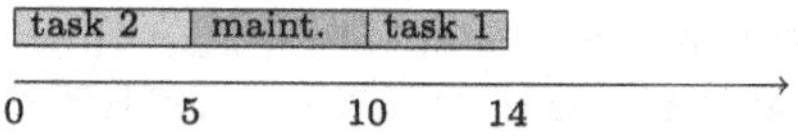

Fig. 1. Deterministic opt. schedule

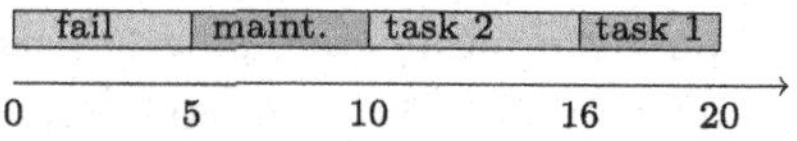

Fig. 2. Repair action when $y_2 = 6$

Decision-focused learning (DFL), extensively reviewed by [13], introduces a novel paradigm for stochastic optimization. This approach embeds an optimization model, like Constraint Programming (CP), in a training procedure to minimize a regret loss [6]. Challenges arise in backpropagation through combinatorial optimization problems, where solutions may change discontinuously. Recent research, including the score-function approach by [17], shows promising directions for handling uncertainty in constraints. Both exploring DFL with uncertainty in constraints, and analyzing the applicability of the score-function method are highlighted as valuable directions for further research [13].

In this research, we explore various scenario-based approaches for stochastic scheduling. The contribution is threefold: 1) we apply DFL for the first time to a repairable stochastic scheduling problem with stochastic processing times, 2) we demonstrate how an existing DFL technique that uses stochastic smoothing can be used to serve a stochastic scheduling problem where historical realizations of processing times are used, and 3) we include an extensive experimental evaluation in which we assess differences in performance between deterministic, stochastic programming, and a DFL approach.

2   Scheduling with Repair

We illustrate the effect of uncertainty in constraints with an example of scheduling two tasks on a single machine, where the average task lengths are $\bar{y}_1 = 4$, and $\bar{y}_2 = 5$. The machine is not available from $t = 5$ to $t = 10$. The task is to minimize makespan. Using the mean values, the optimal decision is to schedule first task 2, and then task 1, which gives us a makespan of 14 (see Figure 1), while scheduling first task 1, and then task 2 results in a makespan of 15.

Now suppose that task 1 is deterministic, and task 2 is stochastic, following the discrete uniform distribution $y_2 \sim U(\{3, 4, 5, 6, 7\})$. It still holds that the expected task lengths are $\bar{y}_1 = 4$, and $\bar{y}_2 = 5$. When we have $y_2 = 6$ and we schedule task 2 first, the effect of the repair strategy can be seen in Figure 2 and leads to a makespan of 20. Considering this repair, we can compute the expected values of the two alternative decisions and find that $\mathbb{E}[\text{first task 1, then task 2}] = 15$ and $\mathbb{E}[\text{first task 2, then task 1}] = 16.6$. So, considering the underlying distributions, it is better to first schedule task 1, instead of task 2. We observe that just using the expected values to come to a decision is not always a good idea when processing times are uncertain.

3   Decision-Focused Learning

Problem Setting. The goal is to optimize an optimization (e.g. scheduling) problem $z^*(y) = \arg\min_z f(z, y)$ s.t. $z \in C(y, z)$, where $f(z, y)$ is the objective function given parameters y and decision z and the constraint set $C(y, z)$. However, the parameters (e.g. processing times) y are unknown at the time of solving. We are given a data set $\mathcal{D} = \{y_i\}_{i=1}^n$ with historical data on y. A common approach is to use the sample averages of $\bar{y}$ to solve the deterministic model and obtain $z^*(\bar{y})$ (which possibly requires reparations when the true values become known). Alternatively, we could take inspiration from the literature on DFL.

Decision-Focused Learning. The idea is to predict the unknowns $\widehat{y} = h_\theta(\mathcal{D})$ based on the data such that the task loss is minimized. Since the unknown parameters occur in the constraints, predicted decisions must sometimes be corrected using a repair function (such as illustrated in Sect. 2). A common task loss for problems with unknown parameters in the constraints is the so-called post-hoc regret $PRegret$ loss [9], defined as:

$$PRegret(\widehat{y}, y) = f(z_{corr}(\widehat{y}, y), y) - f(z^*(y), y) + pen(z^*(\widehat{y}), z_{corr}(\widehat{y}, y)), \quad (1)$$

where y are the true coefficient values, $\widehat{y}$ are the predicted values, $z^*(\widehat{y})$ is the decision based on predicted values, and $z^*(y)$ is the optimal decision with perfect information such as defined by [3]. Then, we have $f(z^*(\widehat{y}), y)$, which are the costs for predicted decisions, and $f(z^*(y), y)$, which are true optimal costs. Due to uncertain parameters in the constraints, predicted decisions must sometimes be corrected using a repair function such that $z^*(\widehat{y}) \rightarrow z_{corr}(\widehat{y}, y)$. How this reparation is penalized is reflected in $pen(z^*(\widehat{y}), z_{corr}(\widehat{y}, y))$.

Zero-Gradient Problem. DFL procedures minimize the post-hoc regret loss by gradient-based optimization with respect to θ to optimize the prediction $\widehat{y} = h_\theta(\mathcal{D})$. However, this loss gives a zero-gradient problem because a combinatorial optimization solver is embedded in the loss computation [6], which is the $\frac{\delta z_{corr}(\widehat{y}, y)}{\delta \widehat{y}}$ term in (2).

$$\frac{\delta PRegret(\widehat{y}, y)}{\delta \theta} = \frac{\delta PRegret(z_{corr}(\widehat{y}, y), y)}{\delta z_{corr}(\widehat{y}, y)} \frac{\delta z_{corr}(\widehat{y}, y)}{\delta \widehat{y}} \frac{\delta \widehat{y}}{\delta \theta} \quad (2)$$

Stochastic Smoothing. A novel approach by Silvestri et al. [17] shows that this zero-gradient problem can be solved with a stochastic smoothing trick. The crux is to use a stochastic estimator $\widehat{y} \sim p_\theta(y)$ (where $p_\theta(y)$ is a parameterized distribution) instead of the point estimator $\widehat{y} = h_\theta(\mathcal{D})$. Using a stochastic estimator makes the loss function an expectation, for which the gradient can be approximated with the score-function gradient estimator (also known as likelihood ratio gradient estimator [7]) that uses:

$$\nabla_\theta \mathbb{E}_{\widehat{y} \sim p_\theta(y)}[PRegret(\widehat{y}, y)] = \mathbb{E}_{\widehat{y} \sim p_\theta(y)}[PRegret(\widehat{y}, y) \nabla_\theta \log(p_\theta(\widehat{y}))] \quad (3)$$

for which the derivation can be found in [17]. The most important assumption on $p_\theta(y)$ is that the probability density function must be differentiable with respect to θ. The right-hand side can be approximated with a Monte-Carlo method [15]. This score-function gradient estimation approach is also the foundation of the REINFORCE algorithm [19], and various other reinforcement learning algorithms [18]. How we exactly apply these techniques to our stochastic scheduling problem is explained in the next section.

4 From Scenarios to Schedules

Algorithm: We adapt DFL to align with our scheduling problem in Algorithm 1. The data $\mathcal{D} = \{y_i\}_{i=1}^n$ comprises historical examples of processing times y. We aim to learn which predictor $\widehat{y}$ minimizes the post-hoc regret. For gradient computation, we use a stochastic estimator parameterized by θ, for which a common choice is the Normal distribution [18]. During training, we sample $\widehat{y} \sim \mathcal{N}(\mu = \theta_\mu \cdot \bar{y}, \sigma = \theta_\sigma \cdot \bar{\sigma})$, where both μ and σ are trainable, and initially set to the sample average $\bar{y}$ and sample standard deviation $\bar{\sigma}$. In each training step, we sample a point y_i and a prediction $\widehat{y}$, compute schedule $z^*(\widehat{y})$, and update θ using the score-function gradient estimator that is provided in equation (3). After training, the stochastic estimator is treated as a point estimator by using $\widehat{y} = \mu$. Note that the distribution is only needed during training for gradient computation on the regret loss.

Example: We explain the zero-gradient problem and smoothing technique with our example from Sect. 2. Suppose $(y_1, y_2) = (4, 6)$, but y_2 is unknown. Figure 3 shows how $\widehat{y}_2$ affects the regret, which is the line with a discontinuity at $\widehat{y}_2 = 5$, indicating a jump in scheduling priority. The blue curves show different stochastic estimators, and the small circles the expected regret values when we sample $\widehat{y}_2$ from each distribution. The line through the circles represents the smoothened expected regret. We assess the applicability of Algorithm 1 using this example. The training data has an underlying distribution with $y_1 = 4$ and $y_2 \sim U(3, 4, 5, 6, 7)$. We expect the algorithm to find scaling $\theta_2 > 1$ to prioritize scheduling task 1. A small experiment confirms that regret drops when μ_1 is above five as anticipated, see Fig. 4.

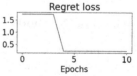

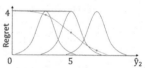

Fig. 3. Smoothing

Algorithm 1: DFL

Require: $\mathcal{D}_{train} = \{y_i\}_{i=1}^{n_{train}}$, $\mathcal{D}_{test} = \{y_i\}_{i=1}^{n_{test}}$
Initialize $\widehat{y} \sim p_\theta(\widehat{y})$ such that
$\widehat{y} \sim \mathcal{N}(\mu = \theta_\mu \cdot \bar{y}, \sigma = \theta_\sigma \cdot \bar{\sigma})$
for each epoch **do**
    for each batch in $\mathcal{D}_{train}$ **do**
        for each instance $(y_i, z^*(y_i))$ in batch **do**
            Sample $\widehat{y}$ from $p_\theta(\widehat{y})$
            Pass $\widehat{y}$ to solver to get schedule
            Compute post-hoc regret$(\widehat{y}, y_i)$
        end for
        Update θ with score-function:
        $\theta = \theta - \mathrm{lr} \cdot \nabla_\theta PRegret(\widehat{y}, y_i) \nabla_\theta \log(p_\theta(\widehat{y}))$
    end for
end for
Pass $\widehat{y} = \mu$ to solver to get schedule
Evaluate post-hoc regret on $\mathcal{D}_{test}$

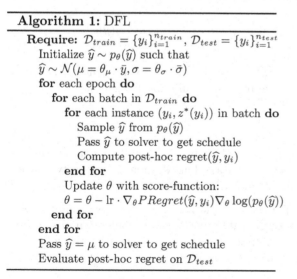

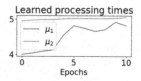

Fig. 4. Training curves

5   Experimental Evaluation

To understand the potential of DFL, we compare performance to deterministic and stochastic programming formulations (Sect. 5.2). We hypothesize that both stochastic programming and DFL outperform the more naive deterministic approach. Furthermore, we explore when DFL or stochastic programming performs better; we expect that DFL has better scalability to larger instances.

5.1   Problem Instances and Evaluation

We study a variant of the **Stochastic Resource Constraint Project Scheduling Problem** (RCPSP). We use two subsets of the PSPLib instances (being j301_1 to j303_10, and j901_1 -to j903_10) [11]. Furthermore, we use two sets of industry-inspired problem instances (small and large instances[1]), related to the factory of our industrial partner DSM-Firmenich. The data describing these instances is provided in our repository [8]. Originally deterministic, these instances are transformed into stochastic versions by sampling task durations from Normal distributions with mean d_j and standard deviation $\sqrt{d_j}$, where d_j is the deterministic processing time of task j from the original instance. We define a scenario as a processing time vector realization for one problem instance. For each instance, three datasets are created: one with 100 training scenarios, another with 50 for validation and tuning, and a final set of 50 for evaluation.

The **evaluation** approach evaluates first-stage start times z_j decisions based on the schedule makespan. Tasks unable to start due to resource constraints or precedence relations undergo a repair policy with (multiple) one-unit time postponements resulting in corrected start times z_j^{corr}. A penalty function measures the sum of start time deviations for all tasks j:

$$\text{pen}(z^*(\widehat{y}), z_{corr}^*(\widehat{y}, y)) = \rho \cdot \sum_{j \in J} z_j^{corr}(\widehat{y}, y) - z_j(\widehat{y}). \tag{4}$$

Here, ρ is the penalty coefficient, which we vary across experiments. Evaluation is conducted using the SimPy discrete-event simulation Python package [14].

5.2   Baseline Methods

We study problem cases where historical realizations of stochastic processing times are available (without feature data). This section describes two other scenario-based methods that are included in the experiments.

Deterministic Approach. This is a simple baseline, where we compute scenario averages of the unknown optimization coefficients. The deterministic constraint programming (CP) model that uses these averages uses the following nomenclature: J: set of all tasks, R: set of all resources, S_j: set of successors

---

[1] 25 instances with 40 to 480 tasks, 13 resource groups with different capacities.

of task j, j: subscript for tasks, r: subscript for resources, *parameters:* y_j: processing time of task j, $r_{r,j}$: resource requirement for task j, b_r: max capacity of resource r, $minLag_{j,i}$: min. difference between start times of tasks j and i, if i is a successor of j and *decision variables:* x_j: interval length for task j. The CP model is:

$$\text{Minimize } Makespan \quad \text{s.t.} \tag{5a}$$

$$\text{Max}(\text{end\_of}(x_j)) \leq Makespan \quad j \in J \tag{5b}$$

$$\text{startOf}(x_i) \geq \text{endOf}(x_j); \quad \forall j \in J \; \forall i \in S_j \quad or$$
$$\text{startOf}(x_i) \geq minLag_{j,i} + \text{startOf}(x_j); \quad \forall j \in J \; \forall i \in S_j \tag{5c}$$

$$\sum_{j \in J} \text{Pulse}(x_j, r_{r,j}) \leq b_r \quad \forall r \in R \tag{5d}$$

$$x_j : \text{IntervalVar}(J, y_j) \quad \forall j \in J \tag{5e}$$

In this model, (5b) defines the makespan which should be larger than the finish time of all tasks, and (5c) enforces precedence constraints between two tasks, where the $minLag_{a,b}$ is the minimal time difference needed between task a and b which is used for the industry instances. The CP pulse constraint (5d) models shared resource usage [4].

Stochastic Programming. The second baseline comprises a scenario-based stochastic programming formulation (again CP). The repair action is added to the stochastic model which comprises the possibility to postpone activities, together with a penalty term for the deviations from the earliest-start-time decision that is included in the objective. Note that we use the same nomenclature as for the deterministic model, but we introduce the notion of scenarios $\omega \in \Omega$, and the first-stage earliest-start-time decision variable $z_j \; \forall j \in J$.

$$\text{Min} \frac{1}{|\Omega|} \sum_{\omega \in \Omega} Makespan(\omega) + \rho \cdot \sum_{\omega \in \Omega} \sum_j \text{startOf}(x_j(\omega)) - z_j \quad \text{s.t} \tag{6a}$$

$$\text{Max}_j(\text{end\_of}(x_j(\omega))) \leq Makespan(\omega) \quad \forall \omega \in \Omega \tag{6b}$$

$$\text{startOf}(x_i(\omega)) \geq \text{endOf}(x_j(\omega)); \quad \forall j \in J \; \forall i \in S_j \quad or$$
$$\text{startOf}(x_i(\omega)) \geq minLag_{j,i} + \text{startOf}(x_j(\omega)); \quad \forall j \in J \; \forall i \in S_j \quad \forall \omega \in \Omega \tag{6c}$$

$$\sum_{j \in J} \text{Pulse}(x_j(\omega), r_{r,j}) \leq b_r \quad \forall r \in R \quad \forall \omega \in \Omega \tag{6d}$$

$$x_j(\omega) : \text{IntervalVar}(J, y_j(\omega)) \quad \forall j \in J \quad \forall \omega \in \Omega \tag{6e}$$

$$z_j \leq \text{startOf}(x_j(\omega)) \quad \forall j \in J \quad \forall \omega \in \Omega \tag{6f}$$

5.3   Results

All experiments are done on a virtual server that uses an Intel(R) Xeon(R) Gold 6148 CPU with two 2.39 GHz processors, and 16.0 GB RAM. All CP models are solved with single thread IBM CP solver [4]. The runtime limits are set per

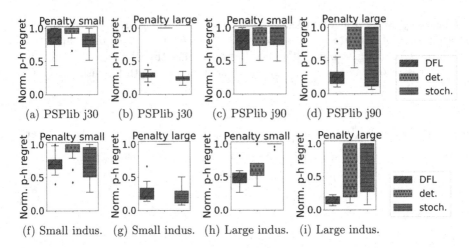

Fig. 5. Normalized post-hoc regret per instance set - penalty setting (smaller regret is better). The box spans from the 25th to the 75th percentile, visualizing the median and interquartile range.

problem size (max. 60 min.) and provided in the README of the repository [8], together with the tuned hyperparameters[2] for deterministic, DFL, and stochastic. Each boxplot in Figure 5 presents the results for the three methods on a single combination of instance set and penalty setting. The y-axis shows the distribution of the normalized post-hoc regret among the test instances of that specific set. We use $\rho = \frac{1}{size}$ (small), where *size* indicates the number of tasks, and $\rho = 1$ (large). We tested the significance of the performance difference of the different algorithms for each setup (a-i) using a paired t-test with $\alpha = 0.05$, and the p-values are included in the repository [8].

On smaller instances with small penalties, both DFL and stochastic methods perform well, with no significant difference. For the large penalty, stochastic tends to outperform both deterministic and DFL methods significantly in PSPlib $j30$ instances. Notably, under $\rho = 1$ for $j30$ instances, no repairs were needed across all instances, emphasizing the robust performance of stochastic when the instances are small enough. For larger instances like PSPlib $j90$, DFL becomes better, even significantly for the small penalty. For the large penalty, there is still a subset of the instances for which stochastic finds very robust solutions that do not need repairs, but because for some of the instances the stochastic model performs much worse than DFL (which is also visible in Fig. 5d), we observe no significant difference between DFL and stochastic looking at all $j90$ instances. We see a somewhat similar pattern in the industrial instances, where again stochastic is most advantageous for the smaller instances and with a high penalty, although not significantly better than DFL. For the larger instances, stochastic performs even worse, especially for the small penalty and DFL is significantly better. We

---

[2] Such as the number of scenarios.

investigated optimality gaps of the outputs of the stochastic model and found that even with a time limit of three hours the gaps are on average approximately around 30%, with outliers of more than 90% (for the largest industry instances) which shows the scalability issue of stochastic programming.

6   Related Work

Previous studies [2, 6, 13] focused mainly on comparing prediction-focused versus DFL approaches for problems with uncertainty in (linear) objectives. The knapsack problem is the most prominent [5, 12]. As far as we know, the set-up without feature data is not studied in earlier work [13]. However, the crux of our problem setting is that uncertain parameters occur in constraints, which can lead to infeasibilities. Hu et al. [10] were the first who introduced a post-hoc regret that penalizes for infeasibilities. The methods introduced in their work rely on specific conditions, such as being recursively and iteratively solvable [9, 10]. This work applies a DFL approach [17] to a pure stochastic repairable scheduling problem, for which both the repairable scheduling setting and the context without features are novel application domains for DFL. It is important to highlight that before this study we did not know if DFL could work for repairable scheduling.

7   To Conclude, and Continue

This study explores a novel application of DFL to stochastic resource-constrained scheduling with repairs, where uncertainty is in the constraints, and the derivative is not smooth by itself. Results indicate that stochastic programming is dominant when it can find the optimal solution, most prominently when the penalty factor is high and the instances are small enough to find robust solutions that do not need reparation. In contrast, we have shown that DFL scales better and is a promising alternative to stochastic programming, even in this pure stochastic scheduling setup. Furthermore, we highlight the potential of DFL because of its flexibility across various settings with different repair strategies, providing a distinct advantage over stochastic programming, in which modeling the exact repair functions is not always possible. We hypothesize that in a setting with features related to stochastic processing times the benefits of DFL for stochastic scheduling are further enhanced, such as shown in earlier research with uncertainty in a (linear) objective [13]. Further interesting directions are investigating alternative gradient estimators or reinforcement learning-inspired algorithms.

Acknowledgments. We acknowledge Mattia Silvestri and Michele Lombardi whose valuable insights improved the quality of this work. We are grateful that Léon Planken was available for questions related to the simulator. This work is supported by the AI4b.io program, a collaboration between TU Delft and dsm-firmenich, and is fully funded by dsm-firmenich and the RVO (Rijksdienst voor Ondernemend Nederland).

Disclosure of Interests. The authors have no competing interests to declare that are relevant to the content of this article.

References

1. Ben-Tal, A., Ghaoui, L.E., Nemirovski, A.: Robust Optimization, 1st edn. Princeton University Press, Princeton (2009)
2. Berthet, Q., Blondel, M., Teboul, O., Cuturi, M., Vert, J.P., Bach, F.: Learning with differentiable pertubed optimizers. In: Larochelle, H., Ranzato, M., Hadsell, R., Balcan, M., Lin, H. (eds.) Advances in neural information processing systems 2020, vol. 33, pp. 9508–9519. The MIT Press (2020)
3. Bertsimas, D., Kallus, N.: From predictive to prescriptive analytics. Manage. Sci. **66**(3), 1025–1044 (2019)
4. Cplex, IBM ILOG: V12. 1: User's manual for cplex. International Business Machines Corporation **46**(53), 157 (2009)
5. Demirović, E., et al.: An investigation into prediction + optimisation for the Knapsack problem. In: Rousseau, L.-M., Stergiou, K. (eds.) CPAIOR 2019. LNCS, vol. 11494, pp. 241–257. Springer, Cham (2019). https://doi.org/10.1007/978-3-030-19212-9_16
6. Elmachtoub, A., Grigas, P.: Smart "predict, then optimize". Manage. Sci. **68**(1), 9–26 (2022)
7. Glynn, P.W.: Likelihood ratio gradient estimation for stochastic systems. Commun. ACM **33**(10), 75–84 (1990). https://doi.org/10.1145/84537.84552
8. van den Houten, K.: Learning from scenarios for repairable stochastic scheduling (2023). https://github.com/kimvandenhouten/Learning-From-Scenarios-for-Repairable-Stochastic-Scheduling
9. Hu, X., Lee, J.C.H., Lee, J.H.M.: Branch and learn with post-hoc correction for predict+optimize with unknown parameters in constraints. In: Cire, A.A. (ed.) Integration of Constraint Programming, Artificial Intelligence, and Operations Research. LNCS, vol. 13884, pp. 264–280. Springer, Cham (2023). https://doi.org/10.1007/978-3-031-33271-5_18
10. Hu, X., Lee, J.C.H., Lee, J.H.M.: Predict+optimize for packing and covering LPs with unknown parameters in constraints. arXiv **2209.03668** (2022)
11. Kolisch, R., Sprecher, A.: PSPLIB - a project scheduling problem library. Eur. J. Oper. Res. **96**, 205–216 (1996)
12. Mandi, J., Demirović, E., Stuckey, P., Guns, T.: Smart predict-and-optimize for hard combinatorial optimization problems. In: Thirty-Fourth AAAI Conference on Artificial Intelligence, AAAI-20 (2020)
13. Mandi, J., et al.: Decision-focused learning: Foundations, state of the art, benchmark and future opportunities. arXiv **2307.13565** (2023)
14. Matloff, N.: Introduction to Discrete-Event Simulation and the SimPy Language (2008)
15. Mohamed, S., Rosca, M., Figurnov, M., Mnih, A.: Monte Carlo gradient estimation in machine learning. J. Mach. Learn. Res. **21**(1), 5183–5244 (2020)
16. Ruszczyński, A., Shapiro, A.: Stochastic Programming, Handbook in Operations Research and Management Science. publisher (2003)
17. Silvestri, M., et al.: Score function gradient estimation to widen the applicability of decision-focused learning. arXiv **2307.05213** (2023)
18. Sutton, R., Barto, A.: Reinforcement Learning: An Introduction, 2nd edn. The MIT Press, Cambridge (2018)
19. Williams, R.J.: Simple statistical gradient-following algorithms for connectionist reinforcement learning. Mach. Learn. **8**, 229–256 (1992)

Explainable Algorithm Selection for the Capacitated Lot Sizing Problem

Andrea Visentin[1,2]($\boxtimes$) , Aodh Ó Gallchóir[1], Jens Kärcher[3] ,
and Herbert Meyr[3]

[1] School of Computer Science and IT, University College Cork, Cork, Ireland
`andrea.visentin@ucc.ie,120323406@umail.ucc.ie`
[2] Insight SFI Research Centre for Data Analytics, Cork, Ireland
[3] Department of Supply Chain Management, University of Hohenheim,
Stuttgart, Germany
{`jens.kaercher,h.meyr`}`@uni-hohenheim.de`

Abstract. Algorithm selection is a class of meta-algorithms that has emerged as a crucial approach for solving complex combinatorial optimization problems. Successful algorithm selection involves navigating a diverse landscape of solvers, each designed with distinct heuristics and search strategies. It is a classification problem in which statistical features of a problem instance are used to select the algorithm that should tackle it most efficiently. However, minimal attention has been given to investigating algorithm selection decisions. This work presents a framework for iterative feature selection and explainable multi-class classification in Algorithm Selection for the Capacitated Lot Sizing Problem (CLSP). The CLSP is a combinatorial optimization problem widely studied with important industrial applications. The framework reduces the features considered by the machine learning approach and uses SHAP analysis to investigate their contribution to the selection. The analysis shows which instance type characteristics positively affect the relative performance of a heuristic. The approach can be used to improve the algorithm selection's transparency and inform the developer of an algorithm's weak and strong points. The experimental analysis shows that the framework selector provides valuable insights with a narrow optimality gap close to a parallel deployment of the heuristic set that generalises well to instances considerably bigger than the training ones.

Keywords: Algorithm Selection · xAI · Capacitated Lot Sizing · Feature Selection · SHAP Analysis

1 Introduction

An algorithm selection approach consists of dynamically choosing the most suitable algorithm from a diverse set based on problem characteristics. It is a prominent class of meta-algorithms based on the *No Free Lunch Theorem* [23] that no individual optimisation algorithm can outperform the competitor in all problem

B. Dilkina (Ed.): CPAIOR 2024, LNCS 14743, pp. 243–252, 2024.
https://doi.org/10.1007/978-3-031-60599-4_16

classes, and aims to leverage the diversity of available algorithms and to dynamically choose the most suitable one for a given problem instance. This approach is particularly noteworthy in solving combinatorial optimization problems, where selecting an appropriate algorithm significantly influences computational efficiency and/or the solution quality [14]. The applicability of algorithm selection spans a wide range of domains, showcasing its versatility and impact. It has been successfully applied in areas such as Boolean Satisfiability (SAT) [21,24], Constraint Programming [7,18], and combinations of the two [9]. These approaches quickly dominated the respected fields, and their success led to the creation of dedicated competitions [16] and problem libraries [1]. Beyond SAT and Constraint Programming, algorithm selection has been successfully employed in various fields, including operations research, artificial intelligence, and data science [12].

Algorithm selection is an artificial intelligence (AI) supervised multi-class classification task. Currently, most machine learning (ML) models trained on large amounts of data are black-box. However, their increased relevance prompts an effort to reach higher transparency and accountability. For this reason, there has been a surge in the explainable AI (xAI) approaches. While these have been recently applied to algorithm selection for ML approaches [13,22], the combinatorial optimisation selection has not yet been investigated from this perspective. In that perspective, being able to explain the decision of successful selectors can lead to useful insights into the strengths and weaknesses of the algorithms. Providing additional information to the developers and transparency to practitioners.

This work focuses on the Capacitated Lot Sizing Problem (CLSP), an NP-hard optimization problem with many practical applications [20]. Due to its complexity, instances relevant to the practitioners are hardly solvable to optimality in a reasonable time. Many different heuristics have been proposed over the years; these algorithms can provide solutions to very large instances with a close cost optimality gap at a reduced time compared to an optimal formulation. In the following setting, the candidates are selected based on the optimality cost gap with an optimal solution and not on the expected computational time. This metric is replicable for deterministic algorithms and does not depend on the hardware in which the algorithms are deployed. Our work is based on the first algorithm selector for CLSP recently introduced by Kärcher and Meyr [10].

In this paper, we introduce an xAI algorithm selection framework for the well-known CLSP problem. To the best of our knowledge, this is the first algorithm selection approach for combinatorial optimisation that uses SHAP analysis to understand the differences between solvers. We present a tailored iterative feature selection approach based on random forest feature importance. In an extensive computational analysis, we show that the performance of the portfolio approach is close to that of an oracle one. The algorithm selection is then applied to instances bigger than the ones in the training data. To analyse if the performance and the explanation can generalise to unseen instances of arbitrary size.

2   Capacitated Lot Sizing Problem (CLSP)

The Capacitated Lotsizing Problem (CLSP) is a dynamic, single-stage, multi-product challenge involving limited production capacity. In each period ($t \in T$), production lots for multiple products ($j \in J$) can be placed, subject to a production line's time capacity constraint (K_t). The production time for one unit of a product (a_j) and the demand for each product in a given period (d_{jt}) are considered. The demand must be met at all periods, and additional items that are not used to satisfy the demand are carried over, incurring per unit holding costs (hc_j). Setup costs are associated with initiating production for each product in a period (sc_j). Continuous non-negative variables represent production quantity (X_{jt}) and inventory level for each product in each period(I_{jt}), along with a binary setup variable (Y_{jt}). Conservation of the setup state across periods is not feasible, and the sequence of lots within a period is not predetermined. The CLSP is recognized as an NP-hard problem [2].

We consider the formulation presented in [8]:

$$\text{Minimize:} \sum_{t=1}^{T} \sum_{j=1}^{J} (hc_j \cdot I_{jt} + sc_j \cdot Y_{jt})$$

$$\text{Demand Constraint:} \quad I_{j,t-1} + X_{jt} - d_{jt} = I_{jt} \quad \forall j \in J, \forall t \in T$$

$$\text{Capacity Constraint:} \quad \sum_{i=1}^{J} X_{jt} \cdot a_j \leq K_t \quad \forall t \in T$$

$$\text{Linkage Constraint:} \quad X_{jt} \leq Y_{jt} \cdot \sum_{p=t}^{T} d_{jp} \quad \forall j \in J, \forall t \in T$$

$$\text{Non-negativity Constraint:} \quad X_{jt} \geq 0, \quad Y_{jt} \in \{0,1\} \quad \forall j \in J, \forall t \in T$$

The problem can be solved to optimality using a MILP solver. For the algorithm selection approach, we considered the following four heuristics: Lambrecht and Vanderveken (**LV**) [15], Dixon and Silver(**DS**) [5], Drogamaci et al. (**DPA**) [6] and Günther (**G**) [8]. These are the same heuristics used in [10]. A wider heuristic survey and comparison can be found in [3,11].

We considered the dataset presented in [10], comprising a total of 7200 CSLP instances generated from a simulated tire manufacturing process. We split the dataset randomly in 80% training and 20% test. They also provide a dataset of 360 instances that are considerably larger (factor of 2.24). For a more detailed description and analysis of the dataset, we refer to the original paper. For each instance, 17 *Key Performance Indicators* (KPIs) have been computed. These aim to capture the structural features of the instances.

The KPIs are divided into 3 main groups:

- KPIs 1–8 relative to production utilization, which try to express the level of difficulty of a problem instance in terms of feasibility or solution quality
- KPIs 9–12 express the course of the demand curve, e.g. increasing/decreasing patterns or seasonality.

– KPI 13–17 related to lot shifting and product properties, capturing the heterogeneity of the products and statistics considered during the load shifting process.

3   Methodology

We consider the selection approach to be a multi-class classification problem and use random forests (RF) for feature and algorithm selection. We opted for this model after benchmarking against various ML techniques for its accuracy and simplicity in extracting feature importance. A high number of features can hinder the model's explainability. Inputs that are strongly correlated or encode the same characteristic divide among them the importance, hiding important patterns. Moreover, a smaller set of features reduces the computational overhead of their computation and of the classifier.

RF provides the impact of each feature on the classifier's decision. However, this rank is strongly affected by correlations and attributes affecting the decisions in a similar way. We designed an iterative method for feature selection that is particularly effective for this problem. We iteratively train a RF on the dataset and compute the features' importance. Then evaluate through cross-validation an RF trained on the single most important feature, then add the following feature and repeat until all the features are selected. Finally, we select the minimum subset of features that gives a similar (or better) accuracy to the complete set. If the selected subset does not include all the current features, we iterate the procedure. The goal is to select enough features to preserve information and make decisions that are as good as the ones taken over the full dataset. However, some of these features could still be redundant. So, by iterating the overall process, it is possible to reduce them further. The experimental analysis shows the efficacy of this technique.

To improve the transparency of the model, we use SHAP (SHapley Additive exPlanations, [17]). SHAP is a technique in the interpretability and explainability of ML models. It is based on cooperative game theory and provides a way to fairly allocate a value (or contribution) to each feature in a prediction. It helps to understand the reasons for a decision by attributing the model's output to each input feature in a fair and comprehensive manner. It is based on RFs, so the proxy model used for the explainability is closer to the original one.

4   Experimental Results

In an extensive numerical analysis, we assess the efficacy of the algorithm selection and the clarity of its explanations and if they are able to generalise to larger instances.

4.1   Feature Importance and Selection

Figure 1 shows the plots generated during the iterative feature selection process. The X-axis label contains the KPIs ordered by importance. In the first iteration, the maximum is reached after including 7 KPIs. This is reduced to 3 after the second iteration, and the last one confirms that further reductions would compromise the accuracy. The final cross-validation accuracy on the training dataset of a model trained on KPI 2, 12 and 17 is equivalent to the accuracy reached using all KPIs. The KPIs selected cover the main aspects of an instance, and they are one for each KPI group described above:

- KPI 2 is related to utilization. It is the maximum utilization of a period.
- KPI 12 is the slope of the regression line that approximates the demand curve. It captures the demand trend.
- KPI 17 is the average ratio of setup and holding costs for a lot shift of the size of an average period demand. The smaller ratio of setup and holding costs implies that fewer lot shifts are necessary to obtain a high-quality solution.

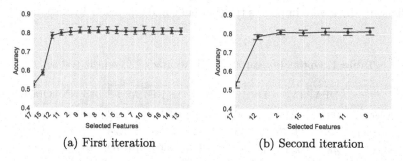

(a) First iteration   (b) Second iteration

Fig. 1. Accuracy of the heuristic selector over increasing subsets of features. In the third iteration, the order of importance did not change.

It is interesting to see how their relevance changes over different iterations. The plots clarify the importance of the iterative approach; just selecting the top 2 features in the first iteration led to an accuracy of less than 60%, while in the last one, it was almost 80% with the same amount of features. This is because multiple KPIs are correlated or contribute to the decision in a similar way, making them redundant. Overall, we managed to reduce the number of KPIs computed to one-fifth of the original ones with an equivalent accuracy.

4.2 Cost Optimality Gap Performance

The aim of the portfolio is to compute competitive solutions to the CLSP problem. We compare the plans provided by the different approaches based on their cost, in particular considering the optimality gap with the optimal solution computed by an MILP solver. The approaches compared are:

- The four heuristics (**DPA**, **LV**, **DS** and **G**).
- The Algorithm Selector based on Random Forest (RF) trained using all the KPIs (**AS RF**).

- The previous approach deployed on the selected features (**AS FS**).
- The deep neural network CLSP-Net presented in [10] (**AS NN**). It is a feed-forward fully connected network with dropout as a regularizer. We can consider this approach state-of-the-art in AS for CLSP.
- An **Oracle** approach that always selects the best heuristic. We consider this as a baseline of what could be achievable by having an AS that always selects the best heuristic.

The comparison results are presented in Table 1. We can see that the AS algorithms clearly outperform all the individual heuristics, reaching an optimality gap close to the oracle. The feature reduction minimally affects the accuracy of the RF. Compared to the deep learning approach, the RF selectors fail to choose the best heuristic for 2 strong outliers. For this reason, their average optimality gap is higher, while the median is really close. A reduced set of features leads to a faster computation of such a set and a quicker classification time. Moreover, the simplicity of the model improves transparency and has faster hyperparametrisation.

Table 1. Optimality gap (%) and Accuracy (%) on the test set.

	DPA	LV	DS	G	AS RF	AS FS	AS NN	Oracle
Average	4.91	9.00	3.60	5.21	2.29	2.32	2.28	2.02
Median	1.70	6.76	1.55	1.82	1.26	1.28	1.27	1.15
Max	48.46	99.04	188.38	188.38	108.02	108.02	108.02	28.22
Accuracy	43.9	2.2	35.7	18.2	79.8	78.5	80.5	100.0

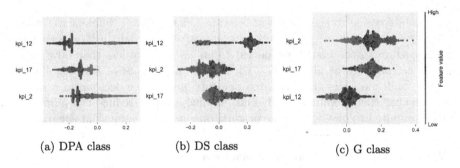

(a) DPA class   (b) DS class   (c) G class

Fig. 2. Feature impact on the classifier decision for the test set. The X-axis contains the SHAP value; positive values are linked to the selection of the specific class.

4.3 Explainability of the Results

We focused on the 3 most relevant heuristics; the classifier hardly selected LV due to poor performance. Figure 2 shows the SHAP analysis of the test set results;

this provides information on why the classifier is selecting a specific heuristic. The KPIs are ordered according to their relevance for the specific class, from top to bottom. Each dot is an instance of the test set; their colour represents if the KPI in that instance is low (blue) or high (red). A positive SHAP value, e.g. points further to the right of the plot, means that KPI is positively contributing to selecting the specific heuristic. We can see that the slope of the regression line of the demand (KPI 12) is the most relevant for selecting the DPA or the DS heuristic. In particular, DS deals better with increasing demand patterns, while decreasing demand leads to the selection of DPA. G works particularly well on instances with high maximum utilization (KPI 2). This means that it can better shift the production to early periods and not with demand-driven production. In a problem instance with a low ratio of setup and holding costs (KPI 17), lot shifts (moving a production lot to a period where the setup is already used) are hardly convenient. These are generally considered easier instances and are those in which DS does not perform well.

Table 2. Optimality gap (%) and Accuracy (%) on the large dataset.

	DPA	LV	DS	G	AS RF	AS FS	AS NN	Oracle
Average	4.37	9.74	3.09	5.97	2.08	2.11	2.04	1.91
Median	1.25	7.95	1.04	1.20	0.73	0.74	0.78	0.72
Max	42.25	49.18	84.32	84.32	19.48	19.48	19.48	19.26
Accuracy	36.9	0.6	38.9	23.6	75.8	73.9	75.6	100.0

4.4   Generalisation on Bigger Instances

A heuristic is designed to solve big instances that are generally not tackleable by optimal solvers. For this reason, we want to investigate if the accuracy of the algorithm selection scales on problems considerably larger than the training set. Table 2 shows the analysis results. We can see that all selectors perform well on the larger dataset, closing the gap with the Oracle approach. The algorithms based on RF exhibit a better median but a higher average than the AS NN. The number of outliers is considerably reduced, and the worst case is the same for all selectors.

Figure 3 shows how the decision patterns described in the previous sections generalise to the larger dataset. This consistency, alongside the good performance of AS FS, means that an increase in the instance size preserves the relationship among solvers inside the specific instance type.

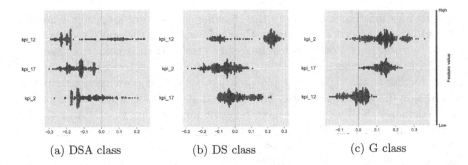

(a) DSA class   (b) DS class   (c) G class

Fig. 3. Feature impact on the classifier decision for the large dataset. The X-axis contains the SHAP value.

5   Conclusions

This paper introduces an xAI algorithm selection approach for the Capacitated Lot Sizing Problem (CLSP). The proposed framework, integrating iterative feature selection and explainable multi-class classification, contributes to enhancing the transparency of algorithm selection decisions. The experimental analysis demonstrates the framework's effectiveness, providing valuable insights with a narrow optimality gap, closely aligning with a parallel deployment of the heuristic set. Moreover, the framework exhibits robust generalization capabilities on a test set considerably larger than the training data, showcasing its potential for informing developers about algorithm strengths and weaknesses in diverse, real-world scenarios. This work emphasizes the importance of understanding and explicability in algorithm selection decisions, paving the way for more informed and transparent approaches to tackling complex optimization challenges.

Several promising avenues emerge from this preliminary work. Concerning the CLSP, the additional KPIs can improve the model accuracy, for example, features learned by machine learning in an unsupervised way [4] or extracted by the heuristics' binaries [19]. In particular, introducing explainability to the last one could lead to more informative feedback for the heuristics developers. Additionally, extending the framework to accommodate a broader array of combinatorial optimization problems beyond CLSP would contribute to its versatility and applicability across diverse domains like SAT.

Acknowledgments. This publication has emanated from research supported by Science Foundation Ireland under Grant No. 12/RC/2289-P2 at Insight, the SFI Research Centre for Data Analytics, which is co-funded under the European Regional Development Fund, and by the EU Horizon project TAILOR (952215). The authors gratefully acknowledge useful inputs from Lars Kotthoff.

References

1. Bischl, B., et al.: ASlib: a benchmark library for algorithm selection. Artif. Intell. **237**, 41–58 (2016)
2. Chen, W.H., Thizy, J.M.: Analysis of relaxations for the multi-item capacitated lot-sizing problem. Ann. Oper. Res. **26**, 29–72 (1990). https://doi.org/10.1007/BF02248584
3. Copil, K., Wörbelauer, M., Meyr, H., Tempelmeier, H.: Simultaneous lotsizing and scheduling problems: a classification and review of models. OR Spect. **39**(1), 1–64 (2017). https://doi.org/10.1007/s00291-015-0429-4
4. Dalla, M., Visentin, A., O'Sullivan, B.: Automated SAT problem feature extraction using convolutional autoencoders. In: IEEE International Conference on Tools with Artificial Intelligence (ICTAI) (2021)
5. Dixon, P.S., Silver, E.A.: A heuristic solution procedure for the multi-item, single-level, limited capacity, lot-sizing problem. J. Oper. Manag. **2**(1), 23–39 (1981). https://doi.org/10.1016/0272-6963(81)90033-4
6. Dogramaci, A., Panayiotopoulos, J.C., Adam, N.R.: The dynamic lot-sizing problem for multiple items under limited capacity. AIIE Trans. **13**(4), 294–303 (1981). https://doi.org/10.1080/05695558108974565
7. Gebruers, C., Hnich, B., Bridge, D., Freuder, E.: Using cbr to select solution strategies in constraint programming. In: Munoz-Avila, H., Ricci, F. (eds.) ICCBR 2005. LNCS, vol. 3620, pp. 222–236. Springer, Heidelberg (2005). https://doi.org/10.1007/11536406_19
8. Günther, H.O.: Planning lot sizes and capacity requirements in a single stage production system. Eur. J. Oper. Res. **31**(2), 223–231 (1987). https://doi.org/10.1016/0377-2217(87)90026-9
9. Hurley, B., Kotthoff, L., Malitsky, Y., O'Sullivan, B.: Proteus: a hierarchical portfolio of solvers and transformations. In: Simonis, H. (ed.) CPAIOR 2014. LNCS, vol. 8451, pp. 301–317. Springer, Heidelberg (2014). https://doi.org/10.1007/978-3-319-07046-9_22
10. Kärcher, J., Meyr, H.: A machine learning approach for identifying the best solution heuristic for a large scaled capacitated lotsizing problem. In: Preprint - Research Square (2023). https://doi.org/10.21203/rs.3.rs-3709286/v1
11. Karimi, B., Fatemi Ghomi, S., Wilson, J.: The capacitated lot sizing problem: a review of models and algorithms. Omega **31**(5), 365–378 (2003). https://doi.org/10.1016/S0305-0483(03)00059-8
12. Kerschke, P., Hoos, H.H., Neumann, F., Trautmann, H.: Automated algorithm selection: survey and perspectives. Evol. Comput. **27**(1), 3–45 (2019)
13. Kostovska, A., Doerr, C., Džeroski, S., Kocev, D., Panov, P., Eftimov, T.: Explainable model-specific algorithm selection for multi-label classification. In: 2022 IEEE Symposium Series on Computational Intelligence (SSCI), pp. 39–46. IEEE (2022)
14. Kotthoff, L.: Algorithm selection for combinatorial search problems: a survey. In: Data Mining and Constraint Programming: Foundations of a Cross-Disciplinary Approach, pp. 149–190 (2016)
15. Lambrecht, M.R., Vanderveken, H.: Heuristic procedures for the single operation, multi-item loading problem. AIIE Trans. **11**(4), 319–326 (1979). https://doi.org/10.1080/05695557908974478
16. Lindauer, M., van Rijn, J.N., Kotthoff, L.: The algorithm selection competitions 2015 and 2017. Artif. Intell. **272**, 86–100 (2019)

17. Lundberg, S.M., Lee, S.I.: A unified approach to interpreting model predictions. Adv. Neural Inf. Process. Syst. **30**, 1–10 (2017)
18. Müller, D., Müller, M.G., Kress, D., Pesch, E.: An algorithm selection approach for the flexible job shop scheduling problem: choosing constraint programming solvers through machine learning. Eur. J. Oper. Res. **302**(3), 874–891 (2022)
19. Pulatov, D., Anastacio, M., Kotthoff, L., Hoos, H.: Opening the black box: automated software analysis for algorithm selection. In: International Conference on Automated Machine Learning, pp. 6–1. PMLR (2022)
20. Ramya, R., Rajendran, C., Ziegler, H., Mohapatra, S., Ganesh, K., et al.: Capacitated Lot Sizing Problems in Process Industries. Springer, Heidelberg (2019). https://doi.org/10.1007/978-3-030-01222-9
21. Sadreddin, A., Mouhoub, M., Sadaoui, S.: Portfolio selection for sat instances. In: 2022 IEEE International Conference on Systems, Man, and Cybernetics (SMC), pp. 2962–2967. IEEE (2022)
22. Shao, X., Wang, H., Zhu, X., Xiong, F., Mu, T., Zhang, Y.: EFFECT: explainable framework for meta-learning in automatic classification algorithm selection. Inf. Sci. **622**, 211–234 (2023)
23. Wolpert, D.H., Macready, W.G.: No free lunch theorems for optimization. IEEE Trans. Evol. Comput. **1**(1), 67–82 (1997)
24. Xu, L., Hutter, F., Hoos, H.H., Leyton-Brown, K.: SATzilla: portfolio-based algorithm selection for sat. J. Artif. Intell. Res. **32**, 565–606 (2008)

An Efficient Structured Perceptron for NP-Hard Combinatorial Optimization Problems

Bastián Véjar[1(✉)], Gaël Aglin[2], Ali İrfan Mahmutoğulları[1], Siegfried Nijssen[2], Pierre Schaus[2], and Tias Guns[1]

[1] Department of Computer Science, KU Leuven, Leuven, Belgium
`Bastian.vejar@kuleuven.be`
[2] ICTEAM, UCLouvain, Louvain-la-Neuve, Belgium

Abstract. A fundamental challenge when modeling combinatorial optimization problems is that often multiple sub-objectives need to be weighted against each other, but it is not clear how much weight each sub-objective should be given: consider routing problems that trade off distance and duration where the relative importance of the two is not known a priori. In recent work, it has been proposed to use machine learning algorithms from the domain of structured output prediction to learn such weights from examples of desirable solutions. However, until now such techniques were only evaluated on fast-to-solve optimization problems. We propose and evaluate three techniques that make it feasible to apply the structured perceptron on NP-hard optimization problems: 1) using heuristic solving methods during the learning process, 2) solving well-chosen satisfaction variants of the problems, 3) caching solutions computed during the learning process and reusing them. Experiments confirm the validity and speed-ups of these techniques, enabling structured output learning on larger combinatorial problems than before.

Keywords: Structured output prediction · Combinatorial optimization · Structured Perceptron

1 Introduction

Combinatorial Optimization (CO) deals with solving optimization problems such as scheduling and planning problems. These problems are formalized as maximizing an objective function over a set of feasible solutions defined through a set of constraints. In many applications, however, multiple objectives can be considered. Notable examples of this appear in engineering [1,6], economics [13], and logistics [7]. As an example, in a logistics application, one may wish to trade off total distance, total travel time, driver familiarity and fairness among drivers.

Since the objectives of a Multi-Objective Combinatorial Optimization Problem (MOCOP) are possibly conflicting, a nontrivial problem is how to formalize MOCOP problems using a single objective function. The classical methods

B. Dilkina (Ed.): CPAIOR 2024, LNCS 14743, pp. 253–262, 2024.
https://doi.org/10.1007/978-3-031-60599-4_17

either try to enumerate all non-dominated solutions (see, for example, [5]) or use a weighted linearization of the objectives –turning them into sub-objectives– to transform the problem into a single-objective optimization problem, for example, [10]. However, the weights in the latter approach may not be explicitly known to the decision-maker.

The motivation for this paper comes from the scheduling and production process of a steel mill company. The company tries to set the weights of many sub-objectives of their scheduling process, e.g. the difference in height and width of adjacent products, based on a perception of their importance. This process requires several iterations of tuning and observing the production process with these updated parameters. The company, however, has a set of historical preferred solutions that a machine learning approach can use to learn from.

Structured Output Prediction (SOP) is a family of machine learning techniques used for learning to predict structured objects. When used to learn the weights of a MOCOP, SOP could be seen as a type of Inverse Optimization (IO) for mixed-integer linear problems [3,14]. However, unlike in IO, in SOP there is typically no initial weight vector given and it considers multiple (noisy) solutions each with different feasible spaces, and the learning focuses on statistical properties that generalize to unseen instances rather than pure reconstruction. The SOP methods are known to be effective if calculating the output structure can be done in polynomial time. Although SOP can also be used over non-polynomial solvers, the training part of SOP, such as done by the Structured Perceptron (SP), requires solving a large number of instances with different weights. In the steel mill company example, the corresponding MOCOP is NP-hard, and finding an optimal solution requires a large amount of computational power, and hence training time for SOP.

We hence propose a number of techniques that make it feasible to apply the SP to hard MOCOP problems, where the goal is to learn the weights of the sub-objectives. A core idea is that we replace the need for an exact solution of the MOCOP during the learning process with less demanding alternatives in terms of solution time and power. Our main observation is that during the learning process, an optimal solution of a MOCOP can be replaced by any solution that shows that the current choice of weights does not yet favor the desired solution according to the training data. Based on this observation we introduce computational enhancements such as timeouts, a reformulation of the problem based on the better-score condition, and scanning a cache of historical solutions. Finally, we conduct computational experiments to observe the benefit of the proposed methods on multi-objective versions of two NP-hard combinatorial optimization problems: the Knapsack Problem (KP) and the Price Collecting Traveling Salesman Problem (PCTSP).

The contributions of our work are:

- We propose a novel valid update approach for the training of the SP. This approach exploits any solution that, for the current choice of weights, has a better score than the preferred solution, to avoid having to solve a MOCOP optimally for each instance at each training epoch.

- We also present computational improvements to speed up the training of the SP. We use a heuristic approach, adding the score criterion as a constraint and solving a SAT, and solution caching instead of solving an NP-hard optimization problem optimally.
- We test our approach on synthetic data for two different combinatorial problems, namely, the knapsack (KP) and prize-collecting traveling salesman problems (PCTSP), of different sizes. The results show that our approach is an improvement over the standard SP not only in terms of computation time but also in terms of the quality of learning.

2 Background: Structured Output Prediction

The general goal of SOP is to predict the most relevant structured output given a description of an input problem. Formally, let $\mathcal{D} = \{(\boldsymbol{x}, \boldsymbol{y})\}$ be a dataset, where $\boldsymbol{x}$ is an input structure (e.g. properties and structural constraints) and $\boldsymbol{y}$ is its corresponding structured output. $\mathcal{X}$ is the set of all possible input structures and $\mathcal{Y}$ is the structured output space consisting of all possible outputs that fulfill the required structure. The objective of SOP is to learn a hypothesis function $h \in \mathcal{H}$ such that for any instance $\boldsymbol{x}$, it produces an output $\boldsymbol{y}' = h(\boldsymbol{x})$ such that $\boldsymbol{y}' \in \mathcal{Y}$ and the distance $dist(\boldsymbol{y}, \boldsymbol{y}')$ between the desired and the predicted outputs is as small as possible. The definition of $dist(\boldsymbol{y}, \boldsymbol{y}')$ is problem specific, e.g. the number of non-overlapping elements in a set prediction problem.

Multiple algorithms for SOP exist, including the structured perceptron [4], stochastic sub-gradient [12], and cutting-plane algorithms [8]. We will build on the seminal SP algorithm of Collins (see [4]). The SP is limited to learning a linear function over its input. However, this setting exactly matches our case of learning the weights of a linear objective function (see Sect. 3).

The key issue in structured output prediction is that $\boldsymbol{x}$ and $\boldsymbol{y}$ are structured. To be able to learn a linear function with weights $\boldsymbol{w}$, the SP assumes that we can define a problem-specific representation function Φ that maps every valid input/output pair to a feature vector: $\Phi : \mathcal{X} \times \mathcal{Y} \mapsto \mathbb{R}^{|w|}$. Given an input structure $\boldsymbol{x}$ and a (learned) weight vector $\boldsymbol{w}$, decoding an output structure (called the *inference* task) then becomes: $h(\boldsymbol{x}) = \arg\max_{\boldsymbol{y}' \in \mathcal{Y}} \boldsymbol{w}^\top \Phi(\boldsymbol{x}, \boldsymbol{y}')$.

Structured Perceptron Algorithm. Algorithm 1 presents its pseudocode. It starts with an arbitrary initialization of the weights on line 1, e.g. using unit weights. As long as a global time limit T has not been reached, it goes over each $(\boldsymbol{x}, \boldsymbol{y})$ instance in the data, and calls the inference algorithm to find the output structure that maximizes $\boldsymbol{w}^\top \Phi(\boldsymbol{x}, \boldsymbol{y}')$ given the current weight vector $\boldsymbol{w}$. The condition on line 5 then checks whether the score of $\boldsymbol{y}' : \boldsymbol{w}^\top \Phi(\boldsymbol{x}, \boldsymbol{y}')$ is higher than the score of the desired solution $\boldsymbol{y}$. If so $\boldsymbol{y} \notin h(\boldsymbol{x})$; hence there is an error and we should update the weights to make $\boldsymbol{y}$ more favorable. This is done in line 6, where the coefficients for features where $\boldsymbol{y}$ has a higher value are increased and others are decreased (scaled by some learning rate η).

Algorithm 1 Structured Perceptron

Require: Training dataset $\mathcal{D} = \{(\boldsymbol{x}, \boldsymbol{y})\}$, maximum time limit T, learning rate η
1: $\boldsymbol{w} \leftarrow \boldsymbol{1}$ ▷ *Parameter initialization*
2: **while** time limit T has not been reached **do**
3:     **for** each instance $(\boldsymbol{x}, \boldsymbol{y})$ in $\mathcal{D}$ **do**
4:        Find $\boldsymbol{y}' \in \mathcal{Y}$ that maximizes $\boldsymbol{w}^\top \boldsymbol{\Phi}(\boldsymbol{x}, \boldsymbol{y}')$ ▷ *Call inference algorithm*
5:        **if** $\boldsymbol{w}^\top \boldsymbol{\Phi}(\boldsymbol{x}, \boldsymbol{y}') > \boldsymbol{w}^\top \boldsymbol{\Phi}(\boldsymbol{x}, \boldsymbol{y})$ **then**
6:          $\boldsymbol{w} \leftarrow \boldsymbol{w} + \eta(\boldsymbol{\Phi}(\boldsymbol{x}, \boldsymbol{y}) - \boldsymbol{\Phi}(\boldsymbol{x}, \boldsymbol{y}'))$ ▷ *Parameter update*
7: **return** $\boldsymbol{w}$

---

3 SP for Multi-Objective Combinatorial Optimization

The goal of SP for MOCOP is to learn a weight vector w that minimizes the distance between the weighted linearization of the predicted and true solutions based on a given data set of features (that fully characterize the optimization problem) and solution pairs. The learned weights can be used in decision-making for unseen features in the future.

Formally, let $z_1(y), \ldots, z_p(y)$ be the p objective functions of a MOCOP and let $\mathcal{C}(y)$ represent the set of constraints it must satisfy. A formulation with a weighted linearization of the sub-objectives can be written as:

$$\max_{y'} \quad w_1 z_1(y') + \ldots + w_p z_p(y') \tag{1}$$
$$\text{s.t.} \quad \mathcal{C}(y')$$

where the weights $w_1, \ldots, w_p$ represent the importance of sub-objectives $1, \ldots, p$, respectively. The formulation (1) is commonly used in industrial applications where the different sub-objectives are typically penalties or bonuses, such as cost, energy use, profitability, overtime penalties, smoothness penalties, etc.

3.1 Input Structure and Representation Function $\boldsymbol{\Phi}$

We now show how the SP algorithm can be used, in case of a set of problem instances and solutions for which the constraint set $\mathcal{C}$ may change, but for which the number of sub-objective functions $(z_1, \ldots, z_p)$ stays the same as well as the nature of what they compute. For example, the first sub-objective might compute the total cost of the solution, the second the total energy consumption, the third the time-to-build, the fourth how much the total overdue time is, etc. More formally we assume a dataset $\{(\boldsymbol{x}, \boldsymbol{y})^i\}$ where $\boldsymbol{x}^i = (\mathcal{C}^i, (z_1^i, \ldots, z_p^i))$ fully characterizes the optimization problem, and $\boldsymbol{y}^i$ is a solution that satisfies $\mathcal{C}^i$.

Given a MOCOP's input structure $\boldsymbol{x} = (\mathcal{C}, (z_1, \ldots, z_p))$ we can now see that the representation function $\boldsymbol{\Phi}$ is simply the vector of the sub-objective values, i.e., $\boldsymbol{\Phi}(\boldsymbol{x}, \boldsymbol{y}) = (z_1(\boldsymbol{y}), \ldots, z_p(\boldsymbol{y}))$ and the inference task in Algorithm 1 line 4 becomes : $h(\boldsymbol{x}) = \arg\max_{y' \in \mathcal{C}} w_1 z_1(y') + \ldots + w_p z_p(y')$, that is exactly our weighted single-objective formulation.

Example 1. For example, assume a bi-objective KP where $\boldsymbol{y}$ is a vector of binary decision variables, and we have two value vectors v_1 and v_2 and a size vector s with a total size limit of l. The mathematical problem formulation is $\max_{\boldsymbol{y}'} w_1(v_1^\top \boldsymbol{y}') + w_2(v_2^\top \boldsymbol{y}')$ s.t. $s^\top \boldsymbol{y}' \leq l$ which corresponds to input structure $\boldsymbol{x} = ((s, l), (v_1, v_2))$ and the representation function $\Phi(\boldsymbol{x}, \boldsymbol{y}) = (v_1^\top \boldsymbol{y}, v_2^\top \boldsymbol{y})$. Note how for different instances $\{(\boldsymbol{x}, \boldsymbol{y})^i\}$ the items might be different, as well as their sizes and values, but the goal is always to balance the first sub-objective (e.g. cost) with the second (e.g. durability).

3.2   Scaling up the SP Algorithm

The main issue with using SOP methods like the SP in this setting is that on line 4 of Algorithm 1, we will have to solve the single-objective CO repeatedly for each instance at each epoch. Solving it for one instance may take minutes to hours, which would make training prohibitively slow.

The key observation that we exploit in this paper is that we do not need to compute (and prove) the optimal solution for the learning to work. In fact, any feasible solution $\boldsymbol{y}'$ for which $\boldsymbol{w}^\top \Phi(\boldsymbol{x}, \boldsymbol{y}') > \boldsymbol{w}^\top \Phi(\boldsymbol{x}, \boldsymbol{y})$ (the update condition on line 5) can be used to do a *valid parameter update* step in line 6.

Based on this observation we take inspiration from computational improvements in related areas such as bilevel optimization [2] and decision-focused learning [9,11]. More specifically we explore three orthogonal techniques for quickly finding a feasible solution $\boldsymbol{y}' \in \mathcal{C}$ that satisfies the update condition:

Heuristic Solving. The first approach is to *replace* the exact solver by a heuristic solver. This could be a greedy algorithm, a local search algorithm, or even the exact solver with a limited timeout. While this is not guaranteed to find a solution that satisfies the valid update condition, if it does find such a solution then it can be used. If not, this instance will be skipped in the hopes of encountering other instances for which the heuristic method does find a usable solution.

Satisfying the Update Condition. An orthogonal approach is based on the realization that an exact solver will spend a lot of time proving optimality. A way to avoid this effort is to stop the solver as soon as it finds a solution that satisfies the update condition. Alternatively, we can remove the objective function from the formulation and instead add the valid update condition directly as a constraint. This turns the problem into a satisfaction problem, which we expect will often be easier to solve than the optimization variant.

Caching Feasible Solutions. A third approach is to *cache* all feasible solutions found in previous solve calls. That is, for any $\mathcal{C}^i$, a feasible solution found for one weight vector w will still be a feasible solution for another weight vector w (this might not be true for different instances with different $\mathcal{C}^i$ hence a separate cache for every $\mathcal{C}^i$ should be maintained). Given such a cache, before calling the solver (on line 4), we can first do a simple linear scan over the solutions in the cache to see if a feasible solution satisfies the valid update condition. If so, no solver needs to be called and an update step can immediately be applied. Otherwise, no solution in the cache can be used and a solver should be called.

4 Experiments

Three metrics are used to evaluate the performance of our proposed techniques: (1) the cosine distance between the real weights and the predicted weights during the learning process (lower is better), (2) the solution accuracy (Score) which measures the similarity of the decisions made with the predicted weight vector compared to the real one (higher is better) and (3) the cost difference (CΔ) when comparing the predicted solutions to the real ones while using the true coefficients (lower is better). The source code, along with the datasets used for experimentation, is available in our public repository https://github.com/BastianVT/EfficientSP.

The research questions we answer in the computational experiments are:

- **RQ1**: How does using a heuristic versus an exact solver impact the speed and quality of learning good weight vectors?
- **RQ2**: Is there an additional benefit to just computing *any* solution that leads to a valid update rule during training?
- **RQ3**: To what extent does the use of a solution cache avoid the need to call the solver, and what is the influence on learning speed and quality?
- **RQ4**: How similar and what is the quality of the *solutions* obtained from solving with the true weight vector and the predicted weight vectors?

Problem Formulations. We will use multi-objective versions of two COPs, namely the Knapsack Problem (KP) and the Price Collecting Traveling Salesman Problem (PCTSP), used in the computational experiments.

Multi-objective Knapsack Problem. This problem is the one introduced in example 1, but generalized to p sub-objectives.

Multi-objective Prize Collecting Traveling Salesman Problem. Let $G = (V, E)$ be a complete undirected graph, where V is the set of cities and E is the set of edges connecting the cities. Each edge $(i, j) \in E$ has p different values $v_1^{ij}, \ldots, v_p^{ij}$ that can be interpreted as different features of that edge such as travel distance, fuel consumption, the total ascent amount, etc., for which smaller values are preferred. Each city $i \in V$ has a penalty γ^i for not being visited and a reward π^i for being visited. Then, the objective of the single-objective formulation (1) for the multi-objective PCTSP is $\min_y \sum_{(i,j) \in E} \left(w_1 v_1^{ij} + \ldots + w_p v_p^{ij} \right) y_{ij} + w_{p+1} \sum_{i \in V} \gamma^i (1 - o_i)$ where y_{ij} is the binary variable indicating whether edge (i, j) is in the tour ($y_{ij} = 1$) or not and o_i is the binary variable indicating whether city is visited ($o_i = 1$) or not. The constraints of the model ensure there is a valid single circuit, link the o_i and y_{ij} variables, and constrain the given minimum reward.

RQ1: Exact Versus Heuristic Solving. Figure 1 shows the cosine distance between the true and predicted weight vectors, using an exact or heuristic solver.

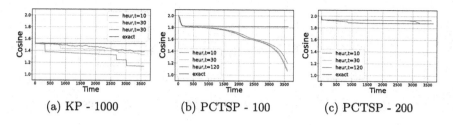

Fig. 1. Cosine distance over time for the heuristic versus an exact solver

Figure 1a shows the results for knapsack (KP) with 1000 items, and Fig. 1b and 1c for the PCTSP with 100/200 stops.

First, for the KP, using an exact solver (shown in purple in Fig. 1a) does not allow for improving the predicted weight vector after the first ten iterations. It spends the rest of the training time unsuccessfully trying to find an optimal solution for one training instance. When using a timeout, leading to heuristic solutions, the algorithm continues to decrease its cosine distance because it can identify valid updates even for instances that are difficult to solve to optimality.

For the PCTSP with 100 stops, the exact solver has a similar issue (shown in purple in Fig. 1b), after a number of successful update steps it encounters a hard-to-solve instance and spends the remaining time on it. On the other hand, the use of timeouts allows the learning to continue. Finally, for the PCTSP with 200 stops, the exact solver behaves analogous to before. But unlike the previous cases, adding a timeout does not lead to better results; in most cases the timeout is too low to find a solution, meaning learning can't progress much either.

RQ2: Valid-Update Solution. Figure 2 presents the cosine distance between the true and predicted weight vectors when using our valid-update condition as a constraint ('any'); we discuss the dotted 'cache' lines later. Based on our previous experiment, we used a timeout of 10 s for KP and 120 s for PCTSP instances. The inverted triangle markers indicate the completion of an epoch (each instance is solved once in an epoch).

We first look at the solid lines (blue and green, 'heur' vs. 'any'). For every dataset tested, our valid-update approach ('any') provides a benefit over heuristic solving as the cosine distance decreases faster. Solving the satisfaction problem allows for a larger number of valid updates compared to solving with a timeout. In Fig. 2c, we observe for PCTSP-200 spikes in the graph corresponding to an *increase* in the cosine distance for the 'any' method. We hypothesize that sometimes the solution found when searching for 'any' valid update, might be very different leading to a direction that is not well aligned with the true cost vector; however, we see that despite these spikes the learning does converge to lower cosine distances.

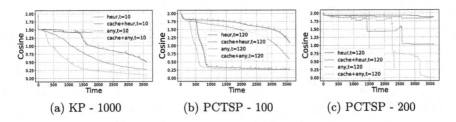

(a) KP - 1000   (b) PCTSP - 100   (c) PCTSP - 200

Fig. 2. Cosine distance over time for the techniques presented (Color figure online)

RQ3: Caching Feasible Solutions. We now focus on the yellow and red lines in Fig. 2 to observe the effect of solution caching. In the first epoch, the performance is identical to not using a cache, as all caches are still empty. We observe that from the second epoch the learning proceeds quicker (more epochs per time unit) and better (faster decreasing cosine distances) when using a cache. Using a cache on top of the 'any' method (yellow lines) leads to the best results overall.

RQ4: Solution Quality of the Predicted Weights. Finally, we examine the effect of the proposed methods on the *solutions* found.

Learning method	KP - 1000		PCTSP - 100		PCTSP - 200	
	Score	CΔ	Score	CΔ	Score	CΔ
Exact	96.8%	1e-3	33.4%	0.64	32.9%	2.08
Cache + heur, $t = 120$	96.4%	1e-3	92.6%	3e-4	33.8%	1.18
Cache + any, $t = 120$	**98.3%**	1e-3	**93.9%**	**1e-4**	**75.7%**	**3e-3**

The above table compares the similarity between the solution obtained with the ground truth weights versus the one obtained with the predicted weights (Score: higher is more similar) and the cost difference of the predicted and real solutions with respect to the true coefficients (CΔ: lower is better). We show the results for 'exact' and caching with heuristic/any valid solution. We see that better cosine distances (previous experiments) also lead to better solutions. We also can see that the 'cache + any' technique leads to the lowest cost difference for the PCTSP, with the most significant improvement for the largest dataset.

The introduction of the techniques presented in this paper achieves the highest score when they are used together.

5   Conclusion

This paper investigates the use of structured output prediction to learn the weights of hard multi-objective combinatorial optimization problems. The key observation is that we don't need to solve to optimality to get a valid update for learning. We propose and evaluate three techniques to speed up the solving: the use of heuristic solving (time limits), the incorporation of a 'valid update' condition as a constraint, and a cache that allows to sometimes skip calling the solver altogether. Our experiments on multi-objective KP and PCTSP problems showed that these techniques speed up the learning compared to using an exact solver, with the combination of caching and solving to find any valid update leading to the best and most scalable results. This opens the door to using structured output prediction on more realistic and hard-to-solve industrial problems.

Future work includes applying these techniques to other SOP methods; deepening algorithm connections to inverse optimisation and bi-level optimization; more datasets and standardized measures for evaluating different methods; and the use and evaluation in real-life applications.

Acknowledgments. This research was partly funded by the European Research Council (ERC) under the EU Horizon 2020 research and innovation program (Grant No. 101002802, CHAT-Opt and Grant No. 101070149, Tuples), and the Institute for the Encouragement of Scientific Research and Innovation of Brussels (Innoviris, 2021-RECONCILE).

References

1. Bemporad, A., de la Peña, D.M.: Multiobjective model predictive control. Automatica **45**(12), 2823–2830 (2009)
2. Camacho-Vallejo, J.F., Corpus, C., Villegas, J.G.: Metaheuristics for bilevel optimization: a comprehensive review. Comput. Oper. Res. 106410 (2023)
3. Chan, T.C., Mahmood, R., Zhu, I.Y.: Inverse optimization: theory and applications. Oper. Res. (2023)
4. Collins, M.: Discriminative training methods for hidden Markov models: theory and experiments with perceptron algorithms. In: Proceedings of the 2002 Conference on Empirical Methods in Natural Language Processing (EMNLP 2002), pp. 1–8 (2002)
5. Deb, K., Sindhya, K., Hakanen, J.: Multi-objective optimization. In: Decision Cciences, pp. 161–200. CRC Press (2016)
6. Ganesan, T., Elamvazuthi, I., Shaari, K.Z.K., Vasant, P.: Hypervolume-driven analytical programming for solar-powered irrigation system optimization. In: Chen, G., Rossler, O., Snasel, V., Abraham, A. (eds.) Nostradamus 2013: Prediction, Modeling and Analysis of Complex Systems. LNCS, vol. 210, pp. 147–154. Springer, Heidelberg (2013). https://doi.org/10.1007/978-3-319-00542-3_15
7. Jayarathna, C.P., Agdas, D., Dawes, L., Yigitcanlar, T.: Multi-objective optimization for sustainable supply chain and logistics: a review. Sustainability **13**(24), 13617 (2021)
8. Joachims, T., Hofmann, T., Yue, Y., Yu, C.N.: Predicting structured objects with support vector machines. Commun. ACM **52**(11), 97–104 (2009)
9. Mandi, J., Stuckey, P.J., Guns, T., et al.: Smart predict-and-optimize for hard combinatorial optimization problems. In: Proceedings of the AAAI Conference on Artificial Intelligence, vol. 34, pp. 1603–1610 (2020)
10. Marler, R.T., Arora, J.S.: The weighted sum method for multi-objective optimization: new insights. Struct. Multidisc. Optim. **41**, 853–862 (2010)

11. Mulamba, M., Mandi, J., Diligenti, M., Lombardi, M., Bucarey, V., Guns, T.: Contrastive losses and solution caching for predict-and-optimize. arXiv preprint arXiv:2011.05354 (2020)
12. Ratliff, N., Bagnell, J.A., Zinkevich, M.: Subgradient methods for maximum margin structured learning. In: ICML Workshop on Learning in Structured Output Spaces, vol. 46 (2006)
13. Tapia, M.G.C., Coello, C.A.C.: Applications of multi-objective evolutionary algorithms in economics and finance: a survey. In: 2007 IEEE Congress on Evolutionary Computation, pp. 532–539. IEEE (2007)
14. Wang, L.: Cutting plane algorithms for the inverse mixed integer linear programming problem. Oper. Res. Lett. $37(2)$, 114–116 (2009)

Robustness Verification in Neural Networks

Adrian Wurm[(✉)]

BTU Cottbus-Senftenberg, Lehrstuhl Theoretische Informatik,
Platz der Deutschen Einheit 1, 03046 Cottbus, Germany
wurm@b-tu.de
https://www.b-tu.de/

Abstract. In this paper we investigate formal verification problems for Neural Network computations. Of central importance will be various robustness and minimization problems such as: Given symbolic specifications of allowed inputs and outputs in form of Linear Programming instances, one question is whether there do exist valid inputs such that the network computes a valid output? And does this property hold for all valid inputs? Do two given networks compute the same function? Is there a smaller network computing the same function?

The complexity of these questions have been investigated recently from a practical point of view and approximated by heuristic algorithms. We complement these achievements by giving a theoretical framework that enables us to interchange security and efficiency questions in neural networks and analyze their computational complexities. We show that the problems are conquerable in a semi-linear setting, meaning that for piecewise linear activation functions and when the sum- or maximum metric is used, most of them are in P or in NP at most.

1 Introduction

Neural networks are widely used in all kinds of data processing, especially on seemingly unfeasible tasks such as image [15] and language recognition [10], as well as applications in medicine [16], and prediction of stock markets [6], just to mention a few. Khan et al. [14] provide a survey of such applications, a mathematically oriented textbook concerning structural issues related to Deep Neural Networks is provided by [3].

Neural networks are nowadays also made use of in safety-critical systems like autonomous driving [8] or power grid management. In such a setting, when security issues become important, aspects of certification come into play [7, 11,17]. If we for example want provable guarantees for certain scenarios to be unreachable, we first need to formulate them as constraints and precisely state for which property of a network we want verification.

In the present paper we are interested in studying certain verification problems for NNs in form of particular robustness and minimization problems such as: How will a network react to a small perturbation of the input [9]? And how

B. Dilkina (Ed.): CPAIOR 2024, LNCS 14743, pp. 263–278, 2024.
https://doi.org/10.1007/978-3-031-60599-4_18

likely is a network to change the classification of an input that is altered a little? These probabilities are crucial when for example a self-driving car is supposed to recognize a speed limit, and they have already been tackled in practical settings by simulations and heuristic algorithms. These approaches however will never guarantee safety, because the conditions under which seemingly correct working nets suddenly tend to decide irrationally often seem arbitrary and unpredictable [2,4,20].

Key results will be that a lot of these questions behave very similar under reasonable assumptions. Note that the network in principle is allowed to compute with real numbers, so the valid inputs we are looking for belong to some space $\mathbb{R}^n$, but the network itself is specified by its discrete structure and rational weights defining the linear combinations computed by its single neurons.

A huge variety of networks arises when changing the underlying activation functions. There are of course many activations frequently used in NN frameworks, and in addition we could extend verification questions to nets using all kinds of activation. One issue to be discussed is the computational model in which one argues. If, for example, the typical sigmoid activation $f(x) = 1/(1 + e^{-x})$ is used, it has to be specified in which sense it is computed by the net: For example exactly or approximately, and at which costs these operations are being performed. In this paper we will cover the most common activation functions, especially *ReLU*.

The paper is organized as follows: In Sect. 2 we collect basic notions, recall the definition of feedforward neural nets as used in this paper as well as for the metrics that we use. Section 3 studies various robustness properties and their comparison to each other when using different network structures and metrics. In Sect. 4, we give criteria on whether a network is as small as possible, or if it can be replaced by a smaller network that is easier to evaluate.

The paper ends with some open questions.

2   Preliminaries and Network Decision Problems

We start by defining the problems we are interested in; here, we follow the definitions and notions of [21] and [19] for everything related to neural networks. The networks considered are exclusively feedforward. In their most general form, they can process real numbers and contain rational weights. This will later on be restricted when necessary.

Definition 1. *A (feedforward) neural network N is a layered graph that represents a function of type $\mathbb{R}^n \to \mathbb{R}^m$, for some $n, m \in \mathbb{N}$. The first layer with label $\ell = 0$ is called the* input *layer and consists of n nodes called input nodes. The input value x_i of the i-th node is also taken as its output $y_{0i} := x_i$. A layer $1 \leq \ell \leq L - 2$ is called* hidden *and consists of $k(\ell)$ nodes called computation nodes. The i-th node of layer ℓ computes the output $y_{\ell i} = \sigma_{\ell i}(\sum_j c_{ji}^{(\ell-1)} y_{(\ell-1)j} + b_{\ell i})$.*

Here, the $\sigma_{\ell i}$ are (typically nonlinear) activation functions (to be specified later on) and the sum runs over all output neurons of the previous layer. The $c_{ji}^{(\ell-1)}$

are real constants which are called weights, *and* $b_{\ell i}$ *is a real constant called* bias. *The outputs of all nodes of layer ℓ combined gives the output* $(y_{l0}, ..., y_{l(k-1)})$ *of the hidden layer. The final layer $L - 1$ is called* output layer *and consists of m nodes called* output nodes. *The i-th node computes an output $y_{(L-1)i}$ in the same way as a node in a hidden layer. The output* $(y_{(L-1)0}, ..., y_{(L-1)(m-1)})$ *of the output layer is considered the output $N(x)$ of the network N.*

Note that above we allow several different activation functions in a single network. This basically is because for some results technically the identity is necessary as a second activation function beside the 'main' activation function used. All of our results hold for this scenario already.

Since we want to study its complexity in the Turing model, we restrict all weights and biases in a NN to be rational numbers. The problem NNREACH involves two Linear Programming LP instances in a decision version, therefore recall that such an instance consists of a system of (componentwise) linear inequalities $A \cdot x \leq b$ for rational matrix A and vector b of suitable dimensions. The decision problem asks for the existence of a real solution vector x. As usual, we refer to this problem as Linear Program Feasibility LPF. By abuse of notation, we denote by A also the set described by the pair (A, b).

As usual for neural networks, we consider different choices for the activation functions used, but concentrate the most frequently used one, namely $ReLU(x) = max\{0, x\}$. We name nodes after their internal activation function, so we call nodes with activation function $\sigma(x) = x$ identity nodes and nodes with activation function $\sigma(x) = ReLU(x)$ ReLU-nodes for example.

We next recall from [21] the definition of the decision problems NNREACH, VIP and NE. These problems have been investigated in [11] and [1], specifically NNREACH in [19,21] and [13].

Definition 2. *a) Let F be a set of activation functions from $\mathbb{R}$ to $\mathbb{R}$. An instance of the* reachability problem for neural networks NNREACH(F) *consists of a (feedforward) neural network N with all its activation functions belonging to F, rational data as weights and biases, and two instances A and B of LP in decision version with rational data, one with the input variables of N as variables, and the other with the output variables of N as variables. These instances are also called* input *and* output specifications, *respectively. The problem is to decide if there exists an $x \in \mathbb{R}^n$ that satisfies the input specifications such that the output $N(x)$ satisfies the output specifications.*

b) The problem verification of interval property VIP(F) *consists of the same instances. The question is whether for all $x \in \mathbb{R}^n$ satisfying the input specifications, $N(x)$ will satisfy the output specifications.*

c) The problem network equivalence NE(F) *is the question whether two F-networks N_1 and N_2 compute the same function or not.*

d) The size *of a network is $T \cdot L$; here, T denotes the number of neurons in the net N and L is the maximal bit-size of any of the weights and biases. The size of an instance of NNREACH, VIP or NE is sum of the network size and the usual bit-sizes of the LP-instances.*

We omit F in the notation if it is obvious from the context and write VIP(N, A, B) if VIP(F) holds for the instance (N, A, B). The notations for the other problems are similar.

We additionally define certain robustness properties that were examined for example in [18] and use them to assess the computed function in a more metric way similar to [5].

Definition 3. *Let N be a NN, N_i the projection of N on the i-th output dimension, $1 \leq j \leq m$ an output dimension, $\bar{x} \in \mathbb{R}^n$, $\varepsilon, \delta \in \mathbb{Q}_{\geq 0} \cup \{\infty\}$ and d a metric on $\mathbb{R}^n$ as well as $\mathbb{R}^m$.*

a) The classification of a network input x is the output dimension $1 \leq i \leq m$ with the biggest value $N_i(x)$, which is interpreted as the attribute that the input is most likely to have. We say that N has ε-classification robustness in $\bar{x}$ with respect to d iff the classification of every input x that is ε-close to $\bar{x}$ is the same as for $\bar{x}$, meaning

$$\mathrm{CR}_d(N, \varepsilon, \bar{x}, j) :\Leftrightarrow \forall x : d(x, \bar{x}) \leq \varepsilon \Rightarrow \arg \max_i N_i(x) = j$$

and arbitrary ε-classification robustness iff

$$\mathrm{aCR}_d(N, \varepsilon, \bar{x}) :\Leftrightarrow \exists j \in \{1, ..., m\} : \mathrm{CR}_d(N, \varepsilon, \bar{x}, j)$$

b) We say that N has ε-δ-standard robustness in $\bar{x}$ with respect to d_λ iff N interpreted as a function is ε-δ-continuous in $\bar{x}$, meaning

$$\mathrm{SR}_d(N, \varepsilon, \delta, \bar{x}) :\Leftrightarrow \forall x : d(x, \bar{x}) \leq \varepsilon \Rightarrow d(N(\bar{x}), N(x)) \leq \delta$$

c) We say that N has ε-L-Lipschitz robustness in $\bar{x}$ with respect to d iff

$$\mathrm{LR}_d(N, \varepsilon, L, \bar{x}) :\Leftrightarrow \forall x : d(x, \bar{x}) \leq \varepsilon \Rightarrow d(N(\bar{x}), N(x)) \leq L \cdot d(\bar{x}, x)$$

d) For a specific set F of activation functions, by abuse of notation we denote by $\mathrm{CR}_d(F)$ the decision problem whether $\mathrm{CR}_d(N, \varepsilon, \bar{x}, j)$ holds where we allow as input only networks N that use functions from F as activations, the notations for the other problems are defined similarly.

We denote by d_p the metric induced by the p-Norm, meaning $d_p(x, y) = (\sum_{i=1}^n (x_i - y_i)^p)^{\frac{1}{p}}$.

Note that d_1 and d_∞ share the property that all ε-balls $B_{\varepsilon,d}(x)$ are semilinear sets and that they can also be described as instances of LPF.

The property aCR seems redundant and unnatural at first, it is however not equivalent to CR at an extent that one could hastily assume. Note that it is in general already very hard to calculate the solution of just one network computation $N(x)$ for given N and x for certain activation functions, the result does not even have to be rational or algebraic any more. It might even be impossible to determine the biggest output dimension, i.e., to tell whether $N(x)_i \geq N(x)_j$, for in the Turing model, one is not capable of precisely computing certain sigmoidal functions, which is why seemingly obvious reductions from CR to aCR fail. Assume for example that an activation similar to the exponential function is included, in this case the problem might become hard for Tarskis exponential problem, which is not even known to be decidable. Checking the properties aCR and CR is therefore not necessarily of the same computational complexity. We also need aCR in reduction proofs later on.

3   Complexity Results for Robustness

In network decision problems, d_1 and d_∞ behave fundamentally different than the other p-metrics, for they are the only ones that lead to linear constraints. General polynomial systems tend have a higher computational complexity than linear ones, compare for example $\exists \mathbb{R}$ (NP-hard) with LP (in P) or Hilberts tenth problem (undecidable) with integer linear systems (NP-complete). The following proposition partially covers the linear part of our network decision problems.

Proposition 1. *Let F be a set of piecewise linear activation functions.*

a) *The problems* VIP(F) *and* NE(F) *are in co-NP*
b) *Let $d \in \{d_1, d_\infty\}$. Then* SR$_d$(F)*,* CR$_d$(F) *and* LR$_d$(F) *are in co-NP.*

This is specifically the case for $F = \{ReLU\}$.

Proof. Finding a witness for violation of the conditions works essentially the same for all the above problems, we will perform it for VIP.

The idea of the proof is that an instance is not satisfiable iff there exists an $x \in \mathbb{R}^n$ fulfilling the input specifications so that $N(x)$ does not fulfill the output specifications, meaning at least one of the equations does not hold. We then guess this equation along with some data that narrows down how the computation on such an x could behave.

For a no-instance, let x be an input violating the specifications. The certificate for the instance not to be solvable is not x itself, because we would need to guarantee that it has a short representation. Instead, the certificate provides the information for every computation node, in which of the linear intervals its input is (for ReLU-nodes for example whether it returns zero or a positive value) when the network computes $N(x)$ together with the index of the violated output specification. This leaves us with a system like Linear Programming of linear inequalities that arise in the following way:

All of the input specifications $\sum_{i=1}^{n} a_i x_i \leq b$ and the violated output specification $\sum_{i=1}^{m} a_i N(x)_i > b$ are already linear. For each σ-node computing

$$y_{\ell i} = \sigma\left(\sum_j c_{ji}^{(\ell-1)} y_{(\ell-1)j} + b_{\ell i}\right)$$

that is by the certificate on an interval $[\alpha, \beta]$ where $\sigma(x)$ coincides with $a \cdot x + b$, we add the linear conditions

$$\alpha \leq \sum_j c_{ji}^{(\ell-1)} y_{(\ell-1)j} + b_{\ell i} \leq \beta$$

and

$$y_{\ell i} = a \cdot \sum_j c_{ji}^{(\ell-1)} y_{(\ell-1)j} + b_{\ell i} + b$$

The only difference to linear programming is that some of the equations are strict and some are not. This decision problem is also well known to still be in P by [12], Lemma 14.

The proof works in the same way for the other problems, as the falsehood of their instances can also formulated as a linear programming instance. For network equivalence for example, one would guess the output dimension that is unequal at some point, encode the computation of both networks, and demand that the outputs in the guessed dimensions are not the same, which is a linear inequality.   ∎

The proof fails for non-linear metrics such as d_2, for the resulting equation system is not linear any more.

Lemma 1. *Let F be a set of activations containing ReLU. For any $\lambda \in [1, \infty]$, there are polynomial time reductions from $d = d_\infty$ or $d = d_1$ to $d = d_\lambda$ for $CR_d(F), aCR_d(F)$ and $SR_d(F)$.*

Proof. We will show how to reduce from d_1 for a network N by constructing a network N' that shows the same behavior in $\mathbb{B}_{\varepsilon,1}(\bar{x})$ as the given network and a sufficiently similar behavior on $\mathbb{B}_{\varepsilon,\lambda}(\bar{x})$. The proof for d_∞ is analogous, as it only relies on the ε-ball to be a convex semi-linear set.

The idea for all three problems is to first retract a superset of the ε-ball to it so that $N' = N \circ T$, where $T(\mathbb{B}_{\varepsilon,\lambda}(\bar{x})) = \mathbb{B}_{\varepsilon,1}(\bar{x})$.

Let $(N, \varepsilon, \bar{x}, j)$ be an instance of $CR_1(F)$. The equivalent instance $(N', \varepsilon, \bar{x}, j)$ of $CR_\lambda(F)$ is constructed as follows:

We first insert n new layers before the initial input layer, the first of them becomes the new input layer. These layers perform a transformation T that has the following properties:

$\forall x \in \mathbb{B}_{\varepsilon,1}(\bar{x}) : T(x) = x$. This ensures that if the initial instance failed the property $CR_1(F)$, the new one fails $CR_\infty(F)$.

$\forall x \in \mathbb{B}_{\varepsilon,\lambda}(\bar{x}) : T(x) \in \mathbb{B}_{\varepsilon,1}(\bar{x})$. This ensures that if the initial instance fulfilled the property $CR_1(F)$, the new one fulfills $CR_\infty(F)$.

Such a function is for example

$$T(x)_i = ReLU(x_i) - ReLU(x_i + \sum_{j=1}^{i-1} |T(x)_j| - \varepsilon)$$

$$-(ReLU(-x_i) + ReLU(-x_i + \sum_{j=1}^{i-1} |T(x)_j| - \varepsilon))$$

because $\mathbb{B}_{\varepsilon,1}(\bar{x}) \subseteq \mathbb{B}_{\varepsilon,\lambda}(\bar{x})$ and $T(x) \in \mathbb{B}_{\varepsilon,1}(\bar{x}) \forall x \in \mathbb{R}^n$.

For CR the reduction is complete, for aCR it is essentially the same. For SR_λ we perform as one further step a transformation P of $N(x)$ into a one-dimensional output, namely $P(N(x)) = \sum_{i=1}^{m} |N(x)_i|$. It is easily seen that $N(x) \in \mathbb{B}_1(\varepsilon) \Leftrightarrow P(N(x)) \in \mathbb{B}_1(\varepsilon) = [0, \varepsilon]$.   ∎

The previous Theorem as well as the following Proposition rely on the semi-linearity of the final sets, a crucial property when using the Ellipsoid Method or the ReLU function. When . This is the reason why we are restricted to metrics where balls are semi-linear sets. When moving on to d_λ for arbitrary λ, we cross the presumably large gap in computational complexity between NP and $\exists\mathbb{R}$.

Proposition 2. *Let $\lambda \in \{1, \infty\}$. Then VIP($\{id\}$), NNReach($id$), NE($id$), $SR_\lambda(\{id\})$, $CR_\lambda(\{id\})$, $aCR_\lambda(\{id\})$ and $LR_\lambda(\{id\})$ are in P.*

Proof. $NNReach(id)$ is a special case of Linear Programming Feasibility. For the other problems, we argue that their complements are in P=co-P. Note that LPF is still in P if we allow strict linear inequalities. The complements of $VIP(\{id\})$, $NE(id)$, $SR_\lambda(\{id\})$, $CR_\lambda(\{id\})$, $aCR_\lambda(\{id\})$ and $LR_\lambda(\{id\})$ are global disjunctions of LP-instances of linear size, so all of these Problems are in P.   ∎

The following Lemma proved in [21] allows us to strengthen a couple of results on complexities in the semi-linear range of network verification.

Lemma 2. *An identity node can be expressed by two ReLU nodes without changing the computed function.*

The Lemma looks somewhat trivial, however for all our decision problems it shows that adding id to the set F of allowed activations already containing ReLU does not change the computational complexity. Furthermore, when reducing one such problem to another one, we can assume id to be among the activations and freely use it in the reductions, as for example in the next Lemma.

Lemma 3. *Let F be a set of activation functions.*
   i) $aCR_d(F)$ is polynomial time reducible to $CR_d(F)$.
   ii) If d is induced by a p-norm and if $ReLU \in F$, then $CR_d(F)$ is polynomial time reducible to $aCR_d(F)$.

Proof. i) To see if $aCR_d(N, \varepsilon, \bar{x})$ holds, check $CR_d(N, \varepsilon, \bar{x}, j)$ for all possible j and see if one of them holds.
   ii) For the reverse direction, let $(N, \varepsilon, \bar{x}, j)$ be an instance of $CR_d(F)$. We will construct an instance $(N_h, \varepsilon, \bar{x})$ of aCR so that

$$CR_d(N, \varepsilon, \bar{x}, j) \Leftrightarrow aCR_d(N_h, \varepsilon, \bar{x})$$

The idea is to apply a function f on the output that collects any possible overlap of all dimensions other than j. This function is then modified to a function h with 3 output dimensions described by a network N_h, where one dimension is the constant zero function, and it is guaranteed to be negative in both remaining dimensions at some point. Also, iff f was at some point positive, which happens exactly if there was an overlap, so should at least one component of h be. Since h is at some point negative in the second and third dimension, the only candidate for the pointwise biggest output dimension is the first, and this is the case iff there was an overlap in the output of the initial network. This means that $\mathrm{CR}_d(N, \varepsilon, \bar{x}, j) \Leftrightarrow \mathrm{CR}_d(N_h, \varepsilon, \bar{x}, 1) \Leftrightarrow \mathrm{aCR}_d(N_h, \varepsilon, \bar{x})$. It remains to show that such functions f, h exist and that they can be described by a network N_h.

Note that the following functions can be computed by neural networks that are modifications of N:

$$f(x) = \sum_{i=1}^{m} ReLU(N(x)_i - N(x)_j)$$

$$g(x) = f(x) - ReLU(f(x) - 1)$$

Note that $f(x) \geq 0$ and $f(x) = 0$ iff j is the index for a maximal component. Furthermore if $f(x) \geq 1$, then $g(x) = 1$ and if $f(x) \in [0,1]$ then $g(x) = f(x)$. Thus, $g \equiv 0$ iff $\mathrm{CR}_d(N, \varepsilon, \bar{x}, j)$. Now define h as

$$h(x) = (0, g(x) - \frac{2}{\delta} ReLU(x_1 - \bar{x}_1), g(x) - \frac{2}{\delta} ReLU(\bar{x}_1 - x_1)).$$

where $\delta \in \mathbb{Q}$ is chosen so that $\delta < \varepsilon$, meaning $\bar{x} + \delta e_1, \bar{x} - \delta e_1 \in \mathbb{B}_{\varepsilon,d}(\bar{x})$ where $e_1 = (1, 0, ..., 0)$.

Let N_h be the network computing h. When asked if $aCR_\lambda(N_h, \varepsilon, \bar{x})$, the second and third dimensions of h cannot be the pointwise biggest, the only possible candidate is 0 (j=1) because $h_2(\bar{x} + \delta e_1), h_3(\bar{x} - \delta e_1) \leq -1$. Also, g is constantly 0 iff $\mathrm{CR}_\lambda(N, \varepsilon, \bar{x}, j)$ holds, on the other hand $g \equiv 0$ holds iff $aCR_\lambda(N_h, \varepsilon, \bar{x})$, so $\mathrm{CR}_\lambda(N, \varepsilon, \bar{x}, j) \Leftrightarrow aCR_\lambda(N_h, \varepsilon, \bar{x})$. ∎

Theorem 1. *Let F be a set of activation functions. We have that*
   i) $\mathrm{SR}_\infty(F)$ is linear time reducible to $VIP(F)$,
   ii) $\mathrm{CR}_\infty(F)$ is linear time reducible to $VIP(F)$,
   iii) if $id \in F$, then $\mathrm{SR}_\infty(F)$ is linear time reducible to $\mathrm{CR}_\infty(F)$,
   iv) if $ReLU \in F$, then $\mathrm{CR}_d(F)$ is linear time reducible to $\mathrm{SR}_d(F)$ for any metric d.

Proof. i) Let $(N, \varepsilon, \delta, \bar{x})$ be an instance of $\mathrm{SR}_\infty(F)$. To construct an equivalent instance of $VIP(F)$, we introduce two copies of the network N and view them as one big network N' with an input dimension twice the input dimension of N. We demand by input specification that the input of the first copy of N is $\bar{x}$ and the input of the second copy is in the ε-ball of $\bar{x}$ with respect to $\| \cdot \|_\infty$, which is equivalent to $\forall i \in \{1, ..., n\} : -\varepsilon \leq \bar{x}_i - x_i \leq \varepsilon$, a linear equation system that we denote A. Similarly, we demand by output specification that $N(x)$ is in the δ-ball of $N(\bar{x})$, which is equivalent to $\forall i \in \{1, ..., m\} : -\delta \leq N(\bar{x})_i - N(x)_i \leq \delta$, we denote this system B. This gives us the $VIP(F)$-instance (N', A, B) for which we have $VIP(N', A, B) \Leftrightarrow SR(N, \varepsilon, \delta, \bar{x})$.

ii) Let $(N, \varepsilon, \bar{x}, j)$ be an instance of $\mathrm{CR}_\infty(F)$. We construct an equivalent instance of $VIP(F)$ as follows. The network remains N, the input specifications are $\forall i \in \{1, ..., n\} : \bar{x}_i - \varepsilon \le x_i \le \bar{x}_i + \varepsilon$ and the output specification is $\bigwedge\limits_{i=1}^{m} N(x)_i \le N(x)_j$.

iii) Let $(N, \varepsilon, \delta, \bar{x})$ be an instance of $\mathrm{SR}_\infty(F)$ with m the output dimension of N. To construct an equivalent instance $(N', \varepsilon, \bar{x}, 2m+1)$ of $\mathrm{CR}_\infty(F)$, we perform the computation of two copies of the network N in parallel, which constitutes a new network N' as follows (Fig. 1):

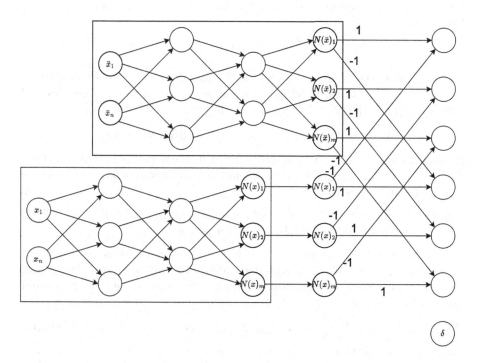

Fig. 1. Idea iii)

The first copy is shifted by one layer, so the layer that was the input layer is now in the first hidden layer with no incoming connections and the bias of node i is $\bar{x}_i$, so that first copy just computes the value $N(\bar{x})$. The input layer of N' is the input layer of the second copy, so the second copy of N computes $N(x)$. We add a layer of id-nodes at the end of it that does not change the output but let it have the same number of layers as the first copy. We add one layer connecting both copies after the old output layer, this one becomes the new output layer, and it computes

$$(N(x)_1 - N(\bar{x})_1, ..., N(x)_m - N(\bar{x})_m, N(\bar{x})_1 - N(x)_1, ..., N(\bar{x})_m - N(x)_m, \delta).$$

The node corresponding to the last value $N'(x)_{2m+1} = \delta$ has no incoming connections and bias δ. We have that $\forall i \in \{1, ..., m\} : |N(\bar{x})_i - N(x)_i| \le \delta$ iff

$\forall i \in \{1, ..., m\} : N(\bar{x})_i - N(x)_i \leq \delta \wedge N(x)_i - N(\bar{x})_i \leq \delta$, so the two instances are in fact equivalent.

iv) Let $(N, \varepsilon, \bar{x}, j)$ be an instance of $\mathrm{CR}_d(F)$. We construct an equivalent instance $(N', \varepsilon, 0, \bar{x})$ of $\mathrm{SR}_d(F)$ as follows: We add two additional layers after the old output layer. The first one computes $\alpha_i = ReLU(N(x)_i - N(x)_j)$ for all i, these values are all 0 everywhere on $B_{d,\varepsilon}(\bar{x})$ iff $\mathrm{CR}_d(N, \varepsilon, \bar{x}, j)$ holds. The next added layer computes $\beta = ReLU(\sum_{i=1}^{m} \alpha_i)$, this value is 0 everywhere on $B_{d,\varepsilon}(\bar{x})$ iff $\mathrm{CR}_d(N, \varepsilon, \bar{x}, j)$ holds. Now we have that $\mathrm{CR}_d(N, \varepsilon, \bar{x}, j) \Leftrightarrow \mathrm{SR}_d(N', \varepsilon, 0, \bar{x}) \wedge N'(\bar{x}) = 0$, so it remains to encode that last condition as a $\mathrm{SR}_d(F)$-instance. The function

$$f(x) := \sum_{i=1}^{n} ReLU(x_i - \bar{x}_i) + ReLU(\bar{x}_i - x_i)$$

is 0 in $\bar{x}$ and strictly positive everywhere else and there exists an F-network N_f computing f. The networks N' and N_f can be combined to an F-network N'' computing $g(x) = ReLU(N' - N_f)(x)$. If $\mathrm{SR}_d(N', \varepsilon, 0, \bar{x})$ holds, then

$$N'(\bar{x}) = 0 \Leftrightarrow g(x) = 0 \forall x \in B_\varepsilon(\bar{x}) \Leftrightarrow \mathrm{SR}_d(N'', \varepsilon, 0, \bar{x}).$$

The conditions $\mathrm{SR}_d(N', \varepsilon, 0, \bar{x})$ and $\mathrm{SR}_d(N'', \varepsilon, 0, \bar{x})$ can be formalized as a single SR_d instance by merging both nets N' and N'' into a single larger one, so the reduction is in fact many-one. ∎

Proposition 3. *Let F be a set of activation functions. If $id \in F$, then $NE(F)$ is linear time reducible to $\mathrm{CR}_d(F)$ for any metric d.*

Proof. Let (N, N') be an instance of $NE(F)$. We merge both networks N and N' into one network N'' by identifying/contracting their corresponding inputs and subtracting their corresponding outputs both ways to obtain the output layer

$$(N(x)_1 - N'(x)_1, ..., N(x)_m - N'(x)_m, N'(x)_1 - N(x)_1, ..., N'(x)_m - N(x)_m, 0).$$

The last output node is always 0, this can be forced by introducing an id-node that has bias 0 and no incoming connections. Now

$$NE(N, N') \Leftrightarrow \mathrm{CR}_d(N'', \infty, 0, 2m + 1),$$

which concludes the proof. ∎

Note that for Lipschitz robustness and standard robustness also global variants make sense:

Definition 4. *Let N be a NN, $\bar{x} \in \mathbb{R}^n$ and $\varepsilon, \delta \in \mathbb{Q}_{\geq 0}$. We say that N has (global) ε-δ-standard robustness wrt the metric d iff it has ε-δ-standard robustness in every $\bar{x} \in \mathbb{R}^n$:*

$$\text{glob-}\mathrm{SR}_d(N, \varepsilon, \delta) := \forall x, \bar{x} : d(x - \bar{x}|) \leq \varepsilon \Rightarrow d(N(\bar{x}) - N(x)) \leq \delta.$$

We define glob-LR_d analogously, note that it coincides with L-Lipschitz continuity independent of ε.

This equivalence does not hold for local Lipschitz robustness: The function $x \cdot \chi_{\mathbb{Q}}(x)$ is ε-1-Lipschitz robust in 0 for every ε, but not Lipschitz-continuous in any ε-ball around 0.

Remark 1. Global arbitrary ε-*classification robustness* glob-aCR$_d$ for a network N, meaning it has arbitrary ε-classification robustness in every input $x \in \mathbb{R}^n$, hardly ever happens at all. We have that only very artificial functions map inputs to different classes and still have global (arbitrary) classification robustness, for between two areas of different classification, there must be a dividing strip of diameter ε in which both classifications must be choosable. The function $f : \mathbb{R} \to \mathbb{R}^2$ where

$$f = \begin{pmatrix} f_1 \\ f_2 \end{pmatrix} \text{ where } f_1(x) = \begin{cases} 0, & \text{for } x \leq -1 \\ x+1, & \text{for } -1 \leq x \leq -\frac{1}{2} \\ \frac{1}{2}, & \text{for } -\frac{1}{2} \leq x \leq \frac{1}{2} \\ x-1, & \text{for } \frac{1}{2} \leq x \leq 1 \\ 1, & \text{for } 1 \leq x \end{cases}, \quad f_2(x) = 1 - f_1(x)$$

for example is globally 1-classification robust. Such functions however are unlikely for a network of "natural origin" for these areas of equal classification do not occur in usual training methods.

Theorem 2. *Let F be a set of activation functions containing ReLU. Then:*
  i) glob-SR$_\infty(F)$ *reduces to* $NE(F)$.
  ii) glob-SR$_\infty(F)$, LR$_\infty(F)$ *and* glob-LR$_\infty(F)$ *are co-NP-hard.*

Proof. i) We reduce an instance (N, ε, δ) of glob-SR$_\infty(F)$ to an instance (N', N'') of $NE(F)$ by using the ReLU-function to check whether certain values are positive or not and by forcing the differences $\|\bar{x}_i - x_i\|_\infty$ to be smaller than ε, again with the use of ReLU-nodes. We construct the network N' as follows: The input $(x, y) = (x_1, ..., x_n, y_1, ..., y_n)$ has twice the dimension of the input of N. On the first half $(x_1, ..., x_n)$ we execute N to obtain $N(x)$. For each of the y_i in the second half, we compute

$$f(y_i) = \varepsilon \cdot (ReLU(y_i) - ReLU(y_i - 1) - \frac{1}{2})$$

by ReLU-nodes in the first hidden layer, a bias of $-\frac{1}{2}$ in the second and a weight of ε in the third. Note that $im(f) = [-\varepsilon, \varepsilon]$, we interpret $f(y_i)$ as $\bar{x}_i - x_i$. We now execute N again, this time on $(x_1 + f(y_1), ..., x_n + f(y_n))$ to obtain $N(x + \bar{x} - x) = N(\bar{x})$. We compute both $N(x) - N(\bar{x})$ and $N(\bar{x}) - N(x)$ by ReLU-nodes with bias $-\delta$ to obtain

$$N'(x, y) = (ReLU(N(\bar{x})_1 - N(x)_1 - \delta), ReLU(N(x)_1 - N(\bar{x})_1 - \delta), ...,$$

$$ReLU(N(\bar{x})_m - N(x)_m - \delta), ReLU(N(x)_m - N(\bar{x})_m - \delta)).$$

SR$_\infty^G(N, \varepsilon, \delta)$ holds iff N' is constantly 0. The second network N'' consists of $2n$ input nodes, no hidden layer and $2m$ output nodes with all weights and biases 0. This network will compute 0 independently of the input, so

$$SR_\infty^G(N, \varepsilon, \delta) \Leftrightarrow NE(N', N'').$$

ii) We reduce an instance φ of 3-SAT with n clauses to an instance $(N, \infty, n - \frac{1}{2})$ of the complement of glob-SR$_\infty(\{ReLU\})$. The idea is to associate pairs of literals and their negations with pairs of nodes with the property that at least one of these nodes must have the value zero. We then interpret the zero-valued node as false and the other one, that may have any value in $[0, 1]$, as true and replace $\wedge$ and $\vee$ by gadgets that work with the function $\Psi : \mathbb{R} \to [0, 1]$ defined as $x \mapsto ReLU(x) - ReLU(x - 1)$. Note that it is easily constructable in a ReLU-network and $\Psi|_{[0,1]} = id|_{[0,1]}$.

For the reduction, introduce for each variable a in φ an input variable α_1 and use the first two hidden layers to compute $\alpha_2 := \Psi(\alpha_1)$. Use the third hidden layer to propagate α_2 as well as to compute $\neg\alpha_2 := 1 - \alpha_2$. In the fourth hidden layer compute $\alpha := 2 \cdot ReLU(\alpha_2 - \frac{1}{2})$ for every atom as well as for the negations. Note that $\alpha = 0$ or $\neg\alpha = 0$ and the other one can be anything in $[0, 1]$. In the fifth and sixth hidden layer compute for every clause (a, b, c) the value $\Psi(\alpha+\beta+\gamma)$. In the seventh layer, add all nodes from the sixth layer, this is the overall output of the network. If the instance of 3-SAT is solvable, then $N(x) = n$ is reachable via setting the true atoms to 1 and the false ones to 0. The value $N(y) = 0$ is always reachable, just assign every variable the value $\frac{1}{2}$. This means that if the 3-SAT instance is solvable, then x and y are witnesses that glob-SR$_\infty(N, \infty, n - \frac{1}{2})$ is false. On the other hand, if the 3-SAT instance is not solvable, then for every input $\Psi(\alpha + \beta + \gamma) = 0$ for at least one of the clauses, so $N(x) \leq n - 1$. Now we always have that $N(y) \geq 0$ because $\Psi \geq 0$, so $|N(x) - N(y)| \leq n - 1$ and glob-SR$_\infty(N, \infty, n - \frac{1}{2})$ is true.

We reduce 3-SAT to LR$_\infty(F)$ and glob-LR$_\infty(F)$ in the same way, the instance for LR$_\infty(F)$ is $(N, \frac{1}{2}, 2n - 1, (\frac{1}{2}, ..., \frac{1}{2}))$ with the same network N as constructed above for glob-SR$_\infty(\{ReLU\})$. We chose $n - 1 < \frac{1}{L} \cdot \varepsilon < n$ so that LR$_\infty(N, \frac{1}{2}, 2n - 1, (\frac{1}{2}, ..., \frac{1}{2}))$ is equivalent to glob-SR$_\infty(N, \infty, n - \frac{1}{2})$. This is true because the minimum of $N(x)$ is 0 attained in $(\frac{1}{2}, ..., \frac{1}{2})$ and the maximum is either n if the 3-SAT instance is satisfiable or at most $n - 1$ if it is not, taken on the boundary of the unit hypercube. The growth between the minimum and the maximum is linear because the intersection points where ReLU is not differentiable are in $(\frac{1}{2}, ..., \frac{1}{2})$ and on the boundary of the unit hypercube, but never in between.

The instance $(N, \frac{1}{2}, 2n - 1)$ constructed by the reduction to glob-LR$_\infty(F)$ is the same, because no other pair of points $x, \bar{x}$ will lead to a steeper descent if the network was derived from a SAT-instance that is not satisfiable. This is because the steepest descent must necessarily point to the minimum if the function is linear along the connecting line and no other local minimum exists. ∎

Corollary 1. glob-SR$_\infty(\{ReLU\})$, $NE(\{ReLU\})$, CR$_\infty(\{ReLU\})$, SR$_\infty(\{ReLU\})$, LR$_\infty(\{ReLU\})$ and $VIP(\{ReLU\})$ are co-NP-complete.

Proof. $VIP(\{ReLU\})$ is in co-NP by Theorem 1 and glob-SR$_\infty(\{ReLU\})$ is co-NP hard by Theorem 2, *ii)*. The chain of reductions in between is provided by Theorem 2 *i)*, Theorem 3, and Theorem 1 *iv), i)*. ∎

4 Network Minimization

In this section, we want to analyze different criteria on whether a network is unnecessarily huge. The evaluation of a function given as a network becomes easier and faster for small networks, it is therefore desirable to look for a tradeoff between accuracy and runtime performance. In the worst case, a long computation in an enormously big network that was assumed to be more accurate due to its size, could lead to rounding errors in both the training and the evaluation, making the overall accuracy worse than in a small network.

Definition 5. *Let F be a set of activations, N an F-network and K the set of nodes of N.*

a) N is called minimal, if no other F-network with less many nodes than N computes the same function. The decision problem $\mathrm{MIN}(F)$ asks whether a given F-network N is minimal, in that case we write $\mathrm{MIN}(N)$.

b) Let $Y \subseteq K$ be a subset of the nodes, G_N the underlying graph of G and G'_N its subgraph induced by $K \backslash Y$. The network N' whose underlying is G'_N and where all activations, biases and weights are the same as in the corresponding places in N, is said to be obtained from N by deleting the nodes in Y. By abuse of notation we denote it by $N \backslash K := N'$.

c) A subset $Y \subseteq K$ of hidden nodes of a network N is called unnecessary, if the network $N \backslash Y$ produced by deleting these nodes still computes the same function and otherwise necessary. The problem to decide whether sets of nodes are necessary is $\mathrm{NECE}(F)$, we write $\mathrm{NECE}(N, Y)$ if Y is necessary.

d) If $\mathrm{NECE}(N, Y)$ for all non-empty subsets $Y \subseteq K$ of nodes, we write $\mathrm{ANECE}(N)$, this also defines a decision problem in the obvious way.

It obviously holds that $\mathrm{MIN}(N) \Rightarrow \mathrm{ANECE}(N)$, the reverse direction however does not hold in general:

Example 1. A network can contain only necessary nodes and still not be minimal. Consider for example the $\{id, ReLU\}$-network N that consists of an input node x, two hidden nodes $y_{1,1} = ReLU(x)$ and $y_{1,2} = ReLU(-x)$ and one output node with weights 1 and -1 and bias 0, so it computes $y_2 = id(y_{1,1} - y_{1,2}) = x$. The network represents the identity map and is not minimal, because the smallest $\{id, ReLU\}$-network M computing the identity consists of an input node, no hidden nodes and one output node with activation id, weight 1 and bias 0. However every non-empty subset of hidden nodes in N is necessary; if we delete $y_{1,1}$, the remaining network computes the negative part $x^- = min\{0, x\}$ of the input, if we delete $y_{1,2}$ it computes the positive part and if we delete both it computes the zero-function for the last node has no more incoming nodes at all and thus returns the empty sum. We call this network Q.[1]

Also consider the net K computing the zero-function in the following way: In the first hidden layer, compute $y_{1,1} = y_{1,2} = ReLU(x)$, the output-node is

---

[1] This also illustrates that the smallest network M that is equivalent to N is in general not isomorphic to a substructure of N.

connected to both with weights 1 and -1, respectively, and bias 0, so it computes $0 = ReLU(x) - ReLU(x)$. In K, both hidden nodes on their own are necessary, the subset containing both however is not, because when deleting those two, we remain with the network Q that also computes the zero-function.

Theorem 3. *Let* $id \in F$, *then the complement of* NECE(F) *and* NE(F) *are linear-time equivalent.*

Proof. The first reduction is trivial,

$$\text{NECE}(N, \{y_1, ..., y_k\}) \Leftrightarrow \neg \text{NE}(N, N \backslash \{y_1, ..., y_k\})$$

For the other direction, let M, N be two F-networks. We construct an instance of NECE(F) that is equivalent to $\neg$NE(N, M) as follows:

First, we merge N and M into a network P that computes $N(x)$ and $M(x)$, then computes $y = N(x) - M(x)$ in the last hidden layer and then propagates the value $P(x) = id(y)$ to the output layer. Now if y is deleted, $P \backslash \{y\} \equiv 0$, because then the output node would have no incoming edges at all, so it would compute the identity over the empty sum, which is 0. It follows that y is unnecessary iff $P \equiv 0$, which in turn is the case iff $N \equiv M$ meaning NE(N, M), so we have that $\neg$NE(N, M) $\Leftrightarrow$ NECE($P, \{y\}$). ∎

Theorem 4. *Let F only contain semi-linear functions. Then*

i) MIN(F) *is in* Π_2^P.
ii) ANECE(F) *is in* Π_2^P.

Proof. i) We argue that co-MIN is in Σ_2^P. The algorithm non-deterministically guesses a smaller representation M of the function/network N and then proves that NE(N, M), which is co-NP-complete, see [21].
ii) In the same way, observe that co-ANECE(F) can be solved by non-deterministically guessing an unnecessary subset Y and then showing that the network remaining after deleting Y is equivalent to the original net. ∎

Theorem 5. *If id is among the activations,* NE *reduces to* ANECE *in polynomial time.*

Proof. Let (N, M) be an instance of NE. First, merge both N and M into one network by subtracting their corresponding output nodes. Then delete all nodes that are constantly zero. These are detected layer-wise starting with the nodes right after the input layer. For each node n, delete all nodes that do not affect it in the sense that they are in subsequent layers, in the same layer, or in a previous layer but with no connection or connection with weights 0 towards n. Then allocate one additional id-node in the following layer and connect it to n with weight 1 and no bias. ∎

5   Conclusion

We examined the computational complexity of various verification problems such as robustness and minimality for neural networks in dependence of the activation function used. We explained the connection between the distance function in use and the complexity of the resulting problem. We provided a framework for minimizing networks and examined the complexities of those questions as well.

Further open questions are:

1.) Is $LR_1(ReLU)$ complete for co-NP? Are $LR_1(F)$ and $LR_\infty(F)$ in general reducible on each other?

2.) Do sets described by functions computed by networks using only $ReLU$ (or only semi-linear activations) have more structure than just any semi-linear set? Can this structure be exploited to calculate or approximate the measure of such a set quicker than with the known algorithms?

3.) Are ANECE and $MIN(F)$ Π_2^P-complete if F only contains semi-linear functions?

4.) Several results rely on the identity activation. Can this be omitted? Or is there a problem that becomes substantially easier when the identity map is dropped?

Acknowledgment. I want to thank Klaus Meer for helpful discussion and the anonymous referees for various hints improving the writing.

References

1. Albarghouthi, A.: Introduction to Neural Network Verification (2021)
2. Athalye, A., Carlini, N., Wagner, D.: Obfuscated gradients give a false sense of security: circumventing defenses to adversarial examples (2018)
3. Calin, O.: Deep Learning Architectures - A Mathematical Approach. Springer, Heidelberg (2020). https://doi.org/10.1007/978-3-030-36721-3
4. Carlini, N., Liu, C., Erlingsson, Ú., Kos, J., Song, D.: The secret sharer: evaluating and testing unintended memorization in neural networks (2019)
5. Casadio, M., Komendantskaya, E., Daggitt, M.L., Kokke, W., Katz, G., Amir, G., Refaeli, I.: Neural network robustness as a verification property: a principled case study (2021)
6. Dixon, M., Klabjan, D., Bang, J.H.: Classification-based financial markets prediction using deep neural networks. Algor. Finan. **6**, 67–77 (2017)
7. Dreossi, T., Ghosh, S., Sangiovanni-Vincentelli, A., Seshia, S.A.: A formalization of robustness for deep neural networks (2019)
8. Grigorescu, S., Trasnea, B., Cocias, T., Macesanu, G.: A survey of deep learning techniques for autonomous driving. J. Field Rob. **37**, 362–386 (2019)
9. Guo, X., Zhou, Z., Zhang, Y., Katz, G., Zhang, M.: Occrob: efficient smt-based occlusion robustness verification of deep neural networks (2023)
10. Hinton, G., et al.: Deep neural networks for acoustic modeling in speech recognition: the shared views of four research groups. IEEE Signal Process **29**, 82–97 (2012)

11. Huang, X., et al.: A survey of safety and trustworthiness of deep neural networks: verification, testing, adversarial attack and defence, and interpretability. Comput. Sci. Rev. **37**, 100270 (2020)
12. Jonsson, P., Bäckström, C.: A unifying approach to temporal constraint reasoning. Artif. Intell. **102**(1), 143–155 (1998)
13. Katz, G., Barrett, C., Dill, D., Julian, K., Kochenderfer, M.: Reluplex: an efficient SMT solver for verifying deep neural networks. Comput. Aided Verificat. **10426**, 97–117 (2017)
14. Khan, A., Sohail, A., Zahoora, U., Qureshi, A.S.: A survey of the recent architectures of deep convolutional neural networks. Artif. Intell. Rev. **53**, 5455–5516 (2020)
15. Krizhevsky, A., Sutskever, I., Hinton, G.E.: ImageNet classification with deep convolutional neural networks. Association for Computing Machinery (2017)
16. Litjens, G., et al.: A survey on deep learning in medical image analysis. Med. Image Anal. **42**, 60–88 (2017)
17. Mahloujifar, S., Mahmoody, M.: Can adversarially robust learning leverage computational hardness? (2018)
18. Ruan, W., Huanga, X., Kwiatkowska, M.: Reachability analysis of deep neural networks with provable guarantees. In: Proceedings of the Twenty-Seventh International Joint Conference on Artificial Intelligence, IJCAI, pp. 2651–2659 (2018)
19. Sälzer, M., Lange, M.: Reachability is NP-complete even for the simplest neural networks. In: International Conference on Reachability Problems, vol. 13035, pp. 149–164 (2021)
20. Uesato, J., O'Donoghue, B., van den Oord, A., Kohli, P.: Adversarial risk and the dangers of evaluating against weak attacks (2018)
21. Wurm, A.: Complexity of reachability problems in neural networks. In: International Conference on Reachability Problems (2023)

An Improved Neuro-Symbolic Architecture to Fine-Tune Generative AI Systems

Chao Yin, Quentin Cappart, and Gilles Pesant[✉]

Department of Computer and Software Engineering, Polytechnique Montréal, Montreal, Canada
{chao.yin,quentin.cappart,gilles.pesant}@polymtl.ca

Abstract. Deep generative models excel at replicating the mechanisms that generate a specific set of sequential data. However, learning the underlying constraints preventing the generation of forbidden sequences poses a challenge. Recently, RL-Tuner, a reinforcement learning framework designed for the *ad hoc* fine-tuning of a neural model to adhere to given constraints, was enhanced to learn from the output of two constraint programming models. The first model computes a score representing the number of constraint violations from the currently generated token while the second model provides the marginal probability of that token being generated if no additional violation is allowed. In this paper, we significantly enhance the latter framework in three ways. First, we propose a simplified architecture that requires only a single constraint programming model. Second, we evaluate constraint violations in a more accurate and consistent manner. Third, we propose a reward signal based on belief propagation on this new model that further improves performance. Our experiments, conducted on the same learning task of music generation, demonstrate that our approach surpasses the previous framework both in terms of convergence speed during training and in post-training accuracy. Additionally, our approach exhibits superior generalization to longer sequences than those used during training.

1 Introduction

Sequential data exists everywhere we look, from text corpus to musical compositions and even stock market time series. They are important as much as they are abundant, as showcased by impressive advances in natural language processing through the development of Large Language Models (LLM). Their impact will only continue to grow as new advances are made in the field of machine learning. As powerful as generative models can be, it is increasingly desirable to control what is generated amid concerns about factual accuracy and potential bias [15], but also when the output is only valid if it exhibits a given (combinatorial) structure. Learning such structure from data can be difficult, may require a prohibitive amount of training data, or suffer from slow convergence. Consider the following approach: given a generative neural network pre-trained for a certain task, experts create a CP model to enforce domain-specific rules. For music it could be classic rules of counterpoint whereas for robotics it could be safety rules when operating around humans or to avoid breaking the robot. This CP model can then be used to fine-tune the neural network using reinforcement learning.

RL-Tuner [13], a reinforcement learning framework to fine-tune a neural model in an *ad hoc* fashion in order to satisfy given structural constraints, was recently enhanced [14] by replacing such *ad hoc* constraint penalties with a pair of constraint programming (CP) models, one which computes a score representing the number of constraint violations from the currently generated token and the other, the marginal probability of that token being generated if no additional violation is allowed, through the use of a CP solver computing such probability distributions over finite domains with belief propagation [20]. Our contribution builds on the idea of leveraging marginal probabilities to craft a reward signal to further increase the performance of the RL agent while simplifying the previous architecture. First, we redesign the CP part to require a single (soft) model. Second, we evaluate constraint violations in a more accurate and consistent manner. Third, we propose a reward signal based on the mathematical expectation of the number of constraint violations represented as a discrete random variable.

We give next some fundamental background in Sect. 2 and then review some related work in Sect. 3. The neuro-symbolic architecture we contribute is detailed in Sect. 4. It is followed by an empirical evaluation in Sect. 5. Finally, Sect. 6 makes some concluding remarks.

2 Background

2.1 Reinforcement Learning

Reinforcement learning is a machine learning paradigm in which an agent tries to maximize rewards collected by interacting with an environment and by learning from experience gained from previous interactions to make better decisions. Each time the agent carries out an action, it moves from a state to another one. The sequence of states encountered by the agent is called an *episode*. One approach to find the optimal policy is to learn the value of an action A_t when taken at a particular state S_t through the reward R_t it receives: such approaches are known as value-based methods. Q-Learning [23] is a specific value-based method that approximates the value of an action at a given state. To use it on problems with large numbers of states and actions, it is paired with a deep neural network, thus giving it the name Deep Q-Learning (DQN) [17], to learn the representation of state-action pairs. Following previous work [14], we will use a DQN to approximate the state-action values and to sample actions for the agent to take.

2.2 Belief Propagation-Based Constraint Programming

Standard constraint propagation, exchanging information between constraints and variables about unsupported variable-value pairs, has been generalized to exchanging information about each variable-value pair's likelihood of being part of a solution (CP-BP) [20], an unsupported pair corresponding to the special case of zero likelihood. Such message passing typically will not reach a fixpoint but will iterate (until some stopping criterion) between constraint-to-variable and variable-to-constraint message phases, following the classic belief propagation message update equations [18]. These enriched messages coming from constraints are computed using weighted counting algorithms specialized for each type of constraint. As a result, this CP-BP framework computes for a given constraint satisfaction problem (CSP) a marginal probability distribution $\hat{\theta}_v$ over the domain of each variable v which approximates the true marginal probability $\theta_v(d)$ of variable v being assigned value d in a solution drawn uniformly at random. Such information has been used to design novel branching heuristics to solve CSPs [3,5] and to build neuro-symbolic AI systems featuring CP [14].

3   Related Work

In recent years there have been many attempts at addressing the difficulty of learning structured output in neural networks. We restrict our attention to approaches able to handle fairly general structure and group them into layer-based and loss-based methods.

Layer-based methods add a differentiable layer in a neural network architecture, through which back-propagation can take place. SATNet [22] designs a differentiable maximum satisfiability (MAXSAT) solver. Neural Logic Machines [10] adopt a tensor representation of logic predicates to build a neural architecture. Semantic Probabilistic Layers [1] translate hard symbolic constraints into a differentiable logic circuit.

Loss-based methods are a looser combination proposing regularization terms in the loss function. A Semantic Loss function [24] was defined for an arbitrary sentence in propositional logic. Neuro-Symbolic Entropy Regularization [2] relies on logic circuits. CL-STE [26] combines Straight-Through Estimators with an encoding of logical constraints in conjunctive normal form as a loss function. DRNets [6] and CLR-DRNets [4] use the concept of entropy to define loss functions for discrete constraints. An adaptation of the negative pseudo-loglikelihood loss function is used to learn discrete graphical models [8].

Outside such classification, NeurASP [25] and DeepProbLog [16] integrate neural networks into their (probabilistic) logic programming language.

RL-Tuner [13] is a reinforcement learning architecture built to fine-tune a generative neural model using domain knowledge in the form of constraints. Even though it was used on a task in the music domain, in principle it is applicable to an arbitrary generative task in any domain. A trained generative neural network is cloned (blue boxes in Fig. 1) so that its copy can be fine-tuned through reinforcement learning (green boxes in Fig. 1) using combined rewards from the original neural network ($\log p(A_t|S_t)$) to reflect the similarity to the trained network and from the domain knowledge constraints to reflect the satisfaction of the constraints. The contribution from the latter consisted of some manually-derived weighted sum of penalties.

CP-based RL-Tuner [14] replaced the *ad hoc* constraints and penalties with a pair of CP models (pink box in Fig. 1): the Violations model to evaluate the number of constraint violations from the currently generated token, and the Marginals model to evaluate the probability of that token being generated if no additional violation is allowed. On a melody generation task, it was shown that a reward signal combining the number of additional violations and the marginal probability of no additional violation after each action taken increased constraint satisfiability with respect to the original RL-Tuner while maintaining similarity to the output of the pretrained neural network (NN).

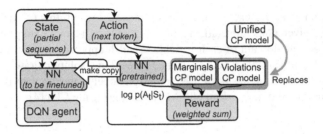

Fig. 1. CP-based RL-Tuner framework and our proposed modification.

4   Proposed Improved Framework

However, the CP-based RL-Tuner framework [14] brought some technical challenges and inefficiencies at the level of the CP models, which we explain below. Because it is in the nature of an RL agent to sometimes take illegal actions in the course of an episode, both CP models must tolerate past constraint violations. To isolate the impact of the next action, regardless of previous violations, a constraint's scope is restricted to future actions, with some adjustments. For example, the upper bound of an ATMOST constraint may be reduced depending on whether the value whose number of occurrences we restrict appeared in previous actions. Or for a REGULAR constraint we play out on the corresponding automaton the sequence of actions so far in order to determine the current "initial" state. And therefore after each new action two new CP models must be created: the *Marginals* model, featuring <u>hard</u> constraints so that the computed marginals correspond to having no future constraint violation, and the *Violations* model, featuring <u>soft</u> constraints (through reification or a soft/optimization version) in order to allow and then count constraint violations. This repeated creation of slightly different models is technically cumbersome and comes at a computational cost. The architecture we propose greatly improves on this.

4.1   Simplified Architecture

Figure 1 gives a schematic representation of our generic architecture, highlighting what we change with respect to the CP-based RL-Tuner architecture. The RL/DQN part driving the learning remains the same and is generic. The neural network (and its clone) serves the generative task in the same way but its nature (e.g. LSTM [11], Transformer [21]) can change depending on the domain. What does change is the CP part, where we replace the two CP models by a single one as explained below.

The design choice made by Lafleur et al. (2022) [14] of using both a hard and a soft CP model was motivated by the desire to reflect in the reward both the number of constraint violations brought about by the current action (requiring the soft model) and the likelihood of that action leading to an episode with no further constraint violation (requiring a hard model so that the corresponding marginal probability $\hat{\theta}_{A_t}(a)$ only reflects violation-free episodes). In other words, a combination of the present and of the future (or promise). It was also remarked by Lafleur et al. (2022) that the two complement each other well during training. In order to simplify this architecture, we give up on the "future" component but, as we will see in Sect. 4.3, replace it by something in the same spirit.

We define a single soft CP model and, even more importantly, use the same one throughout an episode as opposed to a sequence of slightly-different ones. It essentially corresponds to the initial *Violations* model in CP-based RL-Tuner where, instead of restricting the scope and adjusting the arguments of constraints (and create a new model) as an episode proceeds, we simply perform variable assignments on the same model.

The CP model is defined over a sequence of finite-domain variables $\langle A_t \rangle_{t=1}^n$ representing the actions taken by the agent during the episode (or equivalently the tokens generated). It also includes a variable V corresponding to the number of constraint violations. The relevant constraints are formulated in their reified or soft form and to each we associate a violation variable representing its individual number of violations. A SUM constraint relates them to variable V.

4.2 Evaluating Constraint Violations

To count the number of constraint violations from the *Violations* model, Lafleur et al. (2022) [14] use V_{min}, the minimum value in the domain of the variable that sums violations from every constraint. Despite adjusting constraints to ignore violations due to past actions, some inconsistencies about counting violations can still creep in. Consider the following example. We wish to generate a sequence of two tokens in which values a and b should each appear, and count one violation for each missing value. We can use a SOFT-GCC constraint [12] on these values to model this. Suppose the first token generated is c: the constraint reports a minimum of one violation because there is only one other token to generate and two missing required values. So $V_{min} = 1$. A new SOFT-GCC on the remaining token is then posted without adjusting the requirement for a and b since none have been met. Suppose the second token is again c: this time $V_{min} = 2$. So this second action will receive a negative reward twice as large as the first even though they are equally responsible. A fairly simple way to fix this, which we investigate, is to use instead $\Delta V_{min} = V_{min}^t - V_{min}^{t-1}$, where V^t represents the violations variable at time step t (i.e. after the tth token has been generated). This would correct the previous inequity and is in general more consistent.

4.3 Reward Signal

The probability-based reward $\hat{\theta}_{A_t}(a)$ of [14] only takes into account sequences for which no additional violations are allowed. On the one hand, good quality solutions of very low numbers of additional violations are ignored. On the other hand, if constraints are softened then that reward takes into account all solutions good and bad. We address this by considering the marginals for the violations variable instead of that for the current action. We propose a reward signal based on the change in the expected value of V, which we call ΔE. After an action A_t the expected value of the cost variable V is given by $\mathbb{E}[V^t] = \sum_{i=V^t.\min()}^{V^t.\max()} i \cdot \hat{\theta}_V^t(i)$. The difference between it and the expected value at the previous time step gives us our proposed reward signal: $\Delta E = \mathbb{E}[V^t] - \mathbb{E}[V^{t-1}]$.

Note that this is another advantage of having a probability distribution over domains with the CP-BP framework: one can compute expected values.

For example consider generating a binary sequence of length 5 with the rule that each repeated note creates one violation, enforced by a COSTREGULAR constraint on variables $A_1, \ldots, A_5$ with the number of violations represented by variable V. Suppose the first three actions have been taken: $A_1 = 0, A_2 = 1, A_3 = 1$. The remaining sequences are 01100, 01101, 01110, and 01111 which have 2, 1, 2, and 3 violations respectively, so $\mathbb{E}[V^{t=3}] = 2$. If the agent chooses action $A_4 = 1$, $\mathbb{E}[V^{t=4}] = 2.5$ and so $\Delta E = 2.5 - 2 = 0.5$ whereas action $A_4 = 0$ yields $\Delta E = 1.5 - 2 = -0.5$.

5   Empirical Evaluation

We conducted an empirical evaluation[1] of our improved framework, especially its new reward signal but also the way constraint violations are counted, by reusing the learning task on which RL-Tuner and CP-based RL-Tuner were previously evaluated. Given a data set of melodic lines, the goal is to learn to generate melodies that resemble the style of the input while respecting basic rules from music theory. To obtain a more direct comparison with Lafleur et al. (2022) [14], we opt to use the same musical corpus which is the soprano voices of Bach's chorales obtained using the Music21 [7] library and transpose the compositions to C major. The dataset is split into 329 compositions for training and 75 for validation. Following the original paper [13], we use a neural architecture based on long short-term memory (LSTM) cells. This network is pretrained using the music corpus to predict the next musical note, we were able to obtain a final validation accuracy of 54% (similar to [14] due to the small size of the corpus). We then use one copy of this network for our DQN to sample actions (next musical note to generate) and another copy to evaluate the similarity of the musical note generated with respect to the original music corpus.

5.1   CP Model

A melody is represented as a temporal sequence of note variables whose domain is the set of pitches. Auxiliary variables are also defined, such as those representing intervals (difference between consecutive pitches), to ease the expression of the relevant constraints. Many of the music rules are naturally modeled using REGULAR constraints: for our soft model we use COSTREGULAR [9] with a unit cost associated to each transition corresponding to a possible violation. For rules about undesirable pitches or intervals, an AMONG constraint associates a variable to the number of occurrences, which we use as the number of violations. For simpler constraints such as equality, we reify them and use the Boolean variable as the violation variable. We refer the reader to [14] for details about the variety of music rules being enforced. We implement this model using the MiniCPBP solver [19] providing marginal probabilities to compute the reward signals discussed in the previous section. We perform 10 iterations of belief propagation. During training, the solver receives the action taken by the agent (pitch value chosen) at step t and assigns it to variable A_t, constraint propagation and belief propagation are then performed before computing and returning the rewards.

---

[1] https://github.com/ChYinn/RL_Tuner_CPBP.

Table 1. Hyper-parameters for our models

Parameter	Value	Parameter	Value
number of layers (LSTM)	2	learning rate	0.001
hidden size (LSTM)	100	memory size	32000
discount rate	0.5	episodes per epoch	400
batch size	128	number of epochs	8
target network update rate	0.01	random action probability	0.1

5.2 DQN Training and Evaluation

We use the hyper-parameters shown in Table 1 to pre-train our LSTM using musical data and to fine-tune it within the RL-Tuner framework. Training involves generating musical notes one by one until we obtain a melody of 32 notes. After 400 episodes of training, we evaluate our model by randomly sampling a melody of length 32 and 64 to evaluate it by taking the average $\log p(A_t|S_t)$ reward obtained from the pre-trained LSTM and by calculating the total number of violations using the CP model.

5.3 Comparing Reward Signals

There are two metrics when evaluating our models. We would like our agents to learn to respect the constraints imposed on their actions while also keeping a level of similarity to the neural network that was pretrained on the musical corpus. We therefore consider a reward signal to be better when the trained model is able to generate sequences that are equivalent in one metric but superior in another, or superior in both.

The reward signal received by the agent is of the form $\log p(A_t|S_t) + k \times R$ where k is a balance parameter and R is the CP-based reward. Lafleur et al. (2022) [14] found that the combination $k = 2$ and $R = 40 \times \hat{\theta}_{A_t}(a) + V_{min}$ worked best. They also experimented with $k = 2$ and $R = V_{min}$. For our $R = \Delta E$, we chose $k = 4$ after experimenting with a few values. Notice that another advantage of our single-CP-model architecture is that one does not also need to tune the balance between the reward component from each of the two CP models, making tuning much simpler. We also compare $2 \times \Delta V_{min}$ against $2 \times V_{min}$ to illustrate how the *Violations* model can be improved by simply changing how violations are counted (Sect. 4.2).

5.4 Experimental Results

Looking at the top two graphs at Fig. 2, we notice that $4 \times \Delta E$ has the lowest number of violations (and fastest convergence) when generating sequences of the training length (32) while also obtaining higher $\log p(A_t|S_t)$ based rewards when compared to $2 \times (40 \times \hat{\theta}_{A_t}(a) + V)$. It is only slightly surpassed by $2 \times V_{min}$ and $2 \times \Delta V_{min}$ in terms of $\log p(A_t|S_t)$ but at the expense of much higher numbers of violations. This means the agent trained using our new reward is able to generate sequences that are competitively similar with respect to the pretrained LSTM while also obtaining the lowest number

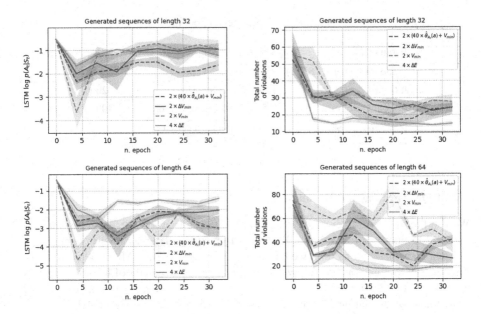

Fig. 2. Evaluation of agents (using four different reward signals) during training after each epoch. Agents generate 10 melodies of length 32 and 64 which are then evaluated using our two metrics: LSTM similarity and constraint satisfaction. The curves and shaded areas respectively represent the mean and standard deviation.

of violations. Looking at the bottom graphs, we observe that the gain in performance increases for both metrics when generating sequences twice as long as those seen during training: this exemplifies better generalization.

Comparing $2 \times \Delta V_{min}$ and $2 \times V_{min}$, we observe comparable results on both metrics for sequences of length 32 but a marked advantage of the former for length 64. This illustrates the improvement made by simply correcting how violations are counted.

6   Conclusion

We improved a reinforcement learning framework designed to fine-tune a neural model to adhere to given structural constraints. We greatly simplified its architecture with respect to the CP part, reducing it to a single CP model throughout an episode and making its implementation considerably more straightforward. We introduced a reward signal that improves the resulting model's ability to generate sequences respecting constraints while slightly diminishing its similarity to the original model. The results also show that agents trained using our reward signal exhibit better generalization.

Acknowledgements. Financial support for this research was provided by NSERC Discovery Grant 05705/2023.

References

1. Ahmed, K., Teso, S., Chang, K., den Broeck, G.V., Vergari, A.: Semantic probabilistic layers for neuro-symbolic learning. In: NeurIPS (2022). http://papers.nips.cc/paper_files/paper/2022/hash/c182ec594f38926b7fcb827635b9a8f4-Abstract-Conference.html
2. Ahmed, K., Wang, E., Chang, K., den Broeck, G.V.: Neuro-symbolic entropy regularization. In: Cussens, J., Zhang, K. (eds.) Uncertainty in Artificial Intelligence, Proceedings of the Thirty-Eighth Conference on Uncertainty in Artificial Intelligence, UAI 2022, 1–5 August 2022, Eindhoven, The Netherlands. Proceedings of Machine Learning Research, vol. 180, pp. 43–53. PMLR (2022). https://proceedings.mlr.press/v180/ahmed22a.html
3. Babaki, B., Omrani, B., Pesant, G.: Combinatorial search in CP-based iterated belief propagation. In: Simonis, H. (ed.) CP 2020. LNCS, vol. 12333, pp. 21–36. Springer, Cham (2020). https://doi.org/10.1007/978-3-030-58475-7_2
4. Bai, Y., Chen, D., Gomes, C.P.: CLR-DRNets: curriculum learning with restarts to solve visual combinatorial games. In: 27th International Conference on Principles and Practice of Constraint Programming (CP 2021). Schloss Dagstuhl-Leibniz-Zentrum für Informatik (2021)
5. Burlats, A., Pesant, G.: Exploiting entropy in constraint programming. In: Ciré, A.A. (ed.) CPAIOR 2023. LNCS, vol. 13884, pp. 320–335. Springer, Cham (2023). https://doi.org/10.1007/978-3-031-33271-5_21
6. Chen, D., Bai, Y., Zhao, W., Ament, S., Gregoire, J., Gomes, C.: Deep reasoning networks for unsupervised pattern de-mixing with constraint reasoning. In: International Conference on Machine Learning, pp. 1500–1509. PMLR (2020)
7. Cuthbert, M.S., Ariza, C.: Music21: a toolkit for computer-aided musicology and symbolic music data. In: Downie, J.S., Veltkamp, R.C. (eds.) ISMIR, pp. 637–642. International Society for Music Information Retrieval (2010). http://dblp.uni-trier.de/db/conf/ismir/ismir2010.html#CuthbertA10
8. Defresne, M., Barbe, S., Schiex, T.: Scalable coupling of deep learning with logical reasoning. In: Proceedings of the Thirty-Second International Joint Conference on Artificial Intelligence, IJCAI 2023, 19–25 August 2023, Macao, SAR, China, pp. 3615–3623. ijcai.org (2023). https://doi.org/10.24963/IJCAI.2023/402
9. Demassey, S., Pesant, G., Rousseau, L.M.: A cost-regular based hybrid column generation approach. Constraints 11(4), 315–333 (2006). https://doi.org/10.1007/s10601-006-9003-7
10. Dong, H., Mao, J., Lin, T., Wang, C., Li, L., Zhou, D.: Neural logic machines. In: 7th International Conference on Learning Representations, ICLR 2019, New Orleans, LA, USA, 6–9 May 2019. OpenReview.net (2019). https://openreview.net/forum?id=B1xY-hRctX
11. Hochreiter, S., Schmidhuber, J.: Long short-term memory. Neural Comput. 9(8), 1735–1780 (1997)
12. van Hoeve, W.J., Pesant, G., Rousseau, L.: On global warming: flow-based soft global constraints. J. Heuristics 12(4–5), 347–373 (2006). https://doi.org/10.1007/s10732-006-6550-4
13. Jaques, N., Gu, S., Turner, R.E., Eck, D.: Tuning recurrent neural networks with reinforcement learning. In: 5th International Conference on Learning Representations, ICLR 2017, Toulon, France, 24–26 April 2017, Workshop Track Proceedings. OpenReview.net (2017). https://openreview.net/forum?id=Syyv2e-Kx
14. Lafleur, D., Chandar, S., Pesant, G.: Combining reinforcement learning and constraint programming for sequence-generation tasks with hard constraints. In: Solnon, C. (ed.) 28th International Conference on Principles and Practice of Constraint Programming (CP 2022). Leibniz International Proceedings in Informatics (LIPIcs), vol. 235, pp. 30:1–30:16. Schloss Dagstuhl – Leibniz-Zentrum für Informatik, Dagstuhl (2022). https://doi.org/10.4230/LIPIcs.CP.2022.30

15. Leslie, D., Rossi, F.: ACM TechBrief: generative artificial intelligence. Technical report, Association for Computing Machinery, New York (2023).https://doi.org/10.1145/3626110

16. Manhaeve, R., Dumancic, S., Kimmig, A., Demeester, T., Raedt, L.D.: Neural probabilistic logic programming in deepproblog. Artif. Intell. **298**, 103504 (2021). https://doi.org/10.1016/j.artint.2021.103504

17. Mnih, V., et al.: Playing Atari with deep reinforcement learning. arXiv preprint arXiv:1312.5602 (2013)

18. Pearl, J.: Reverend bayes on inference engines: a distributed hierarchical approach. In: Waltz, D.L. (ed.) Proceedings of the National Conference on Artificial Intelligence, Pittsburgh, PA, USA, 18–20 August 1982, pp. 133–136. AAAI Press (1982). http://www.aaai.org/Library/AAAI/1982/aaai82-032.php

19. Pesant, G.: The MiniCPBP solver. https://github.com/PesantGilles/MiniCPBP

20. Pesant, G.: From support propagation to belief propagation in constraint programming. J. Artif. Intell. Res. **66** (2019). https://doi.org/10.1613/jair.1.11487

21. Vaswani, A., et al.: Attention is all you need. In: Advances in neural information processing systems, vol. 30 (2017)

22. Wang, P., Donti, P.L., Wilder, B., Kolter, J.Z.: SATNet: bridging deep learning and logical reasoning using a differentiable satisfiability solver. In: Chaudhuri, K., Salakhutdinov, R. (eds.) Proceedings of the 36th International Conference on Machine Learning, ICML 2019, 9–15 June 2019, Long Beach, California, USA. Proceedings of Machine Learning Research, vol. 97, pp. 6545–6554. PMLR (2019). http://proceedings.mlr.press/v97/wang19e.html

23. Watkins, C.J., Dayan, P.: Q-learning. Mach. Learn. **8**, 279–292 (1992)

24. Xu, J., Zhang, Z., Friedman, T., Liang, Y., den Broeck, G.V.: A semantic loss function for deep learning with symbolic knowledge. In: Dy, J.G., Krause, A. (eds.) Proceedings of the 35th International Conference on Machine Learning, ICML 2018, Stockholmsmässan, Stockholm, Sweden, 10–15 July 2018. Proceedings of Machine Learning Research, vol. 80, pp. 5498–5507. PMLR (2018). http://proceedings.mlr.press/v80/xu18h.html

25. Yang, Z., Ishay, A., Lee, J.: NeurASP: embracing neural networks into answer set programming. In: Bessiere, C. (ed.) Proceedings of the Twenty-Ninth International Joint Conference on Artificial Intelligence, IJCAI 2020, pp. 1755–1762. ijcai.org (2020). https://doi.org/10.24963/ijcai.2020/243

26. Yang, Z., Lee, J., Park, C.: Injecting logical constraints into neural networks via straight-through estimators. In: Chaudhuri, K., Jegelka, S., Song, L., Szepesvári, C., Niu, G., Sabato, S. (eds.) International Conference on Machine Learning, ICML 2022, 17–23 July 2022, Baltimore, Maryland, USA. Proceedings of Machine Learning Research, vol. 162, pp. 25096–25122. PMLR (2022). https://proceedings.mlr.press/v162/yang22h.html

Bound Tightening Using Rolling-Horizon Decomposition for Neural Network Verification

Haoruo Zhao[1,2（✉)], Hassan Hijazi[2], Haydn Jones[2,3], Juston Moore[2], Mathieu Tanneau[1], and Pascal Van Hentenryck[1]

[1] Georgia Institute of Technology, Atlanta, GA 30308, USA
{hzhao306,mathieu.tanneau}@gatech.edu,
pascal.vanhentenryck@isye.gatech.edu
[2] Los Alamos National Laboratory, Los Alamos, NM 87545, USA
{hlh,hjones}@lanl.gov
[3] University of Pennsylvania, Philadelphia, PA 19104, USA

Abstract. Neural network verification aims at providing formal guarantees on the output of trained neural networks, to ensure their robustness against adversarial examples and enable deployment in safety-critical applications. This paper introduces a new approach to neural network verification using a novel mixed-integer programming (MIP) rolling-horizon decomposition method. The algorithm leverages the layered structure of neural networks by employing optimization-based bound tightening (OBBT) on smaller sub-graphs of the original network in a rolling-horizon fashion and tightening the bounds in parallel. This strategy strikes a balance between achieving tighter bounds and ensuring the tractability of the underlying mixed-integer programs. Extensive numerical experiments, conducted on instances from the VNN-COMP benchmark library, demonstrate that the proposed approach yields significantly improved bounds compared to existing efficient bound propagation methods. Notably, the proposed method proves effective in solving open verification problems. Our code is built and released as part of the open-source mathematical modeling tool Gravity (https://github.com/coin-or/Gravity), which is extended to support generic neural network models.

Keywords: Neural Network Verification · Optimization-Based Bound Tightening · Mixed-Integer Programming · Decomposition

1 Introduction

Neural networks are being applied in critical systems and high-consequence decision-making settings, e.g., power systems [2] and autonomous driving [4]. How can we trust these models when the stakes are too high, when the price of failure is prohibitive or even life-threatening? To justify trust, these models need to provide robustness guarantees. For example, in the context of power grid applications, a guarantee that a slight change in input (power system state) will

B. Dilkina (Ed.): CPAIOR 2024, LNCS 14743, pp. 289–303, 2024.
https://doi.org/10.1007/978-3-031-60599-4_20

not lead to unreasonable fluctuations in output (control actions predicted by the network). Mathematical optimization can provide such guarantees. While local optimization methods are acceptable for training these models, global optimality—i.e., a formal certificate that no better solution exists—is needed to provide robustness guarantees. There has been significant recent interest in providing verifiable properties for neural networks using global methods such as mixed-integer programming (MIP) [6,7,10]. MIP solvers are appealing because they can, in principle, perform *complete* verification for many network architectures. Unfortunately, global optimization methods come with a hefty computational price tag, and using off-the-shelve solvers is not a scalable approach.

Improving the scalability of global methods for neural network verification is an active research area, with approaches such as mathematical reformulations [15], cutting planes [1], zonotopes [11], custom-built branch-and-bound algorithms and relaxations [12,16–22], to name a few. In 2020, a community-driven effort led to the creation of the VNN competition [3], which has been held yearly since. The competition's goal is to "allow researchers to compare their neural network verifiers on a wide set of benchmarks". The α, β-CROWN team has consistently won this competition since its inception [16–21]. In this paper, we are hoping to bring the use of mixed-integer programming for neural network (NN) verification one step closer to viability. For this purpose, we tackle the verification problem as defined in [3], using a mixed-integer programming decomposition approach combined with optimization-based bound tightening (OBBT) [5]. OBBT is extensively used in global optimization solvers to reduce variable domains, especially for nonconvex mixed-integer nonlinear programs (MINLPs) [8]. The OBBT algorithm solves two auxiliary optimization problems for each decision variable. In its initial form, OBBT relies on convexifying the feasible region and using variables' bounds as objective functions. In this work, we propose to preserve the mixed-integer nature of the subproblems, leveraging instead our proposed rolling horizon decomposition. One nice property of OBBT is its amenability to parallelization, since each auxiliary problem can be run independently. We take advantage of parallelization along with other speedup methods such as early termination using cutoff values as outlined in our approach below.

2   Problem Statement

In the domain of neural network verification, a "white-box" model is given, granting full visibility into the network's architecture and parameters [3]. Our verification challenge, based on the framework established by Bunel et al. [4], is to ascertain whether a neural network, denoted as function f with L layers, produces outputs that satisfy a desired property P for all inputs within a specified range $\mathcal{C}$.

Formally, we verify that for any input $\mathbf{x}_0 \in \mathcal{C}$, the network's output $\mathbf{y}^{(L)}$ adheres to the property $P(\mathbf{y}^{(L)})$, encapsulated by the implication:

$$\mathbf{x}_0 \in \mathcal{C} \Rightarrow P(\mathbf{y}^{(L)}).$$

For instance, in assessing local robustness, we determine whether all inputs within an ϵ-ball around a data point a with label y_a are classified as y_a by the network. This property is widely used in image classification cases, where we assess if a network's output label remains consistent under perturbations within a small tolerance.

2.1   MIP Encoding for Trained Neural Networks

Transitioning neural network architectures into a mixed-integer programming (MIP) format is key for verification. This process, in line with Tjeng et al.'s [14] methodology, enables the application of mathematical programming for thorough network analysis. The MIP model thus becomes a crucial tool for effective verification strategies.

Consider a neural network with an input vector $\mathbf{x}^{(0)} := \mathbf{x}_0 \in \mathbb{R}^{n_0}$. In this network, n_i represents the number of neurons in the i-th layer. The network consists of L layers, where each layer i has an associated weight matrix $W^{(i)} \in \mathbb{R}^{n_i \times n_{i-1}}$ and a bias vector $b^{(i)} \in \mathbb{R}^{n_i}$, for $i \in \{1, \ldots, L\}$. Let $\mathbf{y}^{(i)}$ denote the pre-activation vector and $\mathbf{x}^{(i)}$ the post-activation vector at layer i, with $\mathbf{x}^{(i)} = \sigma(\mathbf{y}^{(i)})$. The output of the network is $\mathbf{y}^{(L)}$. Although σ could be any activation function, we will assume it is the ReLU function throughout this paper. Figure 1 presents a fully connected neural network with ReLU activation functions.

Fig. 1. A fully connected neural network with ReLU activation functions, showcasing the architecture used for MNIST digit classification.

In network verification, we typically define f such that a non-negative outcome of $f(\mathbf{y}^{(L)}) \geq 0$ indicates the satisfaction of the desired property. Therefore, if an adversarial input $\mathbf{x} \in \mathcal{C}$ results in a negative value of f, this is taken as evidence that the verification instance fails to meet the required property, providing a counter-example. In short:

$$\begin{cases} \text{If } \exists \mathbf{x}_{\text{adv}}^{(0)} \in \mathcal{C} \text{ such that } f(\mathbf{y}_{\text{adv}}^{(L)}) < 0, \text{then } P(\mathbf{y}_{\text{adv}}^{(L)}) \text{ does not hold.} \\ \text{If } \forall \mathbf{x}^{(0)} \in \mathcal{C}, f(\mathbf{y}^{(L)}) \geq 0, \text{then } P(\mathbf{y}^{(L)}) \text{ holds.} \end{cases}$$

The optimization problem is as follows:

$$\begin{aligned} \min \quad & f(\mathbf{y}^{(L)}) \\ \text{s.t.} \quad & \mathbf{y}^{(i)} = W^{(i)}\mathbf{x}^{(i-1)} + b^{(i)}, && \forall i \in \{1, \ldots, L\}, \\ & \mathbf{x}^{(i)} = \sigma(\mathbf{y}^{(i)}), && \forall i \in \{1, \ldots, L-1\}, \\ & \mathbf{x}^{(0)} \in \mathcal{C}. \end{aligned} \tag{1}$$

Specifically, if all operators within the trained neural network are piecewise-linear, then the neural network can be linearly represented within the mixed-integer linear programming (MILP) framework. If the activation function is a ReLU function, the MILP formulation of $x = \sigma(y) = ReLU(y) = \max(0, y)$ is given by:

$$x \geq 0, \; x \geq y, \; x \leq y - l(1 - z), \; x \leq u \cdot z, \; z \in \{0, 1\}$$

where l, u are the respective lower and upper bound on y. The binary variable z indicates the activation state of the ReLU.

A ReLU neuron unit can be classified into different categories based on its input domain $[l, u]$: it is deemed "Inactive" if $u \leq 0$, "Active" if $l \geq 0$, and "Unstabilized" when $l < 0$ and $u > 0$. The neuron is called "Stabilized" if it meets either the Active or Inactive condition.

2.2 Bound Tightening

In practice, solving the mixed-integer program (1) can be computationally prohibitive for large neural networks. As a result, it has been proven effective to introduce additional bounds on intermediate layers. Specifically, the problem states

$$\min \quad f(\mathbf{y}^{(L)})$$

$$\text{s.t.} \quad \text{Constraints of model (1)},$$
$$\mathbf{x}_l^{(i)} \leq \mathbf{x}^{(i)} \leq \mathbf{x}_u^{(i)}, \quad \forall i \in \{1, \ldots, L - 1\},$$
$$\mathbf{y}_l^{(i)} \leq \mathbf{y}^{(i)} \leq \mathbf{y}_u^{(i)}, \quad \forall i \in \{1, \ldots, L\}.$$

$$(2)$$

In model (2), $\mathbf{y}_l^{(i)}$ and $\mathbf{y}_u^{(i)}$ denotes the vector lower and upper bound respectively for pre-activation output $\mathbf{y}^{(i)}$ at layer i; we similarly define these bounds for post-activation $\mathbf{x}^{(i)}$. Hence, the additional constraints form hyper-rectangles around the outputs. It is crucial to note that the bounds on $\mathbf{y}^{(i)}$ play a much more important role than those on $\mathbf{x}^{(i)}$, as they serve as the input to the ReLU layer and consequently determine the stabilization of the ReLU neuron. When a ReLU is stabilized, the binary variable indicating its active state becomes fixed to either 0 (inactive) or 1 (active). This stabilization leads to a reduction in the number of binary variables in the problem.

There are several methods to derive bounds for intermediate layers:

- **Interval Bound Propagation (IBP):** This method employs interval bound propagation to establish bounds for each layer [9].
- **DeepPoly:** This method uses a custom polyhedral abstract domain relaxation. It assigns concrete lower and upper bounds to every neuron in a neural network. Symbolic bounds are formulated as linear combinations of the neurons in the network's previous layer [13].
- **CROWN:** This method efficiently leverages linear bound propagation to adaptively determine the lower and upper bounds of neural networks [22].
- α-**CROWN:** This approach builds upon and enhances CROWN, further tightening the linear bounds by utilizing gradients [18].

– **Optimization-Based Bound Tightening (OBBT):** This approach represents the exact bound tightening technique, which involves solving MIP problems [5].

The following remarks can be made about the above methods. First, it is widely accepted that IBP, while simple and fast, yields very weak bounds, especially for deep networks. Second, DeepPoly and CROWN are based on the same polyhedral relaxations, and therefore yield the same bounds [12]. Third, because α-CROWN augments CROWN with a gradient-based procedure, it achieves (both in theory and in practice) tighter bounds than CROWN [12,18]. Finally, as noted in [12], because IBP, CROWN/DeepPoly and α-CROWN are all based on polyhedral (linear) relaxations, they cannot break the convex relaxation barrier described in [12], though α-CROWN matches this theoretical limitation. Note that the convex relaxation barrier presented in [12] is equivalent to LP-based OBBT, wherein bounds are tightened iteratively by solving only the linear relaxation of each OBBT problem. Therefore, in order to further tighten bounds, any procedure must explicitly consider the binary variables associated to each ReLU neuron.

In our work, the primary objective is to obtain tighter bounds. To achieve this, we utilize optimization-based bound tightening, focusing on achieving the tightest box bounds for intermediate layers.

3   Methodology

We introduce the Optimization-Based Bound Tightening with Rolling Horizon (OBBT-RH) here. The method builds upon the MIP-based Optimization-Based Bound Tightening approach, as tighter bounds for intermediate layers in model (2) will in general result in faster solve times. Therefore, applying tighter bounds can speed up the verification process, allowing it to scale to larger models.

OBBT is formulated as two optimization subproblems for each neuron, seeking to find its maximum and minimum bound. Specifically, let $\mathbf{y}_k^{(t)}$ denote the k-th neuron at layer t subject to network constraints. Given $0 \leq s < t \leq L$, the problem states:

$$
\begin{aligned}
\max/\min \quad & \mathbf{y}_k^{(t)} \\
\text{s.t.} \quad & \mathbf{y}^{(i)} = W^{(i)}\mathbf{x}^{(i-1)} + b^{(i)}, && \forall i \in \{s+1, \ldots, t\}, \\
& \mathbf{y}_l^{(i)} \leq \mathbf{y}^{(i)} \leq \mathbf{y}_u^{(i)}, && \forall i \in \{s+1, \ldots, t\}, \\
& \mathbf{x}^{(i)} = \sigma(\mathbf{y}^{(i)}), && \forall i \in \{s+1, \ldots, t-1\}, \\
& \mathbf{x}_l^{(i)} \leq \mathbf{x}^{(i)} \leq \mathbf{x}_u^{(i)} && \forall i \in \{s, \ldots, t-1\}.
\end{aligned}
\tag{3}
$$

However, as the number of neural network layers included in the OBBT instances (3) increases, the OBBT process becomes intractable. To address this issue, we propose a decomposition method in OBBT-RH to reduce the number of layers considered in each OBBT instance.

Algorithm 1. OBBT-RH

Require: Rolling horizon sequence $\mathcal{S} = \{(s_j, t_j) \mid j \in [J]\}$ where J is a positive integer
1: **for** $j = 1, \ldots, J$ **do**
2:     **for** each neuron $k = 1, \ldots, n_i$ **do**
3:         Solve problem (3) to obtain max and min values of $\mathbf{y}_k^{(t_j)}$.
4:     **end for**
5: **end for**
Ensure: Lower/upper bounds $\{\mathbf{y}_l^{(i)}\}, \{\mathbf{y}_u^{(i)}\}$ for all $i = 1, \ldots, L$.

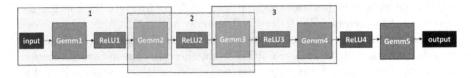

Fig. 2. OBBT-RH with horizon length 2: $\mathcal{S} = \{(0,2), (1,3), (2,4)\}$, meaning that $(s_1, t_1) = (0,2), (s_2, t_2) = (1,3), (s_3, t_3) = (2,4)$.

The proposed OBBT-RH in Algorithm 1 leverages problem (3) to present an effective method for the verification of deep neural networks. OBBT-RH sequentially decomposes the neural network into manageable problem size, focusing on smaller sub-graphs of the original network. This method hence applies optimization-based bound tightening within a rolling horizon framework, effectively balancing the achievement of tighter bounds against maintaining the tractability of the mixed-integer programming problems involved.

3.1   Rolling Horizon Sequence

We introduce a rolling horizon strategy for selecting sequences $\mathcal{S}$ of layers within a neural network for bound tightening. The rolling horizon length, denoted as H, determines the maximum number of General Matrix Multiplications (Gemm) layers to be included in each sub-graph. We begin with an initial pair (s_1, t_1) representing the starting and ending layers of the window, where s_1 is the starting layer and t_1 is the tightening layer. Each subsequent pair (s_j, t_j) in the sequence $\mathcal{S}$ is constructed by shifting the window toward the end layer, with H guiding the span of the window and indicating the layers selected for bound tightening in each iteration. For instance, considering the ReLU neural network, we selected $t_j = \min(j+1, L-1)$ and $s_j = \max(0, t_j - H)$. Indeed, the choice of s_j and t_j can be adjusted for various neural network architectures, highlighting the inherent flexibility of our approach.

The end goal is to tighten the neurons' bounds right before ReLU layers, thus each t in the sequence is a layer that precedes a ReLU layer. Figure 2 shows the OBBT-RH with horizon length $H = 2$ for a neural network with 5 layers. For $H = 2$, the sequence $\mathcal{S} = \{(0,2), (1,3), (2,4)\}$ includes pairs (Input, Gemm2), (Gemm1, Gemm3), and (Gemm2, Gemm4), each ending just before a ReLU layer. As H increases to 3, the sequence $\mathcal{S} = \{(0,2), (0,3), (1,4)\}$ includes pairs (Input, Gemm2), (Input, Gemm3) and (Gemm1, Gemm4), and for $H = 4$, $\mathcal{S} = \{(0,2), (0,3), (0,4)\}$ and it includes $\{(\text{Input, Gemm2}), (\text{Input, Gemm3}),$ (Input, Gemm4)$\}$, now encompassing all Gemm layers leading to a ReLU operator.

The length of the rolling horizon has direct impact on bound tightness and computational time: A longer horizon considers more layers simultaneously, which could help in stabilizing more ReLU neurons and thus simplifying the MIP problem. However, this benefit comes at the cost of increased computational resources. A shorter horizon will generally require less computation, beneficial for efficiency. However, it loses information in the previous layers and produces looser bounds.

One important advantage of this rolling horizon approach is the tightened bounds for neurons in previous layers can speed up the tightening process for neurons in subsequent layers. Note that IBP is equivalent to OBBT for the first ReLU layer if the intermediate layer is linear. We take advantage of this observation to avoid building sub-MIPs leading to this layer.

3.2 Early Termination in the Case of ReLU Activation

Maximizing Neuron Output: When solving the MIP (3) to maximize the output of a neuron $y_k^{(t)}$, if the upper bound reaches zero, it indicates that the neuron will be inactive $(y_k^{(t)} \leq 0)$. Hence, further bound tightening is unnecessary, and the process can be terminated early.

Minimizing Neuron Output: Similarly, when minimizing the output of a neuron, if the lower bound exceeds zero, it implies that the neuron will remain active $(y_k^{(t)} > 0)$. In this case, the OBBT process can also be terminated early.

In either case, the ReLU neuron is stabilized. Therefore, introducing such an early termination criterion speeds up the bound tightening process. In this paper, we simplify our model by assuming a linear structure for the neural network, wherein each layer is dependent solely on its immediate predecessor. However, real-world neural networks often exhibit more complex dependencies, where a given layer may depend not just on its immediate predecessor, but also on layers further back in the sequence. In such cases, the decomposition process becomes more complicated. We propose using Breadth-First Search (BFS) to navigate these complex dependencies. BFS can effectively trace the shortest path from the final layer t to an initial layer s, ensuring that all relevant inter-layer dependencies, including those extending over multiple layers, are appropriately captured for conducting bound tightening of intermediate layers.

3.3 Parallelization of OBBT Sub-MIPs

We take advantage of the graph-structure of neural networks by recognizing that the bounds on each neuron in a given layer can be independently tightened. This observation allows us to distribute the computational effort of solving sub-MIPs corresponding to each neuron using independent threads in parallel. Our implementation uses both the Message Passing Interface (MPI) for multi-machine clusters as well as single-machine multi-threading to parallelize our rolling-horizon algorithm.

4   Numerical Results

4.1   Setup

Benchmark Dataset. Numerical experiments were conducted on all 90 instances from the `mnist_fc` benchmark within the VNN-COMP benchmark library [3].

Comparison Metrics. We primarily focus on the following metrics. First, the number of ReLU neurons stabilized (Table 1). Second, bounds range of ReLU neurons input (Table 2). Third, LP relaxation bounds of the MIPs with bounds on intermediate layers (Table 3). Lastly, the proportion of instances that can be verified and their computing times (Table 4).

Implementation Details. The `auto_LiRPA` package is used to calculate IBP, CROWN, and α-CROWN bounds. A GPU V100 is used for generating these bounds. Note that the code of `auto_LiRPA` returns errors for five instances (51, 54, 60, 76, 85) when computing the α-CROWN bounds. Therefore, for a fair comparison, these instances have been removed. The authors were not able to install DeepPoly and its required dependencies; this method is therefore not included in the results. Nevertheless, recall that DeepPoly and CROWN employ the same relaxations and thus yield numerically identical bounds [12]. The optimization problems were formulated with mathematical modeling tool Gravity (https://github.com/coin-or/Gravity) in C++ and solved using Gurobi 10.0.2. The experiments are carried out on the High-Performance Computing (HPC) platform provided by the Partnership for an Advanced Computing Environment (PACE) at Georgia Institute of Technology, Atlanta, Georgia. Within this benchmark, the setting for OBBT-RH horizon length is dynamically adjusted based on the network architecture. Specifically, for neural networks with three layers, the rolling horizon length H is set to two. In the case of networks with five layers,

Table 1. Average number of neurons (inactive, active, stabilized, unstabilized) for each method in the benchmark.

Method	Inactive	Active	Stabilized	Unstabilized
IBP	118.21	4.21	122.42	633.53
CROWN	435.67	18.72	454.39	301.56
α-CROWN	466.67	19.93	486.60	269.35
OBBT	621.95	31.91	653.86	102.09
OBBT-RH	**622.25**	**31.93**	**654.18**	**101.78**

Table 2. Average bounds range for each method in the benchmark.

Method	Bounds Range
IBP	2028.35
CROWN	96.16
α-CROWN	53.52
OBBT	12.87
OBBT-RH	**12.38**

H is set to three, while for those with seven layers, H is set to five. OBBT can be seen as a special case of OBBT-RH, wherein the horizon length is equal to the depth of the neural network.

Additionally, during the initialization phase of OBBT-RH instances, IBP bounds are applied to intermediate layers. To further enhance efficiency, a parallel approach is employed for tightening the input of each ReLU layer. Given that each ReLU layer in this benchmark comprises 256 neurons, a total of 512 CPU threads are requested, with 2 threads used for each OBBT-RH instance. This strategy of parallelization enables the independent and simultaneous tightening of bounds for each ReLU layer, significantly boosting the process's overall efficiency. In addition, since the goal of each OBBT-RH instance is to tighten the bounds, we have set Gurobi's MIPFocus parameter to value 3 to enhance the quality of the bounds. Moreover, to avoid excessively long solving times for some instances, a 30-second time limit is enforced on each OBBT-RH instance. Therefore, with the need to solve both maximization and minimization problems for unstabilized neuron, the maximum time required for OBBT-RH in one layer is one minute.

4.2   Comparison with Other Bound Tightening Methods

The aim is to compare the bound tightness produced by various methods. Specifically, the goal is to maximize the number of stabilized neurons, thereby reducing the count of binary variables. Additionally, for those unstabilized neurons, the approach seeks to tighten their input bounds as much as possible, aiding in the stabilization of subsequent layer neurons. After determining the bounds preceding the ReLU layers, the LP relaxation bounds of the MIPs will be compared with the intermediate layers' generated bounds. Lastly, the evaluation will focus on the end-to-end comparison of the number of verified instances as well as the time taken across different methods.

In terms of results, Table 1 presents the average number of neurons categorized as inactive, active, stabilized and unstabilized for different bound tightening methods. The OBBT-RH method demonstrates a higher average across all layers from all instances. Table 2 shows the average bounds range for each verification method in a benchmark setting. OBBT-RH outperforms the other methods with the tightest bounds, indicating its effectiveness in bounding neuron activations within the network. Table 3 presents the average LP relaxation bounds for MIPs using the bounds produced by each method. These LP relaxation bounds indicate the tightness of the solution space. OBBT shows the least negative value, implying a closer approximation to the MIP's optimal solution, whereas OBBT-RH has a slightly higher value than OBBT. Table 4 includes the end-to-end verification results by comparing the efficiency of MIP-based verification where variable bounds are computed using different bound tightening methods. BT stands for the bound tightening time and MIP stands for the final MIP solving time with the tightened bounds. A total time limit of 20 min is set for each approach. Overall, MIP with OBBT-RH bounds verifies the most instances (either by proving robustness or finding an adversarial example) than other approaches, and achieves a 2.2x speedup over CROWN-based methods. In addition, OBBT-RH yields 1.25x speedup over OBBT, mostly thanks to shorter bound-tightening times.

4.3   Solving MIP with Tight Bounds

After computing bounds for the intermediate layers, we use these bounds to solve the MIP instances. This is achieved by first obtaining the variable bounds for intermediate layers and then constructing the MIPs using the model as outlined in (2) for each verification instance. In the verification context, tightening the domain of the input for ReLU activations strengthens the MIP formulation by decreasing the big-M coefficients. Furthermore, it can stabilize ReLU units-eliminating binary variables-when a ReLU's input is proven to be always non-negative or always non-positive. This combined effect-yielding stronger relaxations and reducing the number of binary variables-results in substantial improvements when solving the final MIPs corresponding to the verification problems with the tightened bounds. In addition to incorporating the tight bounds into the MIP, we also set the Gurobi cutoff parameters to 0. This parameter setting is critical as it ensures that nodes with a lower bound greater than 0 are pruned, thereby enhancing the efficiency of the verification process.

The end-to-end results are shown in Table 4, with OBBT-RH outperforming the other methods as discussed above. Figure 3 shows the performance of various methods in verifying instances within the benchmark, highlighting the

Table 3. Average LP relaxation bounds of the MIPs for each method in the benchmark.

Method	LP Bounds
IBP	-25232.25
CROWN	-1169.45
α-CROWN	-577.91
OBBT	$\mathbf{-0.64}$
OBBT-RH	-24.77

Table 4. End-to-end comparison on all instances. All time are in seconds.

Method	#Verified	#Timeout	Time (sec)		
			BT	MIP	Total
IBP+MIP	37	48	**3.54**	708.60	712.14
CROWN+MIP	62	23	**3.54**	345.89	349.43
α-CROWN+MIP	61	24	4.93	340.74	345.66
OBBT+MIP	79	6	121.20	81.07	202.27
OBBT-RH+MIP	**80**	**5**	90.19	**64.66**	**154.85**

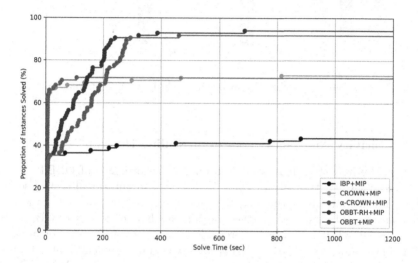

Fig. 3. Performance comparison of the MIP-based verification for different bound-tightening methods.

trade-off between tight bounds and verification time. For low time budgets, fast bound tightening methods such as IBP, CROWN, and α-CROWN are more effective, verifying a greater number of instances quickly. However, as the time budget increases, OBBT-RH and OBBT, which prioritize tighter bounds, begin to outperform the faster methods by verifying a higher proportion of instances. Notably, OBBT-RH is more efficient than OBBT, verifying more instances in less time. Moreover, our approach has successfully closed several challenging instances (48, 78, 83) that previously posed difficulties for state-of-the-art complete verifiers. With our proposed method for generating OBBT-RH bounds, we successfully closed instance 48 within five minutes, a feat not achieved by any complete verifiers from the VNN competition.

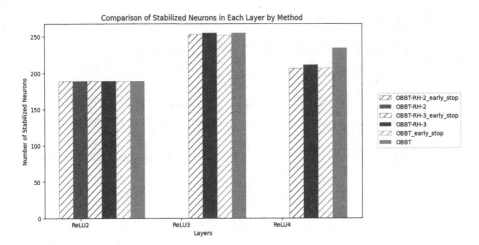

Fig. 4. Comparison of stabilized neurons in each layer by different methods on instance 48.

4.4 Sensitivity Analysis 1: Length of Rolling Horizon

To assess the impact of horizon length on the performance of OBBT-RH, we conducted a sensitivity analysis. Due to the high computational cost of solving each OBBT instance to optimality, the sensitivity analysis was only conducted on instance 48. The analysis involved varying the horizon length of OBBT-RH, denoted as OBBT-RH-H, where H represents the horizon length. For instance, OBBT-RH-3 indicates an OBBT-RH variant with a horizon length of 3. Figure 4 compares the number of stabilized neurons across each layer by OBBT-RH with varying horizon lengths. The analysis reveals a clear trend: OBBT-RH-3 demonstrates a significant increase in the number of stabilized ReLU units compared to its counterpart, OBBT-RH-2, which employs a rolling horizon of length 2. OBBT-RH-2 stabilizes 0 ReLUs in the 3rd and 4th ReLU layers. It should also be emphasized that, considering early-stopping, OBBT-RH-3 stabilizes almost as many neurons as OBBT. This suggests that our proposed MIP-based approach, with an appropriately set rolling horizon, can effectively enhance the tightness of neuron bounds.

4.5 Sensitivity Analysis 2: The Effect of Early Stop

In the process of bound tightening, OBBT-RH is utilized to refine the input bounds of each neuron. This task is computationally intensive, requiring the solution of MIPs for both the upper and lower bounds of potentially hundreds of neurons.

Early stopping is a technique introduced to reduce the computational burden. It achieves this by terminating the bound tightening process before reaching the optimal solution under certain conditions.

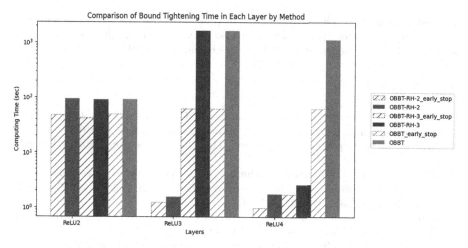

Fig. 5. Comparison of bound tightening time for each layer by different OBBT-based methods on instance 48.

To prevent any single OBBT-RH instance from taking an excessively long time, we have set a 30-second time limit. This decision is informed by empirical evidence indicating that many bounds exhibit only marginal improvement after 30 s. Therefore, this time limit is selected to balance the achievement of high-quality bounds with the need for reasonable computing time.

Figure 5 illustrates the impact of early stopping on bound tightening time across different network layers, presented in a log-scale format. We cease further tightening of a neuron once it stabilizes, regardless of the application of early stopping. To assess the effectiveness of early stopping, we compare the number of stabilized neurons with and without this criterion. As depicted in the figure, early stopping-particularly with a 30-second time limit-does not significantly affect neuron stabilization. This finding supports the use of early stopping as an effective strategy to enhance the scalability of OBBT-RH without significantly compromising the quality of the bounds obtained. Due to the prolonged solving time required for tightening the inputs of the ReLU3 and ReLU4 layers with OBBT-RH-3 and OBBT without early stop (exceeding 4 h), we resorted to using 24 threads to run a single OBBT instance. Therefore, the actual solving time with 2 threads would be even longer.

5   Conclusion and Future Work

We present a new algorithm integrating bound-tightening and mixed-integer programming with a rolling horizon strategy as a promising approach for verifying neural networks. Our method is particularly suited for handling networks where tighter bounds are needed but cannot be obtained from existing effective bound propagation methods. Future work will focus on extending this methodology to other types of nonlinear operators and further improving the computational efficiency of the verification process.

Acknowledgements. This research is partly funded by NSF award 2112533 and supported by the Laboratory Directed Research and Development program of Los Alamos National Laboratory under projects 20230578ER and 20240734DI.

References

1. Anderson, R., Huchette, J., Tjandraatmadja, C., Vielma, J.P.: Strong mixed-integer programming formulations for trained neural networks (2019)
2. ARPA-e, U.S. Department of Energy: The GO Competition (2020). https://gocompetition.energy.gov/
3. Brix, C., Müller, M.N., Bak, S., Johnson, T.T., Liu, C.: First three years of the international verification of neural networks competition (VNN-COMP). Int. J. Softw. Tools Technol. Transf. 1–11 (2023)
4. Bunel, R., Turkaslan, I., Torr, P.H.S., Kohli, P., Kumar, M.P.: A unified view of piecewise linear neural network verification (2018)
5. Caprara, A., Locatelli, M.: Global optimization problems and domain reduction strategies. Math. Program. **125**, 123–137 (2010)
6. Dathathri, S., et al.: Enabling certification of verification-agnostic networks via memory-efficient semidefinite programming. arXiv preprint arXiv:2010.11645 (2020)
7. Dvijotham, K.D., Stanforth, R., Gowal, S., Qin, C., De, S., Kohli, P.: Efficient neural network verification with exactness characterization. In: Uncertainty in Artificial Intelligence, pp. 497–507. PMLR (2020)
8. Gleixner, A.M., Berthold, T., Müller, B., Weltge, S.: Three enhancements for optimization-based bound tightening. J. Global Optim. **67**, 731–757 (2017)
9. Gowal, S., et al.: On the effectiveness of interval bound propagation for training verifiably robust models (2019)
10. Gowal, S., Dvijotham, K., Stanforth, R., Mann, T., Kohli, P.: A dual approach to verify and train deep networks (2019)
11. Kochdumper, N., Schilling, C., Althoff, M., Bak, S.: Open-and closed-loop neural network verification using polynomial zonotopes. In: Rozier, K.Y., Chaudhuri, S. (eds.) NFM 2023. LNCS, vol. 13903, pp. 16–36. Springer, Cham (2023). https://doi.org/10.1007/978-3-031-33170-1_2
12. Salman, H., Yang, G., Zhang, H., Hsieh, C.J., Zhang, P.: A convex relaxation barrier to tight robustness verification of neural networks. In: Advances in Neural Information Processing Systems, vol. 32, pp. 9835–9846 (2019)
13. Singh, G., Gehr, T., Püschel, M., Vechev, M.: An abstract domain for certifying neural networks. Proc. ACM Program. Lang. **3**(POPL) (2019). https://doi.org/10.1145/3290354
14. Tjeng, V., Xiao, K., Tedrake, R.: Evaluating robustness of neural networks with mixed integer programming. arXiv preprint arXiv:1711.07356 (2017)
15. Tsay, C., Kronqvist, J., Thebelt, A., Misener, R.: Partition-based formulations for mixed-integer optimization of trained reLU neural networks. In: Beygelzimer, A., Dauphin, Y., Liang, P., Vaughan, J.W. (eds.) Advances in Neural Information Processing Systems (2021). https://openreview.net/forum?id=jhd62iKzRuj
16. Wang, S., et al.: Beta-CROWN: efficient bound propagation with per-neuron split constraints for complete and incomplete neural network verification. In: Advances in Neural Information Processing Systems, vol. 34 (2021)
17. Xu, K., et al.: Automatic perturbation analysis for scalable certified robustness and beyond. In: Advances in Neural Information Processing Systems, vol. 33 (2020)

18. Xu, K., et al.: Fast and complete: enabling complete neural network verification with rapid and massively parallel incomplete verifiers. In: International Conference on Learning Representations (2021). https://openreview.net/forum?id=nVZtXBI6LNn

19. Zhang, H., et al.: α, β-crown: verified intelligence alpha-beta-crown (2023). https://github.com/Verified-Intelligence/alpha-beta-CROWN. Team from CMU, UCLA, Drexel University, Columbia University, UIUC, RWTH Aachen University, Sun Yat-sen University, University of Michigan. Advisors: Kolter, Zico; Hsieh, Cho-Jui; Jana, Suman; Li, Bo; Lin, Xue

20. Zhang, H., et al.: General cutting planes for bound-propagation-based neural network verification. In: Advances in Neural Information Processing Systems (2022)

21. Zhang, H., et al.: A branch and bound framework for stronger adversarial attacks of ReLU networks. In: Proceedings of the 39th International Conference on Machine Learning, vol. 162, pp. 26591–26604 (2022)

22. Zhang, H., Weng, T.W., Chen, P.Y., Hsieh, C.J., Daniel, L.: Efficient neural network robustness certification with general activation functions. In: Advances in Neural Information Processing Systems, vol. 31, pp. 4939–4948 (2018). https://arxiv.org/pdf/1811.00866.pdf

Learning Heuristics for Combinatorial Optimization Problems on K-Partite Hypergraphs

Mehdi Zouitine[1,2(✉)], Ahmad Berjaoui[1], Agnès Lagnoux[2,3,4],
Clément Pellegrini[2,3,4], and Emmanuel Rachelson[5]

[1] IRT Saint Exupéry, Toulouse, France
{mehdi.zouitine,ahmad.berjaoui}@irt-saintexupery.com
[2] Institut de Mathématiques de Toulouse, UMR5219, Université de Toulouse,
CNRS, Toulouse, France
{agnes.lagnoux,clement.pellegrini}@math.univ-toulouse.fr
[3] UT2J, 31058 Toulouse, France
[4] UT3, 31062 Toulouse, France
[5] ISAE-SUPAERO, Toulouse, France
emmanuel.rachelson@isae-supaero.fr

Abstract. Recently, deep neural networks have demonstrated remarkable performance in addressing combinatorial optimization challenges. The expressive power of graph neural networks combined with Reinforcement Learning (RL) enabled learning heuristics that rival or even surpass conventional methods. Such advancements have paved the way for Neural Combinatorial Optimization (NCO), an emerging paradigm that enables end-to-end heuristic learning without the reliance on expert knowledge. In this paper, we propose an NCO approach to learn heuristics for the vast family of Combinatorial Optimization Problems (COPs) defined on K-partite hypergraphs, including multi-dimensional assignment or scheduling problems. Central to our approach is the ability to represent sophisticated functions on K-partite hypergraphs, using a novel family of neural networks. We show that our heuristic competes with other ones in comparable settings and that our method can also be applied to more complex real-life assignment and scheduling problems.

Keywords: Learning Based-Heuristics · Neural Combinatorial Optimization · Graph Neural Networks · Deep Reinforcement Learning

1 Introduction

Combinatorial optimization, the search for an optimum value item within a discrete set, holds immense practical significance across various application fields and has a long history of research, notably in the Operations Research community. Traditionally, addressing these problems has involved the deployment of either exact or approximate optimization methods. But a substantial fraction of these problems cannot be solved within a reasonable timeframe (e.g., the

B. Dilkina (Ed.): CPAIOR 2024, LNCS 14743, pp. 304–314, 2024.
https://doi.org/10.1007/978-3-031-60599-4_21

knapsack problem, the traveling salesman, or the multi-dimensional assignment problems), rendering their exact solutions computationally infeasible [11]. In such cases, heuristics emerge as a pragmatic alternative, aiming to strike a balance between solution quality and computational efficiency [1,2,5,14,16,19]. However, the design of effective heuristics often demands specialized expert knowledge and can be time and resource consuming.

Recently, a paradigm shift has been observed with the advent of Neural Combinatorial Optimization (NCO) [3]. Approaching combinatorial challenges from a reinforcement learning [26] perspective, NCO enables the swift learning of good heuristics, reducing the need for domain expertise. Despite its potential to learn constructive heuristics, NCO demands meticulous neural network architecture and optimization considerations. Numerous combinatorial problems, particularly those with intricate structures, remain largely unexplored by NCO. In this work, we expand the scope of tackled problems, by addressing the broad family of combinatorial challenges arising from K-partite hypergraphs, extending the Deep Bipartite Assignment (DBA) framework introduced by [12]. We develop Deep K-partite Assignment (DKA), a generic NCO method, based on Transformer architectures [27] and designed to tackle optimization problems expressed via K-partite hypergraphs. In this paper, we make three contributions.

1. To the best of our knowledge, combinatorial problems on K-partite hypergraphs for $K > 2$ is addressed here for the first time with an NCO perspective.
2. We propose a new graph neural network architecture, which generalizes the one proposed in [12] from bipartite graphs to K-partite hypergraphs, and offers a much desired flexibility in NCO, permitting training on one problem size and performing inference on another.
3. We use this architecture to incrementally construct efficient heuristics for solving a set of COPs on K-partite hypergraphs, and demonstrate empirically their relevance and efficiency.

2   Background and Related Work

Hypergraphs (HG). Let V be a finite set where each element is referred to as a node. An undirected HG $H = (V, E)$ is defined by a set V of nodes and a set E where each element $e \in E$ is a non-empty subset of V. Each subset e of E is called a hyperedge (HE). While a regular edge in a graph connects exactly two nodes, a HE in a HG can connect any number of nodes, from one to the size of the entire set V. This flexibility enables HGs to model intricate and multiple types of relationships, going beyond the representational capacity of standard graphs. A standard graph is simply a HG where all HEs have a cardinality of 2. To learn functions defined on the complex structure of HGs, several neural network architectures have recently emerged as HG neural networks [7,10]. A weighted HG is denoted $H = (V, E, W)$, where each scalar value in W is the weight of the corresponding HE in E. In a K-partite weighted HG, the set of vertices V is partitioned into K disjoint subsets $V_1, \ldots, V_K$. Each HE $e \in E$ contains exactly one vertex from each partition V_i (hence its cardinality is K, and K-partite HGs are also K-uniform).

COPs on K-Partite HGs. K-partite HGs efficiently describe a broad spectrum of difficult COP's. For instance, the Multi-dimensional Assignment Problem (MAP) in dimension K consists in grouping together items from K disjoint sets of the same cardinality for a minimal overall cost; e.g. assigning workers (set 1) to tasks (set 2) and time windows (set 3) is a 3-dimensional assignment problem. This problem arises in many real-life situations such as multi-target tracking, manufacturing, course scheduling, or multi-sensor data fusion [6,9,20], and can be represented by a K-partite weighted HG, where each HE stands for the grouping of K items together from the disjoint K partition of nodes. A solution to such a problem is a subset among the set E of HEs and can be represented by a binary vector $x \in \{0,1\}^{|E|}$.

MAPs are NP-hard in general (for $K > 2$), highlighting the necessity to develop heuristic methods to tackle them efficiently. Similarly, K-partite HGs are also significantly useful in resource allocation problems, such as the multi-Level generalized assignment problem [4,8,13]. Additionally, K-partite HGs play a role in scheduling problems. A notable instance is the extension of unrelated parallel machine scheduling to incorporate multiple time windows [29], a generalization of the classical problem defined by [15]. Formally, COPs on K-partite HGs can be expressed as:

$$\min_{x \in \{0,1\}^{|E|}} \sum_{i_1=1}^{|V_1|} \cdots \sum_{i_K=1}^{|V_K|} c\left(W_{i_1 \ldots i_K}, x_{i_1 \ldots i_K}\right) \quad \text{subject to} \quad \mathcal{C}(x) \leq b, \quad (1)$$

where $c(W_{i_1 \ldots i_K}, x_{i_1 \ldots i_K})$ is the cost of choosing HE $(i_1, \ldots, i_K)$ (i.e. setting $x_{i_1 \ldots i_K} = 1$) and $\mathcal{C}(x) \leq b$ is a set of problem-specific constraints.

Reinforcement Learning (RL) [26] addresses the challenge of developing a decision-making strategy for an entity (or agent) that iteratively interacts with a dynamical system. At every iteration, both the entity and the context can be defined using a state s from the state space $\mathcal{S}$. The agent takes an action a from the action space $\mathcal{A}$, which prompts the system to transition to a subsequent state s', determined by the transition probability $\mathcal{P}(s'|s, a)$, while receiving a reward denoted by $\mathcal{R}(s, a)$. The tuple $M = (\mathcal{S}, \mathcal{A}, \mathcal{P}, \mathcal{R})$ constitutes a Markov Decision Process (MDP) [21], where we consider an initial state probability distribution $p_0(s)$ on $\mathcal{S}$. A policy, encapsulated by parameters θ, is portrayed as $\pi_\theta(a|s)$, which translates states into likely actions. The objective in training an RL agent is to uncover the policy that elevates the anticipated cumulative return, expressed as $J(\pi_\theta) = \mathbb{E}[\sum_{t=0}^{\infty} \gamma^t \mathcal{R}(s_t, a_t)]$ where $\gamma \in [0, 1)$ is a discount factor.

Graph Neural Network (GNN) and Message Passing. In recent years, Graph Neural Networks have emerged as powerful tools for tasks defined on graph structures [17]. GNNs excel in processing graph data through message passing, which involves exchanging and processing information based on the neighborhoods of nodes and edges. Various GNN models have been implemented to handle increasingly complex graph structures, such as bipartite graphs [12], heterogeneous graphs [31], or relational graphs [23]. Our contribution aims to develop suitable message passing strategies for K-partite HGs by defining an appropriate neighborhood and function, enabling the efficient learning heuristics.

Learning Based Heuristics. Recently, a new approach known as NCO has emerged as a promising paradigm for solving COPs. In a foundational work, [28] introduced the pointer network, a novel architecture designed to solve permutation-based problems such as convex-hull and travelling salesman problems. Building upon this, [3] employed RL to shift away from label ties, learning heuristics exclusively through reward signals, thereby addressing both travelling salesman and knapsack problems. RL is used to learn constructive heuristics functions that incrementally build solutions step by step, optimizing a cumulative reward based on partial solutions. The efficiency of these constructive heuristics relies significantly on the chosen neural architecture. Finding the right function with the appropriate inductive biases is crucial for the heuristic to effectively navigate the solution space and make smart incremental decisions. [18] proposed an attention model rooted in the transformer architecture [27], demonstrating impressive outcomes on a variety of routing challenges. Recently, the research landscape has witnessed a proliferation of advanced NCO architectures, each meticulously crafted to address distinct combinatorial challenges. Notable among these are solutions for the job shop scheduling problem [31] which leverage directed acyclic GNN, and most similar to our work, the 2D assignment problem [12] employing bipartite GNN.

3   A Generic Family of Neural Policies on K-Partite HG

3.1 K-Partite HG Neural Network

The main contribution of this work is the development of a novel K-partite HG neural network policy, designed for learning solutions to a variety of combinatorial problems. Central to our formalisation is the concept of the dual H^* of a HG $H = (V, E, W)$ [30]. In this dual $H^* = (E^*, V^*, W^*)$, the roles of nodes and HEs are interchanged, where each node in E^* corresponds to a HEs in E and each HEs in V^* corresponds to a node in V. The set W^* represents the weight or the attribute of the nodes E^*. We reinterpret a K-partite HG as a K-relational multigraph (K-multigraph). A K-multigraph is characterized by K different types of edges (relations) and the capability for any two nodes to be connected by multiple edges.

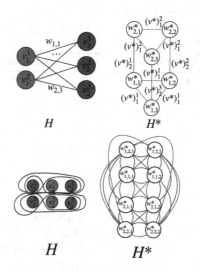

Fig. 1. Complete bipartite and 3-partite (Only a limited number of HEs are represented.) HGs with their respective dual representations

In the dual formulation, nodes are linked by relation i (ranging from 1 to K) if their associated HEs in H intersect in partition V_i. In the upcoming methodological discussions, "node" refers to an element in H^*, (a HE in H).

K-Partite HG Encoder and Practical Implementation. We propose a new HG neural network encoder E_θ, central to which is a new message-passing layer. This layer has the unique ability to learn embeddings from a K-partite HG. This approach is fundamentally based on the dual formulation of HGs (Fig. 1), which facilitates a more conventional representation of message passing between nodes (Fig. 2).

Our encoder first processes the node attributes $w_i^* \in W^*$ through a learned linear projection, $z_i^0 = g_\theta(w_i^*)$. The embeddings update via L layers, each with K distinct sublayers $\theta_1, \ldots, \theta_K$, using different adjacency matrices A_k for various relations. Node embeddings from layer ℓ are denoted by $z_i^{(\ell)}$, for ℓ ranging from 1 to L.

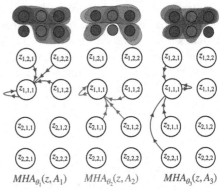

$$MHA_{\theta_1}(z, A_1) \qquad MHA_{\theta_2}(z, A_2) \qquad MHA_{\theta_3}(z, A_3)$$

Fig. 2. Message passing in 3-partite HG: node embeddings in H^* are learned using K-distinct functions, corresponding to each relation. The node embeddings are updated using multi-head attention across each relation.

Message Passing Layer. Following the Transformer architecture [27], each attention layer consists of two sublayers: a multi-head attention (MHA) layer that executes message passing between the nodes and a fully connected feed-forward (FF) layer. Each sublayer adds a skip-connection and batch normalization (BN).

Algorithm 1: Message passing layer

Parameters: $\theta_1, \ldots, \theta_K$
Input: Nodes embeddings $z^{(\ell-1)}$, adjacency matrices $A_1, \ldots, A_K$
$z \leftarrow z^{(\ell-1)}$
for k *in* $1 \ldots, K$ **do**
$\quad$ $z \leftarrow BN(z + MHA_{\theta_k}(z, A_k))$ $\qquad$ `// Message passing across`
$\quad$ `partition` k
$\quad$ $z \leftarrow BN(z + FF_{\theta_k}(z))$ $\qquad$ `// Normalization and projection`

Build on top of this new encoder, a policy head p_θ focuses on the sequential construction of a solution $X \in \{0, 1\}^{|E^*|}$. Mathematically, our policy $\pi_\theta := p_\theta \circ E_\theta$ is a function that accepts a K-multigraph H^*. It outputs a score p_i for each node, representing the probability of including the node in the partial solution. The θ parameters are optimized by Proximal Policy Optimization (PPO) [24]. Drawing inspiration from the method described in [3], the policy employs a mask $\mathcal{M}$ to identify and avoid selecting nodes that would violate the specific constraints of the problem. This ensures that only feasible nodes are considered during the assignment process, maintaining the integrity of the solution with respect to the problem constraints. Following [3], we scale the node scores (before masking) within $[-C, C]$ using tanh action function. Then we compute the final output probability p_i using a softmax:

$$
u_i = \begin{cases} C \cdot \tanh \left(p_\theta(z_i^{(L)}) \right) & \text{if } i \notin \mathcal{M} \\ -\infty & \text{otherwise} \end{cases} \qquad p_i = \frac{e^{u_i}}{\sum_{j=1}^{|E^*|} e^{u_j}}. \qquad (2)
$$

We employ RL to optimize this node selection policy. To achieve this, we delineate the underlying MDP as outlined below and illustrated in (Fig. 3):

The **state** represents the solution to an instance at timestep t. To do so $s_t := \{W^*\} \cup \{X_t\}$ is defined as the union of the HG and the partial solution.

The **action** a_t points to the index of the forthcoming node to be added in the partial solution.

The **transition** function $\mathcal{P}(s_{t+1}|s_t, a_t)$ is influenced by the assignment problem specifics. However, in most cases, the subsequent state s_{t+1} is deterministically derived from s_t through a simple update to the current solution, X_t.

The **reward** function is typically contingent on the nodes weights W^*. The overarching objective is to maximize the reward given by $r(s_t, a_t) = c(W^*, a_t)$. As an example, in specific case of the MAP, the reward function is simply $r(s_t, a_t) = W^*[a_t]$ which is the weight of the node indexed by a_t.

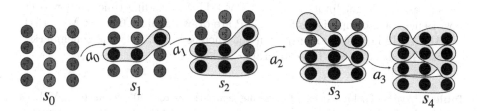

Fig. 3. DKA Policy trained via RL sequentially selects a node to construct a solution to a 3D assignment problem. The agent's current state at each step is represented as s_t encapsulating both HG and the partial solution.

4   Experiments and Discussion

To validate our method, we benchmark DKA against established heuristics on several problems, notably the well-known 3D Assignment Problem (3DAP) [20]. We explore two variants: a random version ('r'), with arbitrary assignment costs and a geometric version ('g'), where costs correspond to the area of the triangle formed by assignments [25]. Instances are denoted as 3(**type**)n, e.g., 3r10 for a 3DAP with random costs and 10 elements in each subset. Our learned heuristics are compared against random and established constructive heuristics: Greedy [16], Max-regret [1,2], ROM [14], and ROM-shift [16] (Table 1).

Table 1. Average solution costs for various heuristics across 500 instances of 3DAP; lower values indicate better performance. Values in parentheses represent the execution time in milliseconds (ms).

	Random	Greedy	Max-regret	ROM	ROM-shift	DKA
3r5 ($\downarrow$)	2.465 (0)	0.833 (0)	0.632 (0)	0.687 (0)	0.533 (0)	**0.452** (33)
3r10 ($\downarrow$)	4.999 (0)	0.921 (0)	0.568 (3)	0.749 (0)	0.619 (2)	**0.330** (69)
3r15 ($\downarrow$)	7.468 (0)	0.945 (0)	0.553 (6)	0.810 (2)	0.683 (5)	**0.269** (121)
3g5 ($\downarrow$)	0.380 (0)	0.160 (0)	0.103 (0)	0.090 (0)	0.0712 (0)	**0.063** (33)
3g10 ($\downarrow$)	0.757 (0)	0.208 (0)	0.096 (3)	0.084 (0)	0.067 (2)	**0.0365** (69)
3g15 ($\downarrow$)	1.152 (0)	0.220 (0)	0.084 (6)	0.077 (2)	0.063 (5)	**0.026** (121)

Our study addresses two correlated problems: the Multi-Level Generalized Assignment Problem (MGAP) [4,13] and the Unrelated Parallel Machine Scheduling with Multiple Time Windows (UPMSMTW) [29]. Both problems entail assigning multi-level tasks to resources, with the MGAP focusing on task weights and the UPMSMTW on time windows, both impacting the profit based on the task level, thus introducing additional complexity. MGAP instances are denoted by MGAP (resource number)r(task number)t(level number)l, such as 3r20t3l. In contrast, UPMSMTW instances are represented as (machine number)m(task number)t(time window number)w, for example, 2m20t3w. The objective is to maximize profit while adhering to capacity constraints-for MGAP, not surpassing weight limits, and for UPMSMTW, avoiding time window overlaps. Benchmark comparisons for MGAP are drawn against Greedy, MGAPH1, and MGAPH2 heuristics described in [8], whereas UPMSMTW is benchmarked against Greedy [22] (See Table 3) (Table 2).

Table 2. Average solution costs for various heuristics across 500 instances of MGAP; higher values indicate better performance. Values in parentheses represent the execution time in milliseconds (ms).

	Random	Greedy	MGAPH1	MGAPH2	DKA
2r10t2l ($\uparrow$)	4.381 (1)	6.603 (2)	6.711 (9)	6.802 (11)	**7.007** (66)
2r20t2l ($\uparrow$)	8.441 (2)	13.594 (3)	13.843 (14)	14.021 (16)	**14.846** (125)
2r40t5l ($\uparrow$)	17.257 (3)	26.960 (5)	27.143 (21)	27.583 (24)	**29.760** (240)

Table 3. Average solution costs for various heuristics across 500 instances of MGAP and UPMSMTW; higher values indicate better performance. Values in parentheses represent the execution time in milliseconds (ms).

UPMSMTW

	Random	Greedy	DKA
2m20t2w (↑)	4.962 (1)	8.266 (2)	**8.630** (13)
3m20t3w (↑)	7.539 (1)	12.987 (3)	**13.360** (17)
4m100t3w (↑)	46.735 (2)	75.003 (4)	**80.821** (111)

In Table 4, we demonstrate that our agent can generalize in a zero-shot manner to larger instances than those encountered during training. This capability underscores the model's adaptability and robustness in handling increased problem complexities.

Table 4. Zero-shot generalization performance of a model trained on 3r15.

	Random	Greedy	Max-regret	ROM	ROM-shift	DKA
3r30 (↓)	14.975 (0)	0.963 (0)	0.519 (30)	0.918 (6)	0.807 (20)	**0.221** (4000)
3r40 (↓)	19.970 (0)	0.969 (2)	0.510 (47)	0.970 (10)	0.865 (30)	**0.244** (22000)
3r50 (↓)	24.919 (0)	0.974 (30)	0.625 (793)	0.979 (173)	0.907 (51)	**0.243** (61000)

A key aspect of our model is the use of K unique parameters $\theta_1, ..., \theta_K$ in the message-passing layer. Our ablation study on small-scale problems highlights the importance of treating the HG as a K-multigraph and using distinct parameters for each relation (Table 5).

Table 5. Results of ablation study on three problems: impact of K distinct parameters.

	$\theta_1 \neq ... \neq \theta_K$	$\theta_1 = ... = \theta_K$
(3r5) (↓)	**0.452**	0.677
(2r20t2l) (↑)	**14.050**	9.076
(2r20t2w) (↑)	**8.630**	8.102

Experimental Settings

All the experiments were run on a desktop machine (Intel i9, 10th generation processor, 64 GB RAM) with a single NVIDIA RTX 3090 GPU, using Python. The parameters in Table 5 were used to optimize our policy (Table 6).

Table 6. Hyperparameters used in the study.

Hyperparameter	Value	Hyperparameter	Value
Number of layers	4	Number of attention heads per layer	8
Node embeddings dimension	128	Number of parallel environments	256
Episode length	256	Discount factor γ	1
Optimizer (θ of PPO)	Adam ($lr = 2e - 5$)	PPO epochs	3
PPO clip	0.2	PPO GAE λ	0.95
PPO value loss coefficient	0.5	PPO entropy loss coefficient	0.01
PPO batch size	256		

The training time of DKA on the 3DAP varies with the size, taking 1 h for 3r5, 8 h for 3r10, and 14 h for 3r15 as an indication. Similarly, for MGAP and UPMSMTW, the training durations align closely, starting at 1 h for the smallest instances and extending up to 14 h for the largest ones.

5 Conclusion, Limitation and Future Work

We have proposed the first GNN architecture that can learn heuristics for K-partite HG without the need for expert knowledge, relying solely on reward signals through deep RL. This expands the application of neural constructive heuristics to these family of problems. Ongoing experiments suggest that our model can outperform several established heuristics and generalize to larger, unseen problems. Our architecture, designed for COPs, could be applicable in other domains where problem can be modeled as K-partite HG. However, our framework is currently limited to relatively small instance sizes due to significant memory requirements, which grow exponentially with the number of partition. We anticipate that DKA could surpass other methods in uncertain settings. Future work on DKA will explore scalability and complex, uncertain problems, such as satellite constellation scheduling under cloud uncertainties.

Acknowledgment. We acknowledge the support of the IRT-MINDS project.

References

1. Balas, E., Saltzman, M.J.: An algorithm for the three-index assignment problem. Oper. Res. **39**(1), 150–161 (1991)
2. Bekker, H., Braad, E.P., Goldengorin, B.: Using bipartite and multidimensional matching to select the roots of a system of polynomial equations. In: Gervasi, O., et al. (eds.) ICCSA 2005, Part IV. LNCS, vol. 3483, pp. 397–406. Springer, Heidelberg (2005). https://doi.org/10.1007/11424925_43
3. Bello, I., Pham, H., Le, Q.V., Norouzi, M., Bengio, S.: Neural combinatorial optimization with reinforcement learning (2017). https://doi.org/10.48550/arXiv.1611.09940, arXiv:1611.09940 [cs, stat]

4. Ceselli, A., Righini, G.: A branch-and-price algorithm for the multilevel generalized assignment problem. Oper. Res. **54**(6), 1172–1184 (2006)
5. Croes, G.A.: A method for solving traveling-salesman problems. Oper. Res. **6**(6), 791–812 (1958)
6. Dang, X., Cheng, Q., Zhu, H.: Indoor multiple sound source localization via multi-dimensional assignment data association. IEEE/ACM Trans. Audio Speech Lang. Process. **27**(12), 1944–1956 (2019). https://doi.org/10.1109/TASLP.2019.2935837
7. Feng, Y., You, H., Zhang, Z., Ji, R., Gao, Y.: Hypergraph neural networks. CoRR abs/1809.09401 (2018). http://arxiv.org/abs/1809.09401
8. French, A.P., Wilson, J.M.: Heuristic solution methods for the multilevel generalized assignment problem. J. Heuristics **8**, 143–153 (2002)
9. Frieze, A., Yadegar, J.: An algorithm for solving 3-dimensional assignment problems with application to scheduling a teaching practice. J. Oper. Res. Soc. **32**, 989–995 (1981)
10. Gao, Y., Feng, Y., Ji, S., Ji, R.: HGNN+: general hypergraph neural networks. IEEE Trans. Pattern Anal. Mach. Intell. **45**(3), 3181–3199 (2022)
11. Garey, M.R., Johnson, D.S.: Computers and Intractability, vol. 174. Freeman, San Francisco (1979)
12. Gibbons, D., Lim, C.C., Shi, P.: Deep learning for bipartite assignment problems. In: 2019 IEEE International Conference on Systems, Man and Cybernetics (SMC), pp. 2318–2325 (2019). https://doi.org/10.1109/SMC.2019.8914228
13. Glover, F., Hultz, J., Klingman, D.: Improved computer-based planning techniques, part 1. Interfaces **8**(4), 16–25 (1978)
14. Gutin, G., Goldengorin, B., Huang, J.: Worst case analysis of max-regret, greedy and other heuristics for multidimensional assignment and traveling salesman problems. In: Erlebach, T., Kaklamanis, C. (eds.) WAOA 2006. LNCS, vol. 4368, pp. 214–225. Springer, Heidelberg (2007). https://doi.org/10.1007/11970125_17
15. Horowitz, E., Sahni, S.: Exact and approximate algorithms for scheduling nonidentical processors. J. ACM (JACM) **23**(2), 317–327 (1976)
16. Karapetyan, D., Gutin, G., Goldengorin, B.: Empirical evaluation of construction heuristics for the multidimensional assignment problem. arXiv preprint arXiv:0906.2960 (2009)
17. Kipf, T.N., Welling, M.: Semi-supervised classification with graph convolutional networks. arXiv preprint arXiv:1609.02907 (2016)
18. Kool, W., van Hoof, H., Welling, M.: Attention, learn to solve routing problems! (2019). https://doi.org/10.48550/arXiv.1803.08475, arXiv:1803.08475 [cs, stat]
19. Lin, S., Kernighan, B.W.: An effective heuristic algorithm for the traveling-salesman problem. Oper. Res. **21**(2), 498–516 (1973)
20. Pierskalla, W.P.: The multidimensional assignment problem. Oper. Res. **16**(2), 422–431 (1968)
21. Puterman, M.L.: Markov Decision Processes: Discrete Stochastic Dynamic Programming. Wiley, Hoboken (2014)
22. Rainjonneau, S., et al.: Quantum algorithms applied to satellite mission planning for earth observation. IEEE J. Sel. Top. Appl. Earth Observ. Remote Sens. **16**, 7062–7075 (2023). https://doi.org/10.1109/jstars.2023.3287154, http://dx.doi.org/10.1109/JSTARS.2023.3287154
23. Schlichtkrull, M., Kipf, T.N., Bloem, P., van den Berg, R., Titov, I., Welling, M.: Modeling relational data with graph convolutional networks. In: Gangemi, A., et al. (eds.) ESWC 2018. LNCS, vol. 10843, pp. 593–607. Springer, Cham (2018). https://doi.org/10.1007/978-3-319-93417-4_38

24. Schulman, J., Wolski, F., Dhariwal, P., Radford, A., Klimov, O.: Proximal policy optimization algorithms. arXiv preprint arXiv:1707.06347 (2017)
25. Spieksma, F.C., Woeginger, G.J.: Geometric three-dimensional assignment problems. Eur. J. Oper. Res. **91**(3), 611–618 (1996)
26. Sutton, R.S., Barto, A.G.: Reinforcement Learning: An Introduction. MIT Press, Cambridge (2018)
27. Vaswani, A., et al.: Attention is all you need. In: Advances in Neural Information Processing Systems, vol. 30 (2017)
28. Vinyals, O., Fortunato, M., Jaitly, N.: Pointer networks (2017). https://doi.org/10.48550/arXiv.1506.03134, arXiv:1506.03134 [cs, stat]
29. Wang, J., Song, G., Liang, Z., Demeulemeester, E., Hu, X., Liu, J.: Unrelated parallel machine scheduling with multiple time windows: an application to earth observation satellite scheduling. Comput. Oper. Res. **149**, 106010 (2023)
30. Whitney, H.: A theorem on graphs. Ann. Math. 378–390 (1931)
31. Zhang, R., et al.: Learning to solve multiple-TSP with time window and rejections via deep reinforcement learning. IEEE Trans. Intell. Transp. Syst. 1–12 (2022). https://doi.org/10.1109/TITS.2022.3207011

Author Index

B. Dilkina (Ed.): CPAIOR 2024, LNCS 14743, pp. 315–317, 2024.
https://doi.org/10.1007/978-3-031-60599-4

Printed in the United States
by Baker & Taylor Publisher Services